Samuel Almond Miller

North American Mesozoic and Cænozoic Geology and Palæontology

An abridged history of our knowledge of the triassic, jurassic, cretaceous and

tertiary formations of this continent

Samuel Almond Miller

North American Mesozoic and Cænozoic Geology and Palæontology
An abridged history of our knowledge of the triassic, jurassic, cretaceous and tertiary formations of this continent

ISBN/EAN: 9783337409708

Printed in Europe, USA, Canada, Australia, Japan

Cover: Foto ©berggeist007 / pixelio.de

More available books at **www.hansebooks.com**

NORTH AMERICAN

MESOZOIC AND CÆNOZOIC

Geology and Palæontology;

OR,

AN ABRIDGED HISTORY OF OUR KNOWLEDGE OF THE

TRIASSIC, JURASSIC, CRETACEOUS AND TERTIARY

FORMATIONS OF THIS CONTINENT.

By S. A. MILLER.

CINCINNATI:

PRINTED BY JAMES BARCLAY, 269 VINE STREET.

1881.

PREFACE.

This work is a historical review of what we know of the Triassic, Jurassic, Cretaceous and Tertiary formations of North America. It is not exhaustive, yet it contains more information in regard to these formations than will be found in any other single publication. In compiling the work, the language of the various authors, whose books are referred to, has been used wherever practicable, and when it has been abridged the substance has not been changed. The author has not had an opportunity to specially study any of these formations beyond that part which is embraced within the period of the drift. The latter he has explored and studied, in its distribution, over many of the States and a considerable part of Canada.

He has undertaken to overthrow the glacial hypothesis, and now submits the facts and the conclusions he has drawn to the learned of this country and of Europe, and asks for their verdict. There is this further fact to be remembered. If there was no glacial period in North America, there was none in Europe.

Only a limited number of copies of the book has been published in this form, though it has appeared, in parts, in the last three volumes of the JOURNAL OF THE CINCINNATI SOCIETY OF NATURAL HISTORY.

From the Journal of the Cincinnati Society of Natural History, October, 1879.

NORTH AMERICAN MESOZOIC AND CÆNOZOIC GEOLOGY AND PALÆONTOLOGY.

By S. A. MILLER, Esq.

The sciences of Geology and Palæontology had not advanced many steps, in Europe, before their growth had commenced in America. Their development, therefore, has been nearly contemporaneous in the two countries, though more rapid in the early part of the century in the Old World than in the New. Europe has had William Smith, J. S. Miller, Sowerby, Murchison, Lyell, Brongniart, D'Orbigny, Goldfuss, Sternberg, Barrande, and many other distinguished authors; while America has had McClure, Morton, Vanuxem, Hitchcock, Conrad, Leidy, Hall, Lesquereux, Logan, Billings, Dawson, and others, original discoverers, who possessed the philosophical learning necessary for the correct application of the discoveries to the advancement and growth of the sciences. The facts, however, upon which these sciences are based, and which constitute the superstructure, as now understood, have been ascertained, so recently, that one would hardly undertake to enumerate a score of the principal fathers of them, in either country, without mentioning the names of some who are still living.

The first society organized for the advancement of science in North America, of which we have any account, is the American Philosophical Society, instituted in 1769, in Philadelphia. The earliest geological papers that seem to be worth mentioning, appeared in the Transactions of this Society, and though its publications have not been rapid, they continue to appear, and to hold a high rank, whether devoted to Geology, Palæontology, or other departments of science. The society is indebted for its organization to Benjamin Franklin. The first volume of the Transactions appeared, in quarto, in 1771.

Belknap wrote, upon the White Mountains, in 1784; Hutchins, on the Rock and Cascade of the Youghiogheny, in 1786; William Dunbar, on large mammalian bones found in Louisiana, a set of human teeth found while digging a well at the depth of 30 to 35 feet; and on the Mississippi river and its delta, in 1804, which was continued in 1809. B. II. Latrobe described the freestone quarries on the Potomac and Rappahannock, in 1809; and William McClure, in the same year, published his Observations on the Geology of the United States, explanatory of a geological map. He divided the formations into four classes, viz: 1st, Primitive rocks; 2d, Transition rocks; 3d, Flœtz or Secondary rocks; and 4th, Alluvial rocks. These classes he separated on their mineralogical characters, and he treated of their dip and extent, as far as his observations permitted. And Thomas Jefferson, who had been President of the United States, described the fossil bones of the Megalonyx, in 1818.

The American Academy of Arts and Sciences was established in Boston, and commenced the publication of its Memoirs in 1780. The Academy of Natural Sciences, of Philadelphia, originated in 1812, but commenced its publications in 1817. It soon collected an extensive library of works upon Natural History, largely owing to the fine donation by the generous and distinguished geologist, William McClure, and at once entered the field as an active society, alive to the importance of the publication of facts, as distinguished from theoretical considerations. Its publications, from the commencement, have occupied the first rank in science, and are now, absolutely, indispensable to every American naturalist, and should occupy a shelf in every public library.

An idea of the absence of geological information, in this country, in 1803, may be formed when it is remembered that geology was not separated as a science from mineralogy, and that so little was known of mineralogy that it could hardly have ranked as a science ; for later in life, Prof. Silliman, speaking of this period, says, "it was a matter of extreme difficulty to obtain, *among ourselves* even, *the names* of the most common stones and minerals; and one might inquire earnestly, and long, before he could find any one to identify even *quartz, feldspar,* or *hornblende,* among the simple minerals; or *granite, porphyry,* or *trap,* among the rocks. *We speak from experience,* and well remember with what impatient, but almost despairing curiosity, we eyed the bleak, naked ridges, which impended over the valleys and plains that were the scenes of our youthful excursions. In vain did we doubt that the glittering spangles of · mica, and the still more alluring brilliancy of pyrites, gave assurance of the existence of the precious metals in those substances; or that the cutting of glass by the garnet, and by quartz, proved that these minerals were the diamond; but if they were not precious metals, and if they were not diamonds, we in vain inquired of our companions, and even our teachers, what they were."

An idea of the low state of Palæontology, in 1809, may be formed from a letter written by Parker Cleveland, Professor of Mathematics and Natural Philosophy, in Bowdoin College, and published in the Memoirs of the American Academy of Arts and Sciences, vol. iii., part 1. He had carefully watched the digging of two wells through sand and into blue clay; one of them was at a distance of about 20 miles from the sea, and three or four miles from the tide, in Cathance river, and had an elevation estimated at 70 or 80 feet above the tide. This well was dug 20 feet deep. The first 10 feet was through sand and gravel. At the depth of 10 feet a stratum of blue clay was found, which had the appearance and smell of that dug on flats, or near salt marshes. In this clay he found shells; one a clam, "two varieties of muscle," and one large conical form, whose generic name he knew not, but the same genus he said "is found on our sea shores." The other well was near Brunswick, 80 feet above tide water, in the Androscoggin, and half a mile from the river. At the depth of 12 feet, a four feet stratum of clay was found having the same smell, and containing shells plentifully interpersed, similar to those found in the well near Cathance river. He thought that important advantages would result from possessing a geographical map, indicating the different species of fossil shells, and the places in which they were found, especially where the country or coast might be thickly inhabited; because, he says, "with such a map before us, we should be better enabled to compare individual facts, and hence to draw several conclusions."

In 1818, Prof. Benjamin Silliman commenced the publication of the American Journal of Science and Arts, which, through his remarkable talent, and unbounded energy, at once took rank with the scientific journals of Europe. It has now reached the 119th volume, and the aid it has rendered the sciences of Geology and Palæontology is unmeasured.

In 1818, William McClure prepared an "Essay on the Formation of Rocks, or an inquiry into the probable origin of their present form and structure," which was published in the Journal of the Academy of Natural Sciences, of Philadelphia, vol. i., part 2. He says:

"Concerning the nature and properties of the great mass, which constitutes the interior of the earth, we are entirely ignorant; few of our mines penetrate deeper than one fifty thousandth part of the earth's diameter, under the surface, and none of them go beyond one twenty-five thousandth part of that diameter: it would appear, therefore, that any mere supposition concerning the actual and present state or the nature of those substances, which form the interior of the earth, is unsupported, as yet, by any reasonable analogy: and that all conjectures, concerning former changes, partial or total, in the nature and structure of those substances, are removed still farther from anything analogous, in our present state of knowledge."

" The earth being flattened, at the poles, does not necessarily imply its former fluidity. We may be permitted to doubt the analogy between our experiments on bodies moving, in our atmosphere, and the earth's motion in space; our total ignorance of the nature of the fluid, which occupies what is usually called space, tends to render the analogy inconclusive."

" May not the mode of casting patent shot be considered as an experiment, on the form which liquid bodies would take by a rotary motion? A drop of melted lead let fall from the height of 200 feet is completely globular, and not flattened at the poles; the lead might be thrown with force from the top of the tower, which would imitate the centrifugal force, as gravitation does the centripetal force, and make the experiment more analogous."

" The supposition that the earth was in a fluid state, when it took its present form, leads to the supposition that it was always so; and that fluidity was the original state of the earth, kept so by all the general laws and order of nature, all of which general order and laws of nature must be totally changed before the earth would take a solid form."

" On the supposition that the earth, previous to its fluid state, had existed always in a solid state, and that some creation or accident produced the fire or water necessary to its liquefaction, we have, in that case, first to suppose that the order and nature of the general laws, which had kept it always in a solid state, were totally changed to produce a fluid state ; and that another change, in the general laws, which produced and kept it in a fluid state, must have taken place previous to its having become again solid."

" It may be doubted whether the uniformity, order and regularity of the general laws of nature, which have at any time come within the limits of our observation, can warrant a supposition founded on such complete changes in the mode of action."

" As we do not comprehend either the creation or annihilation of matter, by the origin of rocks, we mean the last change which produced their present form, and the agents that nature employed to give them that form, or effectuate that change."

He divided the rocks into three classes (not, however, without expressing grave doubts as to the correctness of his conclusions), as follows:

1st Class—of Neptunian origin. 1st Order: Sand beds, Gravel beds, Sea salt, Sandstone, Pudding-stone, Brown coal, Bog Iron ore, Calcareous tufa, Calcareous depositions, and Silex from hot springs.

2d Order, resembling in structure, position or component parts, the 1st order, the evidence of their origin resting on direct and positive analogy: Coal, Gypsum, Chalk, Compact limestone, Sandstone, Puddingstone, Rock-Salt, Old Red Sandstone, Graywacke and Graywacke slate, Transition sandstone, Transition limestone, Transition gypsum, Transition clayslate, Anthracite and Siliceous schist.

2d Class—Volcanic origin. 1st Order, thrown out of active volcanoes, and resting on the evidence of our senses: Compact lava, Porous lava, Porphyritic lava, Scoria, Mud lava, Obsidian or Volcanic glass, Pumice-stone and cinders.

2d Order, resembling the 1st order in structure, position, and component parts, having the remains of craters, with currents of lava diverging from them; though the fire which may have formed them is now extinct, the evidence of their origin resting on direct and positive analogy: Basalt, Trap formation called by Werner the newest Flœts Trap formation, Pitchstone, Pearlstone, Porphyry attending the trap as above, and Clinkstone.

3d Order, where the rocks resemble the second in texture and component parts, but the proof of their origin resting on a more distant analogy: Basalt, Trap, Pitchstone, Porphyry, and Clinkstone.

3d Class—the origin doubtful, resembling a little, the 2d order of the 1st and 2d classes, but the analogy neither direct nor positive, amounting only to probable conjecture. 1st Order: such rocks as probable conjecture would incline to place in the Neptunian origin: Gneiss, Mica slate, Clay slate, Primitive slate and limestone.

2d Order, such rocks as probable conjecture would incline to place in the volcanic origin: Hornblende, Porphyry, Greenstone, Sienite and Granite.

The greatest good that this author accomplished may have resulted from constantly teaching that it is through observation, and not through the imagination, that a knowledge of Geology can be acquired. He said:

"The short period of time that mankind seem to have been capable of correct observation, and the minute segment of the immense circle of nature's operations, that has revolved during the comparatively short period, renders all speculations on the origin of the crust of the earth mere conjectures, founded on distant and obscure analogy. Were it possible to separate this metaphysical part, from the collection and classification of facts, the truth and accuracy of observation would be much augmented, and the progress of knowledge much more certain and uniform; but the pleasure of indulging the imagination is so superior to that derived from the labor and drudgery of observation—the self love of mankind is so flattered by the intoxicating idea of acting a part in the creation—that we can scarcely expect to find any great collection of facts, untinged by the false coloring of systems."

Very few facts, which now constitute the sciences of Geology and Palæontology were, at this time, known, and even later, theories and unwarranted assumptions constituted the greater part of what was taught as Geology, notwithstanding the exhortations of McClure, urging empirical study as against the injurious speculations and pretensions founded upon the imagination, or in the zeal to suppress investigation, because it seemed inimical to the teachings of the clergy. As a sample of what was taught, we may quote from Prof. Amos Eaton's "Index to the Geology of the Northern States," published in 1820.

He says, page 223 :

"I think I may say, with confidence, that the remains of two genera of animals, *Anomia* and *Pecten*, form, at the least, two thirds of all the secondary limerocks in North America. It may be deemed arrogant to include *all* the territory of this vast continent. But it has been my good fortune to see specimens of this rock from Canada to Mexico, and from Hudson's river to the Mississippi, taken from numerous localities. Perhaps I ought not, however, from these examinations, to infer that there may not be compact limestone of a great extent made up of different organic remains west of the Rocky mountain."

Again :

"Moses says, the Lord made 'every herb of the field before it grew,' ----'whose seed is in itself,' etc. This accords with the well known fact, that new plants are still springing up from seeds, probably planted at the creation, wherever forests are cut away and other steps taken to prepare particular patches of earth for giving growth to such particular plants. It is even said that pulverized rocks have been known to afford seeds, and to give growth to new plants. Perhaps this latter fact is not well authenticated."

In short, prior to about 1820, but little was known of North American Geology and Palæontology, and except as a matter of historical curiosity, rather than instruction, we need not seek these sciences in earlier publications.

The Mesozoic and Cænozoic rocks, to which this essay will be confined, constitute the superior one fourth part of the geological column, in the sedimentary strata, of the continent; the other three fourths belong to Palæozoic age. As a striking illustration: the upturned palæozoic strata, in the little state of New Hampshire, reveal a thickness twice that furnished by the Mesozoic and Cænozoic rocks throughout their extensive distribution to the remotest parts of the continent.

It will be observed in the sequel, that I have followed the chronological order of discovery, as near as practicable, with a view of presenting the history, the development and the growth of these sciences, as well as the facts, within the scope considered, upon which they are now supposed to rest.

First, we will pursue the Mesozoic rocks and fossils, and afterward the Cænozoic.

THE MESOZOIC AGE.

The Mesozoic age is divided into three periods, beginning with the earliest, as follows:

1. The Triassic Period.
2. The Jurassic Period.
3. The Cretaceous Period.

The name Triassic was given to the 1st Period in allusion to a three-fold division, which it presents in Germany. The Jurassic derives its name from the Jura mountains of Switzerland; and the name Cretaceous is derived from *creta*, chalk.

It will be convenient to consider the Triassic and Jurassic together, because the line of separation, at many places, still remains a matter of doubt, and because the rocks at one place, at one time, have been considered as Triassic, and at another as Jurassic, and even now great uncertainty exists as to their correct classification.

TRIASSIC AND JURASSIC.

In 1832, Prof. Edward Hitchcock * described the New Red Sandstone which extends across the State of Massachusetts, on both sides of the Connecticut river.

In 1833,* he referred all the sandstone in the valley of the Connecticut to the age of the New Red Sandstone of Europe. The opinion was

---

* Geo. of Mass.

fortified by the organic remains which had been collected at that time, as well as by the mineral character of the rocks. He described the rocks as micaceous sandstone, variegated sandstone, brecciated sandstone, shales, argillaceous slate and limestones. He discussed the dip, direction and thickness of the strata, and the occurrence of valuable minerals.

In 1836,* he described, from Massachusetts, *Ornithichnites giganteus,* now *Brontozoum giganteum, O. tuberosus,* and *O. tuberosus, var. dubius,* now *B. loxonyx,* and *B. sillimanium, O. ingens,* now *Tridentipes ingens, O. diversus,* now *Tridentipes elegans, O. minimus,* now *Argozoum minimum, O. palmatus,* and *O. tetradactylus.*

In 1839, Prof. H. D. Rogers† described the Red Sandstone of Pennsylvania, which stretches through the central and northern portions in a long and irregular tract, from New Jersey to Maryland. It is found in the vicinity of Reading, and near the Potomac river, from which place is quarried the famous Red Sandstone used in Washington city. Prof. Rogers proposed to call this the "Middle Secondary Red Sandstone formation," because it is higher than the Coal Measures, and below the Cretaceous Green Sand of New Jersey.

In 1841, W. C. Redfield‡ described, from the Connecticut Valley, *Palæoniscus macropterus,* now *Ischypterus macropterus, P. agassizi, P. ovatus, Catopterus anguilliformis, C. parvulus,* and *C. macrurus,* now *Dictyopyge macrura.*

In the same year, Prof. Hitchcock§ said the New Red Sandstone, extending from New Haven to the north line of Mass., in Northfield, occupies a narrow synclinal trough, having a width of about 20 miles, from East Hampton, in Massachusetts, to the Sound at New Haven; but from East Hampton to Northfield a width of only 6 or 7 miles. He described *Fucoides connecticutensis, F. shepardi, Sauroidichnites barratti, S. heteroclitus,* now *Ancyropus heteroclitus, S. minitans,* now *Plectropterna minitans, S. longipes, S. palmatus,* and *S. polemarchus,* now *Polemarchus gigas.* He used the word *Sauroidichnites* as a generic name, but described it as the name of a sub-order under the class *Ichnolite.* He also described *Ornithoidichnites* as a sub-order, and used it as a generic name, and described numerous species under it. These names have, however, been abandoned, and the species have also been abandoned or referred to genera properly defined. The

---

* Am. Jour. Sci. and Arts, vol. xxix.
† 3d Ann. Rep., Pa,
‡ Am. Jour. Sci. and Arts, vol. xli.
§ Geo. of Mass.

Ornithoidichnites are *O. giganteus*, *O. tuberosus*, *O. expansus*, *O. cuneatus*, *O. parvulus*, *O. ingens*, *O. elegans*, *O. deani*, *O. tenuis*, *O. macrodactylus*, *O. divaricatus*, *O. isodactylus*, *O. delicatulus*, *O. minimus*, *O. gracilior*, and *O. tetradactylus*. He afterward, before the Association of American Geologists and Naturalists, described some species under these names, which he subsequently referred to other genera.

In 1842, Prof. J. G. Percival* described the existence of these rocks in two places in Connecticut, as follows:

The larger secondary formation extends from Morris Cove, on the east side of New Haven Harbor, on the south, to the north end of Northfield village, in Mass., on the north, a distance of nearly 80 miles. Its greatest width, near the central part of the basin, exceeds 20 miles. This basin is entirely surrounded by Primary rocks, except at New Haven Harbor, where, however, Primary rocks form the two points on the opposite sides of the basin. The smaller secondary formation extends 6 to 7 miles from south to north, and at its widest point scarcely exceeds two miles in breadth, and is about equally included in the towns of Woodbury and Southbury. It forms a small isolated tract, nearly in the center of that part of the Western Primary, within the limits of the State, and nearly 15 miles west of the larger secondary formation. The rocks of both these formations consist of Red Sandstones, Conglomerates and Shales, and the physical characters and organic remains indicate a peculiar relation to the New Red Sandstone of Europe.

In 1843, Prof. W. W. Mather† described these rocks in the State of New York, as follows:

The New Red Sandstone occupies that portion of Rockland county, from Grassy point along the base of the Highlands to New Jersey, and eastward to the Hudson, but a portion of its area is covered over by trap rocks. It has also been found in a small area in Richmond county. In color, it varies from chocolate brown, through brick-red and gray to white; in texture, it varies from pebbly conglomerate, through common sandstone, fissile and micaceous sandstone, to shale; and in composition, from perfectly siliceous to an argillo-calcareous marl. Where the trappean rocks have cut through the strata, or have spread laterally between them, their texture and appearance are much modified, and appear to have been subjected to the action of heat, which has partly melted them, or rendered them more compact and hard, like a hard-burnt brick, or has made them metalliferous.

---

* Geo. of Conn.　　　　　　　　　† Geo. of N. Y.

In the same year, Prof. W. B. Rogers* described, from the Trias of Eastern Virginia, *Equisetum arundiniforme, Calamites planicostatus, Tæniopteris magnifolia, Zamites obtusifolius*, and *Z. tenuistriatus.*

In 1847, Sir Charles Lyell† described the Triassic coal field, on the James river, near Richmond, Virginia, as follows: The tract of country occupied by the crystalline or hypogene rocks, which runs parallel to the Alleghany mountains, and on their eastern side is in this part of Virginia about 70 miles broad; in the midst of this space the coal-field occurs in a depression of the granitic and other hypogene rocks, on which the coal rests, and by which it is surrounded, along its outcrop. The length of the coal-field, from north to south, is about 26 miles, and its breadth varies from 4 to 12 miles. The James river flows through the middle of it, about 15 miles from its northern extremity, while the Appomattox traverses it near its southern borders; on its eastern side it is distant about 13 miles from the city of Richmond; it occupies an elliptical area, the beds lying in a trough, the lowest of them usually highly inclined, where they crop out along the margin of the basin, while the strata higher in the series, which appear in the central part of the basin, are very nearly horizontal. The general strike is about N.N.E. and S.S.W., while that of the nearest ridges of the Appalachian chain is about N.E. and S.W.

A great portion of these coal measures consists of quartzose sandstone, and coarse grit, some of the beds, in the lower part of the series resembling granite or syenite, being entirely composed of the detritus of the neighboring granitic and syenitic rocks. Dark carbonaceous shales and clays, occasionally charged with iron ores, abound in the proximity of the coal seams, and numerous impressions of plants, chiefly ferns and zamites, are met with in shales, together with flattened and prostrate stems of Calamites and Equisetum. These last, however, the Calamites and Equisetum, are very commonly met with in a vertical position, more or less compressed perpendicularly. That the greater number of Calamites standing erect in the beds above and between the seams or beds of coal, which I saw at points many miles distant from each other, have grown in the places where they are now buried in sand and mud, I entertain no doubt. This fact would imply the gradual accumulation of the coal measures during a slow and repeated subsidence of the whole region.

The coal seams have hitherto been all found at or near the bottom of the series, and the plants in beds below or between them, or immediate-

---

* Trans. Ass. Am. Geo. and Nat.   † Quar. Jour. Geo. Sci., vol. iii.

ly overlying. One or two species of shells (Posidonomya?) also occur in the same part of the series, at a small height above the coal-seams, and above these a great number of fossil fish, chiefly referable to two nearly allied species of a genus, very distinct from any ichthyolite hitherto discovered elsewhere. Above these fossiliferous beds, which probably never exceed 400 or 500 feet in thickness, a great succession of grits, sandstone and shales of unknown depth occur. They have yielded no coal, nor as yet any organic remains, and no speculator has been bold enough to sink a shaft through them, as it is feared that toward the central parts of the basin they might have to pass through 2000 or 2500 feet of sterile measures before reaching the fundamental coal seams.

The coal is separated almost everywhere into three distinct beds, and sometimes into five. The upper bed is the thickest, except in a few places where a thin layer of coal is found above it. In some places the main seam of coal is from 30 to 40 feet thick, and at Blackheath it is seen actually to touch the fundamental granite, or is parted from it only by an inch or two of shale.

A section at the Midlothian Pit, half a mile south of Blackheath, on the eastern outcrop of the coal, is as follows: Sandstone and shale, 570 feet; slate with calamites, 1½ feet; sandstone and shale, 43 10.12 feet; sandstone with calamites, 8 feet; sandstone and slaty shale, 48 feet; slate and long vegetable stems, 2½ feet; sandstone, 6½ feet; slate with calamites, 5½ feet; sandstone, 14 feet; black rock, 13 feet; slate, 5 feet; main coal, 36 feet; sandstone not laminated, 5 feet; slate, 4 feet; coal, 1 foot; slate, 3 feet; sandstone or grit, 7 feet. Total, 773 10-12 feet. This rests upon granite of unknown depth. Some deductions must be made for the thickness of the beds on account of the inclination at an angle of 20 degrees.

The unevenness of the granite floor is extremely great, and the thickness of the coal seams quite variable. The disturbances have been extremely great, and dikes of greenstone occur in some places 20 feet in thickness. Some of the upper beds of coal have been reduced to coke, by being deprived of their volatile matter, while others below remain unaltered and bituminous. This is accounted for on the ground that the greenstone, although intrusive, has made its way between the strata like a conformable deposit, and has driven the gaseous matter from the upper coal, while its influence has not extended to lower seams. A remarkable example of coke, in a bed eight feet in thickness, occurs at Edge-hill, a locality between five and six miles north of James river, and ten miles north of Blackheath, being on the eastern outcrop of the basin, and within 500 yards of the granite. The measures passed through above the 8 feet bed of coke, are 110 feet thick, including a conformable bed of blue basalt, 16 feet thick. The shale immediately below the trap is white for 11 feet, and then 25 feet of dark, leafy shale succeed, below which comes the bed of coke, resting on white shale; and lower down, coal-measures with two seams of inferior coal, each about 4 or 5 feet thick. The shale, 47 feet thick, interposed between the basalt and the coke, exhibits so many polished surfaces or slickensides, and is so much jointed and cracked, and in some places disturbed and tilted, that we may probably attribute the change from coal to coke, not so much to the heating agency of the intrusive basalt, as to its mechanical effect in breaking up the integrity of the beds, and rendering them permeable to water or the gases of decomposing coal.   In some places, in the same district, where the upper part of a seam is coke, the lower is coal, and there is sometimes a gradation from the one to the other, and sometimes a somewhat abrupt separation.

In the same year, C. J. F. Bunbury * described, from North Carolina, *Neuropteris linnœifolia, Pecopteris bullata, Filicites fimbriatus,* and *Zamites gramineus.*   And Prof. Hitchcock† described, from Massachusetts, *Brontozoum moodi,* and *B. parallelum.*   He also discussed the Trap Tuff or Volcanic grit of the Connecticut valley, with the bearing of its history upon the Trap Rock and the Red Sandstone.

In 1848, Prof. J. W. Dawson‡ described the New Red Sandstone of Nova Scotia, which extends on the north side of Cobequid bay, from Moose river to the point at the mouth of North river, and on the south side, from the mouth of Shubenacadie to the mouth of North river. It rests upon carboniferous strata, and, in some places, presents cliffs rising to an eminence of 400 feet. It is also extensively developed at Blomidon, in the valley of Cornwallis, on the south side of the Bay of Fundy, and at other places. This sandstone appears to have been deposited in an arm of the sea, somewhat resembling, in its general form, the southern part of the present Bay of Fundy, but rather longer and wider. This ancient bay was bounded by disturbed Carboniferous and Silurian strata. The evidences of volcanic action are numerous, and in some places showing great quantities of melted rock brought to the surface, without altering the soft arenaceous beds through which it has been poured, and whose surface it has

---

* Quar. Jour. Geo. Soc., vol. iii.
† Am. Jour. Sci. and Arts. 2d Ser., vol. iv.
‡ Quar. Jour. Geo. Soc., vol. iv.

overflowed. The Sandstone contains no valuable minerals, and no fossils had then been detected in it.

In 1853, Isaac Lea* described, from the Triassic of Lehigh county, Pennsylvania, *Clepsysaurus pennsylvanicus.*

In 1854, Dr. Joseph Leidy† described, from the Triassic of Prince Edward Island, *Bathygnathus borealis.*

In 1855, Prof. J. W. Dawson described Prince Edward Island, which stretches for 125 miles along the northern coast of Nova Scotia and New Brunswick, has everywhere a low, undulating surface, and consists almost entirely of soft red sandstone and arenaceous shale, much resembling the new red of Nova Scotia, and like it having the component particles of the rock united by a calcareous cement. In some places the calcareous matter has been in sufficient abundance to form bands of impure limestone, usually thin and arenaceous. Over the greater part of the island these beds dip at small angles to the northward, with, however, large undulations to the south, which probably cause the same beds to be repeated in the sections on the opposite sides of the island.

In the same year, Dr. E. Hitchcock, jr.‡ described *Clathropteris rectiusculus,* from the sandstone of Mt. Tom, in Easthampton, Mass., of the age of the lower Jurassic.

In 1856, Prof. E. Emmons§ described, from the Lower Triassic of the Deep and Dan river beds of North Carolina, *Chondrites gracilis, C. interruptus, C. ramosus, Gymnocaulus alternatus, Equisetum columnaroides, Dictuocaulus striatus, Rutiodon carolinensis, Clepsysaurus leai, Palæosaurus carolinensis, P. sulcatus* and *Posidonia ovalis,* now referred to the genus *Estheria,* and from the Upper Triassic of the Deep and Dan river beds, *Strangerites obliquus, Acrostichites oblongus, Pecopteris carolinensis, P. falcata, Pterozamites decussatus, Cycadites acutus, C. longifolius, Zamites graminioides, Podozamites lanceolatus, P. longifolius, Lepacyclotes circularis, L. ellipticus, Walchia diffusa, W. longifolia, Calamites disjunctus, Sphenoglossum quadrifolium,* and *Posidonia multicostata,* and *P. triangularis,* which are now regarded as synonyms or varieties only of *Estheria ovalis.*

And in 1857‖ he described, from North Carolina, *Calamites punctatus, Walchia angustifolia, W. variabilis, W. brevifolia, W. gracilis, Sphenopteris egyptiaca, Cyclopteris obscura, Odontopteris tenuifolia.*

---

* Jour. Acad. Nat. Sci., 2d Ser., vol. ii.
† Jour. Acad. Nat. Sci., 2d Ser. vol. ii.
‡ Am. Jour. Sci. and Arts, 2d Ser., vol. xx.
§ N. Carolina Sur.
‖ Am. Geo., pt. 6.

Pterozamites gracilis, P. obtusus, P. linearis, P. spatulatus, Dioonites linearis, Strangerites planus, Pterophyllum robustum, Noeggerathia striata, Comephyllum, cristatum, Amblypterus ornatus, Rabdiolepis speciosus, Microdus lævis, Palæonornis struthionoides, and *Dromatherium silvestre,* the most ancient mammalian remains yet found upon the continent.

In 1857, T. A. Conrad* described, from the Triassic black shale at Phœnixville, Pennsylvania, *Myacites pennsylvanicus.*

In 1858, Meek and Hayden† described, from the Jurassic of the Black Hills, *Pentacrinus asteriscus, Lingula brevirostra, Avicula tenuicostata, Mytilus pertenuis, Arca inornata,* now *Grammatodon inornatus, Panopæa subelliptica,* now *Myacites subellipticus, Ammonites cordiformis, A. henryi,* and *Belemnites densus.*

Prof. Hitchcock made his report on the Ichnology of New England, being " A report on the Sandstone of the Connecticut valley, especially its Fossil Footmarks, made to the government of the Commonwealth of Massachusetts." This work contains a bibliography of North American Fossil Footmarks; the history of the discoveries of the tracks; a discussion of the geological position of the Connecticut river sandstone, and the evidences tending to prove the Jurassic Age of at least the upper half of the strata, with geological sections across the valley, showing that in general the dip is easterly, varying from 5° to 50°.

The sandstone of the Connecticut valley extends from Northfield, in the Northern part of Massachusetts, across the latter State, and Connecticut to Long Island Sound, a distance of 105 miles. The greatest width is at the mouth of the Farmington river, though Hitchcock's Springfield section was taken where the width is nearly as great. Several ranges of trap rock (greenstone, amygdaloid, and volcanic grit), traverse the sandstone longitudinally, having for the most part a northeasterly trend, and being generally in the form of interstratified beds or masses. Along the west side of the valley, there is a coarse, thick-bedded sandstone, whose prevailing color is red, but which is sometimes mottled, and near the trap and the hypozoic rocks, sometimes nearly white. This sandstone underlies the trap. Immediately above the trap, on the east side of the valley, the rocks consist of interstratified red and black shales, volcanic grit, micaceous sandstone, compact, fetid blue and gray limestone, and in some places coarse sandstone and conglomerate. It is in the shales and sandstones lying im-

---

* Proc. Acad. Nat. Sci., vol. ix.
† Proc. Acad. Nat. Sci.

mediately above the trap, with very few exceptions, that the organic remains—the fishes, the tracks, and plants—are found. His sections show the thickness of the sandstone above and below the trap, as follows:

	Feet.
Turner's Falls section, above	4,190
" " " below	7,788
Mettawampe section, above.	1,584
" " below	5,283
Mount Tom section, above	8,102
" " below	5,115
Agawam and Chicopee or Springfield section, above	11,500
" " " " " " below	8,128

The rock below the trap seems, from the evidences adduced, to be of Triassic Age. He argues that the strata of sandstone were not deposited in their present inclined position, and subsequently elevated, and that the sandstone was not elevated or tilted up by the eruption of the trap rock; but, on the contrary, that the lower beds of sandstone were deposited, and perhaps somewhat tilted up, when the trap was ejected from beneath, and spread over the upper part of the strata, and that afterward the work of depositing the sandstone was resumed, and that which lies above the trap laid down. New outbursts of the trap, however, occurred at subsequent periods, but less in quantity, as if the eruptive force were dying out. This is followed by a very learned essay upon the constant and distinctive characters in the feet of animals, and the application of the rules laid down, to the footmarks, which he described and illustrated. He called these tracks *Lithichnozoa*—stone-track animals; or animals made known by their tracks in stone.

The longest trough, and greatest exposure, in the Eastern States, begins at Stony Point, on the Hudson, and extending across New Jersey, Pennsylvania and Maryland, reaches Culpepper county, in Virginia. It has a length of about 350 miles, and though frequently narrowing to a breadth of four or five miles, it expands, in New Jersey, to a width of about thirty six miles. The character of the deposit very much resembles that of the Connecticut valley. The other Virginia deposit exists in Henry, Pittsylvania, Halifax, Prince Edward and Buckingham counties.

Prof. Emmons first ascertained the extent and general character of the two basins of Triassic strata, in North Carolina. One is in Stokes and Rockingham counties, bordering on Virginia. It begins at Leaksville, and runs about thirty miles southwest to Germantown, and is from four to six miles wide. The other commences six miles south of Oxford, in Granville county, and runs southwest through a part of

Orange, Chatham, Moore, Montgomery, Richmond and Anson counties, and extends about six miles into South Carolina. Its length is about 120 miles, and it has a breadth, in the widest part, of 18 miles, though its width is generally about six miles.

In 1859, Major Hawn* gave a section in Kansas, of rocks 410 feet in thickness, which he referred to the Triassic. But Dr. Mudge has maintained since that time, that the cretaceous rocks rest directly upon the Permian, in that State.

In 1860, Meek & Hayden† described, from the Jurassic, at the south-west base of the Black Hills, *Pholadomya humilis*, *Myacites nebrascensis*, *Thracia arcuata*, *T. sublævis*, *Cardium shumardi*, *Tancredia æquilateralis*, *T. warrenana*, *Astarte fragilis*, *A. inornata*, *Trigonia conradi*, *Pecten extenuatus*, now *Camptonectes extenuatus*, and from Red Buttes, on the North Platte, *Ostrea engelmanni*, *Pecten bellistriata*, now *Camptonectes bellistriatus*, and *Dentalium subquadratum*.

And Wm. M. Gabb‡ described, from the Triassic in Bath county, Virginia, *Ceratites virginianus* and *Rhynchonella halli*.

In 1861, Dr. F. V. Hayden,§ in his reconnoissance of the country about the headwaters of the Missouri and Yellowstone rivers, found the red arenaceous deposits, usually referred to the Triassic age, exposed in outcropping belts, from one to two miles wide, around the margins of the mountain elevations, but not generally otherwise exposed. They occur on the northeastern side of the Big Horn mountains, on the west slope of the Wind River mountains, along the mountains at the source of the Missouri, around the Judith mountains, and at numerous other places. Frequently thick layers of gypsum are found in the deposits. The thickness observed is from 1000 to 1500 feet.

He also found the Jurassic rocks overlying the red arenaceous beds, referred to the Triassic, and possessing an equal geographical extension. They are found along the margins of the Black Hills, along the northeastern slope of the Big Horn mountains; at Red Buttes; along the southwest side of the Big Horn, and the northeast side of the Wind River mountains, sometimes having a thickness of 1000 feet, and containing organic remains in the greatest abundance.

In the same year, Meek & Hayden‖ described, from the Jurassic, at

---

* Proc. Am. Ass. Ad. Sci.
† Proc. Acad. Nat. Sci.
‡ Jour. Acad. Nat. Sci., 2d Ser., vol. iv.
§ Am. Jour. Sci. and Arts, 2d Ser., vol. xxxi.
‖ Proc. Acad. Nat Sci.

the head of Wind River valley, *Gryphæa calceola*, var. *nebrascensis*, *Modiola formosa*, now *Volsella formosa*, *Neritella nebrascensis*, *Melania ceterna*, now *Lioplacodes ceterna*.

In 1864, F. B. Meek* described, from the Jurassic, of California, *Rhynchonella gnathophora*, *Lima sinuata*, *L. recticostata*, *L. cuneata*, *Pecten acutiplicatus*, *Inoceramus obliquus*, *I. rectangulus*, *Trigonia pandicosta*, *Mytilus multistriatus*, *Astarte ventricosa*, *Unicardium gibbosum*, *Myacites depressus*. And W. M. Gabb† described *Lima erringtoni*, now *Aucella erringtoni*, and *Belemnites pacificus*.

And from the Triassic rocks,‡ in the Buena Vista District, and the Humboldt Mining Region of Nevada Territory, *Orthoceras blakei*, *Nautilus whitneyi*, *N. multicameratus*, *Ceratites whitneyi*, *Ammonites blakei*, *A. homfrayi*, *A. billingsanus*, *Myacites humboldtensis*, *Corbula blakei*, *Mytilus homfrayi*, *Avicula homfrayi*, *Halobia dubia*, *Rhynchopterus obesus*, *Posidonomya stella*, *P. daytonensis*, *Myophoria alta*, *Terebratula humboldtensis*, *Rhynchonella lingulata*, *R. æquiplicata*, *Spirifera homfrayi*, from Sonora Mexico, *Panopæa remondi*, from Gifford's Ranch, Plumas county, California, *Avicula mucronata*, *Monotis subcircularis*, *Pecten deformis*.

In 1865, F. B. Meek§ described, from the Jurassic, near the southwest base of the Black Hills, *Viviparus gilli;* from the auriferous slates on the Mariposa estate of California,‖ *Aucella erringtoni*, var. *linguiformis*, and *Amussium aurarium*. J. D. Whitney referred the auriferous rocks of El Dorado, Mariposa, and Toulomne counties, California, to Jurassic or Triassic age. And Bailey and Matthew¶ showed that the Trias of New Brunswick consists of three small patches, on the coast of the Bay of Fundy, one at Quaco Head, another at Gardner's Creek, and the other at Salisbury Cove.

In 1866, Prof. E. D. Cope** described, from the Triassic, at Phœnixville, Pa., *Mastodonsaurus durus*, now *Eupelor durus*, and *Pterodactylus longispinus*.

In 1867, Prof. Swallow†† found, in eastern Kansas, what he supposed to be the Triassic, consisting of a series of buff, red and

---

* Pal. of Cal., vol. i.
† Proc. Cal. Acad. Sci.
‡ Pal. of Cal., vol. i.
§ Pal. Up., Mo.
‖ Geo. Sur. Cal., vol. i.
¶ Rep. on S. N. Brunswick.
** Proc. Acad. Nat. Sci.
†† Proc. Am. Ass. Ad. Sci.

mottled sandstones, red and drab marls, buff, magnesian and black limestones, blue and brown shales and gypsum, 344 feet in thickness. These rocks extend in an irregular belt across the State, from the head waters of the Blue and Fancy, across the Republican and Solomon, and over the Kansas, between Turkey Creek and the Saline; thence south and southeasterly up the Smoky Hill and Gypsum, Holland and Turkey Creeks; along the northern slope of the divide, south of the Kansas, to the heads of Lyon and Diamond Creeks; sweeping thence westward across the Cottonwood and down the divide, south of that stream, to the Walnut and White Water. The gypsum beds vary in thickness from 0 to 50 feet, and crop out on the Blue, the Republican, and the Kansas, and on Turkey Creek; and on the divides between the Gypsum and Holland, and between Turkey Creek and the Cottonwood.

In the same year, Dr. F. V. Hayden * referred the celebrated Pipestone quarry of northeastern Dakota, to the Triassic, and showed that the manufacture of it into pipes commenced by the Indians, at a quite recent date—probably within the last 50 or 100 years. The pipestone is called Catlinite.

The Triassic rocks of New Jersey † are included in a belt of country which has the Highland Range of mountains on its northwest side, and a line almost straight from Staten Island Sound, near Woodbridge, to Trenton, on its southeast; the Hudson river on the northeast, and the Delaware on the southwest. The length of the southern border line is 74 miles; that on the northwest is 68 miles. These measurements are from the Delaware river to the State line. Its greatest breadth is on the Delaware, where it is over 30 miles across. From Mine mountain to the Raritan river, near the mouth of Lawrence Brook, its breadth is 19 miles. On the State line, from the Hudson river to Sufferns, it is 15 miles. The area embraced within these limits, excluding the bays, is about 1500 square miles. Of this about 330 square miles are occupied by trap rock. It consists of red sandstone, and is fossiliferous, at Pompton, Boonton, Milford, Tumble Station, Belleville, Newark, Pluckamin and other places.

The ordinary way of computing the thickness of a rock formation is to take its dip, and also the breadth of country across which this dip is continued, and use them as two parts of a right-angled triangle forgetting the remaining parts, one of which is the perpendicular thick-

---

* Am. Jour. Sci. and Arts, 2 ser., vol. xliii.
† Geo. of N. Jersey, 1868.

ness of the rock. The red sandstone has an average dip along the Delaware river, of at least 10 degrees, for 30 miles. This would give a thickness of 27,000 feet for this formation, or more than five miles. If the mode of computation is accepted, the result must be received as correct. Those who think the strata were once horizontal, and that they have been thrown into their present inclined position at some later period, adopt this conclusion without hesitation. Others who think the strata were deposited on a slope as we now find them, do not consider the above to be the true thickness. They suppose that the strata along the southeast border were first deposited on this northwest slope; and then that the upper edges were worn off, and the material carried farther northwest to be again deposited, and form new strata upon the lower parts of those already deposited. Without any addition of material there would in this way be a multiplication of strata, all having the same dip. And such a process could go on until the formation had widened out to its present extent.

The aqueous rocks of the new red sandstone period, in Nova Scotia and Prince Edward Island,* are principally coarse and soft red sandstones, with a calcareous cement, which causes them to effervesce with acids, and contributes to the fertility of the soils formed from them. In the low part of the formation, there are conglomerates made up of well-worn pebbles of the harder and older rocks.

The volcanic rocks of this period are of that character known to geologists as trap, and are quite analogous to the products of modern volcanoes; and, like them, consist principally of *Augite*, a dark green or blackish mineral, composed of silica, lime and magnesia, with iron as a coloring material. Various kinds of trap are distinguished, corresponding to the varieties of modern lavas. Crystalline or basaltic trap is a black or dark green rock, of a fine crystalline texture, and having on the large scale a strong tendency, to assume a rude columnar or basaltic structure. Amygdaloid or almond-cake trap is full of round or oval cavities or air bubbles, filled with light colored minerals introduced by water after the formation of the rock. This represents the vesicular or porous lava which forms the upper surface of lava currents, just as the basalt trap represents the basaltiform lava which appears in their lower and more central parts. The only difference is, that in the amygdaloid the cavities are filled up, while in the modern lavas they are empty. In some old lavas, however, the cavities are already wholly or partially filled. A third

---

* Acadian Geology, 1868.

kind of trap, very abundant in Nova Scotia, is Tufa or Tuff, or volcanic sandstone, a rock of earthy or sandy appearance, and of gray, greenish or brown color. It consists of fine volcanic dust, and scoriæ, popularly known as the ashes and cinders of volcanoes, cemented together into a somewhat tough rock. Modern tufa, quite analogous to that of the trap, is very abundant in volcanic countries, and sometimes sufficiently hard to be quarried as a stone.

In the valley of the Salmon river, 4½ miles eastward of the village of Truro, the eastern extremity of the New Red Sandstone is seen to rest unconformably on hard, reddish, brown sandstones and shales, belonging to the lower part of the Carboniferous system, and dipping N. 80 degrees, E. at an angle of 40 degrees. At this place the overlying formation is nearly horizontal, and consists of soft and rather coarse, bright, red, silicious sandstones. Southward of Truro, at the distance of less than a mile, the horizontal soft red sandstone is seen in the banks of a brook, to run against hard, brownish grits and shales, dipping to the eastward at angles varying from 45 to 50 degrees. Westward of this place, the red sandstones extend in a narrow band, about a mile in width, to the mouth of the Shubenacadie, ten miles distant. This band is bounded on the North by Cobequid Bay, and on the South by highly inclined sandstone, shale, and limestone of the Lower Carboniferous series. In the coast section, between Truro and the Shubenacadie, the red sandstone presents the same characters as at the former place, except that near the Shubenacadie, some of the beds, which, like most of the red sandstones of Truro, have a calcareous cement, show a tendency to arrangement in large concretionary balls. West of the mouth of the Shubenacadie, the red sandstone ceases to form a continuous belt, but occurs in several patches, especially at Salter's head, Barncote and Walton. At the latter place, it is seen to rest on the edges of sandstones and other rocks of the Lower Carboniferous system, affording a very fine example of that unconformable superposition, which, in Geology, proves the underlying formation to have been elevated and disturbed before the overlying beds were deposited upon it.

Westward of Walton, the estuary of the Avon river and Minas basin make a very wide gap in the new red sandstone. On the western side of Minas basin, however, this formation attains its greatest width and grandest proportions. Blomidon is the eastern extremity of a long band of trappean rocks, forming an elevated ridge, named in the greater part of its length the " North Mountains." This ridge is about 123 miles in length, including two insular portions at its western extremity, and does not exceed five miles in breadth, except near Cape

Blomidon, where a narrow promontory, terminating in Cape Split, extends to the northward. The trap of the North Mountains presents to the Bay of Fundy, a range of high cliffs, and is bounded on the inland side by soft red sandstones, which form a long valley separating the trappean rocks from another and more extensive hilly district, occupied principally by metamorphic slates and granite. The trap has protected the softer sandstones from the waves and tides of the bay, and probably also from older denuding agents; and where it terminates, the shore at once recedes to the southward, forming the western side of the Minas basin, and affording a cross section of the North Mountains and the valley of Cornwallis.

At Cape Blomidon, the cliff, which in some parts is 400 feet in height, is composed of red sandstone surmounted by trap. The sandstone is soft, arranged in beds of various degrees of coarseness, and is variegated by greenish bands and blotches. It contains veins of selenite and fibrous gypsum, the latter usually parallel to the containing beds, but sometimes crossing them obliquely. It dips to the N. W. at an angle of 16 degrees. Resting on the sandstone, and appearing to dip with it to the N. W., is a thick bed of amygdaloidal trap, varying in color from gray to dull red, but in general of grayish tints. It is full of cavities and fissures; and these, as well as its vesicles, are filled or coated with quartz, in different states, and with various zeolites, especially heulandite, analcime, natrolite, stilbite, and apophylite, often in large and beautiful masses of crystals. In its lower part there are some portions which are scarcely vesicular, and often appear to contain quartz sand like that of the subjacent sandstone. Above the beds of amygdaloid is a still thicker stratum of crystalline basaltic trap, having a rude columnar structure.

The columnar trap of Blomidon, in consequence of its hardness and vertical joints, presents a perpendicular wall, extending along the top of the precipice. The amygdaloid beneath, being friable and much fissured, falls away in a slope from the base of this wall, and the sandstone in some places forms a continuation of the slope, or is altogether concealed by the fallen fragments of trap. In other places, the sandstone has been cut into a nearly vertical cliff, above which is a terrace of fragments of amygdaloid.

Northward of Cape Blomidon, the northwesterly dips of the sandstone and trap cause the base of the former to descend to the sea-level, the columnar trap, which here appears to be of increased thickness, still presenting a lofty cliff. Southward of the Cape, on the other hand, the amygdaloid and basalt thin out, until the red sandstones occupy the whole of the cliff. It thus appears that the trap at Blomidon is a comformable bed, resting on the sandstone, exactly as in some places on the opposite shore.

The coast section between Blomidon and Horton, as seen near Pereau river and Bass creek, and at Starr's Point, Long Island and Bout Island, exhibits red sandstones, with northwest dips at angles of about 15 degrees, and precisely similar in mineral character to those of Blomidon, except that near Bass creek some of them contain layers of small pebbles of quartz, slate, granite and trap. The whole of these sandstones underlie those of Blomidon, and resemble those which occupy the long valley of Cornwallis and the Annapolis river, westward of this section. In this valley, the red sandstone, in consequence of its soft and friable nature, is rarely well exposed, but where observed, it has the same dip as on the coast. The comparatively high level of the sandstone, where it underlies the trap, shows that the present form of this valley is in great part due to denudation.

Commencing at Truro, the New Red Sandstone extends with several interruptions, as far as Cape d'Or. It consists of a narrow strip extending only about three miles from the bay, with occasional masses of trap. At Cape d'Or a great mass of trap rests on slightly inclined red sandstone, and this again on disturbed carboniferous rocks, while, behind and from beneath these last, still older slates rise into mountain ridges. Cape d'Or forms a great salient mass standing out into the bay, and separated from the old slate hills behind, by a valley occupied by the red sandstone and carboniferous shales. It differs from most of the trappean masses in the arrangement of its component parts. The upper part of the cliff consists of amygdaloid and tufa, often of a brownish color, while beneath is a more compact trap, showing a tendency to a columnar structure.

The small patches of New Red Sandstone on the New Brunswick side of the Bay of Fundy, with the deposits in Nova Scotia, show that the depression occupied by the Triassic Bay was of similar form (though somewhat enlarged probably) to that occupied by the present Bay of Fundy.

Dr Joseph Leidy* described, from the Triassic rocks of Star Canon. Humboldt county, and from Toiyabe Range northeast of Austin, Nevada, *Cymbospondylus petrinus, C. piscosus* and *Chonespondylus grandis.* And Prof. E. D. Cope, from Chatham county, North Carolina. the batrachian *Pariostegus myops.*

In 1869, T. A. Conrad† referred the clays on the Raritan river, in New Jersey, which are found at the base of the Cretaceous, to the Triassic, and described *Podozamites proximus* and *Palœocypris trinodiferus.* He described from South river,‡ New Jersey, *Astarte veta,* and *A. annosa,* and from Perkiomen Creek, Pa.,§ *Solenomya triasina.*

Prof. E. D. Cope,‖ from the Triassic at Phoenixville, on the Schuylkill river, Pa., the Saurian *Belodon lepturus.*

In 1870, W. M. Gabb¶ described, from New Pass, near Austin, and in the slates of Star Canon, *Cardinia ponderosa, Posidonomya blatchleyi, Cassianella lingulata* and *Monotis circularis.* And from the Jurassic of the mining district of Volcano. in Nevada, *Ammonites neradensis. Turbo regius, T. elevatus, Pholadomya multilineata, P. nevadana, Goniomya aperta, Cardium arciformis, Astarte appressa, Plicatula perimbricata,* and *Spirifera obtusa;* and from the slates on the west slope of the Sierra Nevada, near Colfax, *Ammonites colfaxi.*

In 1871, Prof. J. W. Dawson** described from the Trias of Prince Edward Island, *Dadoxylon edwardanum* and *Cycadoidea abequidensis.*

In 1872,†† he said the Trias of Prince Edward Island is represented principally by bright red sandstone, sometimes mottled with white and associated occasionally with beds of gray and white sandstone. Subordinate to these sandstones are beds of red and mottled clay, of reddish concretionary and conglomerate limestone, sometimes dolomitic and of reddish conglomerate with quartz pebbles and arenaceous

---

* Proc. Acad. Nat Sci.
† Am. Journal Sci. and Arts, 2d series, vol. 47,
‡ Am, Jour. Conch., vol. 4.
§ Am. Jour. Conch., vol. 5.
   Trans. Am. Phil. Soc.
‖ Am. Jour. Conch., vol. 5.
** Report on Prince Edward Island.
†† Lond. Geo. Mag., vol. 9.

cement.   These beds undulate in low synclinals and anticlinals, having in general a northeast and southwest direction, and rise in some places to an elevation of 400 feet above the sea.   They are probably about 500 feet in thickness.   The lower half of this thickness, which contains the limestone beds, and also certain hard beds of conglomerate and concretionary calcareous sandstone, may be regarded as an equivalent of the Bunter Sandstone ; while the upper portion, consisting principally of soft red sandstone, with some beds of fine grained conglomerate may be regarded as corresponding to the Keuper.

These beds rest conformably upon the newer coal measures without the intervention of the Permian.   They appear to have been deposited in a shallow sea area, not improbably coincident with the Southern Bay of the Gulf of St. Lawrence, limited to the north by the Magdalen Islands and the banks in their vicinity, which represent an old Lower Carboniferous outcrop.   Their materials were derived from the waste of red sandstones and marls of the Carboniferous, and have been thrown down with sufficient rapidity to prevent the coating of red oxide of iron from being removed by abrasion, or by the chemical action of organic matter.   The dolomitic character of some of the coarse limestones may either indicate the occurrence of occasional isolated basins and depositions of magnesia from sea water, or may have been connected with the outburst of igneous matter in magnesia, like the dolerite of Hog Island, near to which place the beds richest in magnesia were observed.

In 1872, F. B. Meek[*] described from the Jurassic, at Lincoln Valley, near Fort Hall, Idaho, *Aviculopecten idahoensis.*

In 1873, Dr. F. V. Hayden estimated the thickness of the Jurassic, on the Missouri, below the Canon at the Three Forks at 1,500 feet.   A section, in Spring Canon, on the headwaters of the Gallatin river in Montana, of limestones, sandstones, quartzites and conglomerates, displays a thickness of 425 feet, followed below by 65 feet of Triassic age.[†] And F. B. Meek described from Montana *Gervillia montanensis, Goniomya montanensis, Myacites subcompressus, Pholadomya kingi, Trigonia americana, T. montanensis,* and *Volsella subimbricata.*

In 1874, Dr. Hayden[‡] estimated the thickness of the Triassic on Eagle river, consisting of brick-red sandstones and clays at from 1,200 to 1,500 feet, and above them 200 feet or more of Jurassic rocks, suc-

---

[*] 5th Rep. Hayden's U. S. Geo. Sur. Terr.
[†] 6th Rep. Hayden's U. S. Geo. Sur. Terr.
[‡] 7th Rep U. S. Geo. Sur. Terr.

ceeded by a quartzite belonging to the Dakota Group, having a thickness of 150 feet. At Little Thompson's creek it consists of soft granite sandstones and conglomerates below, followed by red, shaly and massive sandstones above, and reposes upon the smoothed and often irregular surface of Archaean rocks. It has a thickness of 750 feet. It thins out north of Golden City, where it has a thickness of scarcely 400 feet, but rapidly thickens in its extension southward to where the South Platte debouches from the mountains; here it attains a thickness of 1,600 feet. Dr. A. C. Peale's section through Pleasant Park represents its thickness at 1,580 feet, and from Glen Eyrie eastward to Camp creek, 1,280 feet.

A section of the Jurassic rocks, taken by Wm. H. Holmes near Saint Vrain's Creek, gave a thickness of between 400 and 500 feet; and another in Bear Cañon, 870 feet; another near Ralston creek, 660 feet; and another near Bear Creek, 770 feet. Dr. A. C. Peale's section through Pleasant Park furnishes a thickness of about 461 feet, and from Glen Eyrie eastward to Camp Creek, 405 feet.

Prof. E. D. Cope[*] detected the first vertebrate fossils in the Trias of the Rocky Mountains of New Mexico, including carnivorous Saurians and Unionidæ, the latter indicating a lacustrine deposit.

In North Carolina [†] there are two narrow fringes of an eroded and obliterated anticlinal which belong to the Triassic; the smaller or Dan river belt, from 2 to 4 miles wide, following the trough-like valley of that stream, about N. 65° E. for more than 30 miles, to the Virginia line, and then extending into Virginia about 10 miles; the other, the Deep river belt, extending in a similar trough 5 to 15 miles wide (and depressed 100 to 200 feet below the general level of the country), from the southern boundary of the State in Anson county, in a N. E. direction, to the middle of Granville county, within 15 miles of the Virginia line. They are separated, therefore, by a swell of country 75 to 100 miles wide, which rises along its topographical axis to 800 or 900 feet above the sea, the troughs themselves having respectively an elevation of 500 to 600 feet, and 200 to 300 feet. The belts are convergent in the direction of the Triassic beds of Virginia, with which they were doubtless once connected (as well as with some small intervening outliers) in one continous formation.

The dip of the Dan river beds is about 35° N. W. (20° to 70°) and those of Deep river, 20° S. E. (10° to 35°). The rocks are sandstones,

---

[*] Proc. Am. Phil. Soc.
[†] Kerr's Geo. of N. Carolina, 1875. Emmons Geo. Sur. 1856.

clay slates, shales and conglomerates, generally ferruginous and brick red, but often gray and drab. The shales are occasionally marly, and these and the sandstones are sometimes saliferous. Many of the beds consist of loose and uncompacted materials, and are therefore easily abraded.

The most important and conspicuous member of the series, is a large body of black shales, which enclose seams of bituminous coal 2 to 6 feet. This coal lies near the base of the system in both belts, and is underlaid on Dan river by shales; and on Deep river by sandstones and conglomerates; the latter constituting the lowest member of the series, and being in places very coarse. And near the eastern margin in Wake county, where the belt reaches its greatest breadth (some 15 miles), the conglomerates are of great thickness and very coarse, uncompacted and rudely stratified, resembling somewhat the half stratified drift of the mountain slopes, the fragments often little worn, and sometimes 10 to 12 inches in diameter, and evidently derived from the Huronian rocks of the hills to the eastward. The conglomerates of the Dan river belt are among the upper members of the series, and are mostly fine and graduating to grits and sandstones.

The black shales near the base of the system contain beds of fire clay and black band iron ore, interstratified with the coal. They are also highly fossiliferous, especially on Deep river. Silicified trunks of trees are very abundant in the lower sandstones, as may be seen conspicuously near Germantown, in Stokes county, the public road being in a measure obstructed by the multitude of fragments and entire trunks and projecting stumps of a petrified Triassic forest; and similar petrifactions are abundant in the Deep river belt, occurring in this, as in the other, among the sandstones near the horizon of the coal.

The actual vertical depth to the underlying Archæan rocks on Dan river may not exceed 1000 feet, but what was the original thickness of the strata before denudation began can only be conjectured. The beds on Dan river, however, measured at right angles to the dip, gives a minimum thickness for that side of the formation of near 10,000 feet. In the section of the Deep river belt, which is exposed in the valley of the Yadkin, not only is there a width of six miles with the usual dip of 20°, but there is an additional outcrop more than a mile in breadth, *ten miles south* of the principal belt, which preserves the southeasterly dip of nearly 20°, and hence the calculation for a minimum thickness, at this margin, must be based on a breadth of 16 miles, which gives a thickness of more than 25,000 feet.

There is no way of accounting for the present position of these beds

with their opposite and considerable dips, but by supposing an uplift
of the intervening tract, such and so great, that if the movement were
now reversed, it would carry this swell of nearly 100 miles breadth
into a depression much below the present level of the troughs in which
these remnant fringes lie, so that there has been an erosion not only of
10,000 to 20,000 feet of the broken arch of Triassic beds over this area
but also of a considerable thickness of the underlying rocks on which
they had been deposited.

The present area of Triassic in North Carolina is about 1,000 square
miles, about one third of which, it is estimated, is underlaid with coal.

Prof. G. K. Gilbert* found a section of the Trias exposed by the
North fork of Virgin river, from the vicinity of Mountain Lakelet to
Rockville, in Southern Utah, 3,250 feet in thickness, and the Jurassic
at the same place 350 feet. The Triassic on the West Fork of Paria
Creek, 2,575 feet, and the Jurassic 710 feet. And the Triassic at
Jacob's Pool, Northern Arizona, 2,150 feet in thickness. E. E. Ho-
well estimated the Trias at Rock Canon, near Provo in the Wahsatch
Range, at from 4,000 to 5,000 feet, and the Jurassic from 6,000 to
8,000 feet. On Pine Mountain, the Trias at 4,650 feet, and the Jurassic
at 1,200 feet.

On the Dirty Devil river in Northern Utah, the Jurassic is about
800 feet thick, on the southwest side of Escalante river, 60 miles farther south, from 1,000 to 1,200 feet. The thickness of the Triassic in
New Mexico and Eastern Arizona is from 1,200 to 1,800 feet. This
gradually increases to the westward until near Paria, it is 2,250 feet
Ninety miles to the northeast, on the Dirty Devil river, 1,700 to 1,900
feet, is found, while near St. George, farther west, the thickness is estimated between 5,000 and 6,000 feet.

J. J. Stevenson found the Triassic on Beaver Creek, a few miles
northeast of Canon City, 2,700 feet in thickness, and unconformable
with the Jurassic above, wherever it is observed in this region.

Prof. G. M. Dawson,† separated the Triassic or Jurassic of the Rocky
Mountains, near the boundary monument, in descending order, into—
1st. Fawn-colored flaggy beds, 100 feet. 2d. Beds characterized by a
predominant red color, and chiefly red sandstone, but including some
thin greyish beds, and magnesian sandstones, the whole generally thin
bedded, though sometimes rather massive. Ripple marks, etc., weathers to a steep rocky talus, where exposed on the mountain sides ; and
passes gradually down into the next series, 300 feet.

* Geo. Sur. W. 100th Meridian, vol. 3.
† Rep. Geo. 49th Parallel.

Theo. B. Comstock* found the Jurassic limestones outcropping in many places, in the Wind river country, particularly in the neighborhood of the mountains, upon both sides of the plateau, and having a thickness of about 1,000 feet.

And Prof. E. D. Cope† described from the Trias of the Rocky Mountains, in New Mexico, *Typothorax coccinarum.*

In 1876, Prof. J. W. Powell‡ separated the Jurassic and Triassic rocks of the Plateau Province of the west in descending order, as follows—

1. Flaming Gorge Group,	1200 feet.
2. White Cliff Group, .	1100 "
3. Vermilion Cliff Group,	1100 "
4. Shinarump Group, ...	1800 "

The Flaming Gorge Group is of Jurassic age, the other three are situated above the carboniferous, but whether they should be referred to the Jurassic or the Triassic has not been determined.

The Flaming Gorge Group consists of bad-land sandstones, sometimes argillaceous with much gypsum, massive sandstones and limestones. A bed of limestone at the base is from 10 to 200 feet in thickness. In Southern Utah it caps an extensive escarpment which is called the white cliff limestone. It can be well studied at Flaming Gorge, the type locality. Commencing at the conglomerate of the Henry's Fork Group, and going southward, you pass over the upturned edges of the beds, crossing the bad-land sandstones, then the mid-group limestones, and then the bad-land indurated sandstones until the limestone is reached. The bad-land sandstones both above and below the mid-group limestone are of fresh water origin.

The White Cliff Group is a massive, obliquely laminated sandstone, often a beautiful white or golden color, sometimes red. In a few places there are heavily bedded sandstones. The typical locality is in Southern Utah. The Paria, Kanab, and Rio Virgen with their many tributaries that head in the Pink Cliffs above and to the north, have cut many canons and canon valleys through these escarpments plainly revealing the structural geology and stratigraphy.

The Vermilion Cliff Group consists of massive sandstones with ferruginous layers, and often with thin, irregular beds of cherty limestone ; the massive beds sometimes broken into thinner strata. It is also well exposed in the Paria, Kanab, Rio Virgen and their tributa-

---

* Jones' Report on Northwestern Wyoming, etc.
† Proc. Acad. Nat. Sci.
‡ Geo. of Uinta Mountains.

ries. The wagon road from Toquerville to Paria, a little town on the Paria river, soon after climbing the Hurricane Ledge, reaches the foot of the Vermilion Cliffs, and continues at this geological horizon until it commences to descend into the valley of the Paria. For seventy-five miles the road lies under this great ledge, whose salient buttes, deep alcoves, terraced and buttressed walls, towering pinnacles, all brightly colored in orange, vermilion and purple, and dotted here and there with straggling cedars and nut pines, constitute a grand panorama to the passing traveler. Flaming Gorge, on Green river, is cut through beds of this group, and received its name from the bright colors of the sandstone. Labyrinth Canon and Glen Canon present fine exposures, and fine exposures may also be seen along the Colorado Chiquito.

The Shinarump Group is separable into the Upper Shinarump consisting of bad-land sandstones with much gypsum; often argillaceous; sometimes indurated sandstones. 2d, the Shinarump conglomerate, consisting of a fine conglomerate, not easily recognized toward the north, about 20 feet in thickness, but increasing southward until it attains 200 feet. It is found capping an extensive escarpment, known as the Shinarump Cliffs. And 3d, the Lower Shinarump, consisting of bad-land sandstones with much gypsum; sometimes argillaceous; in a few places they are indurated sandstones; sometimes unconformable by erosion with the next. In such places a conglomerate is found at the base, composed of rounded and angular fragments of carboniferous rocks.

The variegated beds above and below the conglomerate are seen in many places on either flank of the Uinta Mountains, and from time to time this horizon is brought up by faults or flexures in all the stretch of country which intervenes between the Shinarump Cliffs and the Uinta Mountains. This group may be seen at the foot of the cliff on the south side of Flaming Gorge, and throughout the valley of Sheep Creek. Outcrops are found in Po Canon district, at the foot of the Yampa plateau to the east, south and west, from the foot of Whirlpool Canon, through the Island Park district, and south of Echo Park, at the foot of the Yampa plateau.

The Shinarump Conglomerate is characterized by the occurrence of silicified wood in large quantities. Sometimes trunks of trees occur, from 50 to 100 feet in length. Shinarump means literally "Shin-au-av's Rock." Shinauav is one of the Gods of the Indians of that country, and they believed these trees to have been his arrows.

The plane of demarkation between the Shinarump and the summit of the Carboniferous is always well marked.

Prof. C. A. White described from the Flaming Gorge Group, Green river, near the northern boundary line of Utah, *Unio stewardi*, and from the mouth of Thistle Creek, Spanish Fork Canon, Utah, *Neritina powelli*.

R. P. Whitfield [*] described from the Jurassic in the Bridger Mountains, Montana, *Gryphaea planoconvexa, Gervillia sparsalirata* and *Myalina perplana*.

Dr. F. V. Hayden,[†] speaking of the Triassic Group of Colorado and the West, as late as 1876, says:

The Red Beds or Triassic Group is very persistent, and if absent at all, only at very short intervals. No organic remains have yet been found in this group, by the members of the survey under my charge, yet, for various reasons, we have assumed the red sandstones to be of Triassic Age. It is barely possible that a portion or all of the Group is of Jurassic Age. Yet Prof. Cope is of the opinion that he has discovered evidence in New Mexico of its Triassic Age. The history of this Group is still obscure, and remains as one of the problems to be solved by more extended and more thorough explorations. Geographically it is one of the most widely distributed formations in the west. From the northern boundary to the southern line and east of the Wasatch range, in Utah, this red formation makes its appearance wherever a mountain range is elevated so as to expose the various sedimentary groups. The evidence indicates that it extends without any important interruption over the broad area as defined above. These red sandstones have always attracted much attention, on account of their peculiar color, but nowhere have I ever observed them performing such a conspicuous part in giving form to the scenery of the country, as along the eastern base of the Rocky Mountains in Colorado. This feature is more marked from a point about fifty miles north of Denver to Colorado Springs, than in any other portion of the continent. Along this belt the sandstones are more compact, with every variety of red, from a pale, dull tint to a deep purple color. There is also every variety of texture, from a rather coarse conglomerate to a fine sandstone. It varies much in thickness, ranging from 400 to 2,000 feet. These sandstones in Pleasant Park, the " Garden of the Gods," and other places have been weathered into the most fantastic shapes, and stand up in immense walls or columns from 50 to 250 feet in height.

Dr. A. C. Peale found Permian fossils in the beds below the red sandstones referred to the Triassic, and as Dr. Hayden and others had

---

[*] Carroll to Yellow Stone, Nat. Park.
[†] U. S. Geo. Sur. Terr

found Jurassic fossils, in the beds above, which are referred to Jurassic age, it left the conclusion with him that the red sandstones are of Triassic age. The credit, however, of first announcing the age of these sandstones is due to M. Jules Marcou, who, as early as 1853, in his "geological map," etc., "with an explanatory text," referred the beds of conglomerate, described by Capt. Stansbury, in the environs of the Devil's Gate, Rocky Mountains, and the conglomerate and sandstone described by Prof. Dana, on the Shaste river, and the boundary between Oregon and California, to the Trias. The reader may also be referred to his "Resume and Field Notes," in Vol. 3, Pacific R. R. Survey, where he identified these rocks at numerous places near the 35th parallel.

Dr. Peale found a section of Jurassic rocks, at the head of Second Canon, Eagle river, about 940 feet in thickness, and consisting of marls, sandstones and limestones. Another on Roaring Fork below station No. 14, 440 feet thick, and another in the lower Canon of Gunnison river, near station 60, representing 242 feet in thickness. It occurs usually only as a narrow belt outcropping beneath the Dakota Group.

In 1877, Arnold Hague[*] estimated the thickness of the Triassic on the outlying ridges and foot hills of the east side of the Colorado Range at 800 feet. The group is found immediately overlying the Coal Measures all along the foot-hills of the range, the continuity of the out crop being broken in only a few places, and in most cases, simply by being concealed below the uncomformable Tertiary beds. The rocks are characterized by a prevailing brilliant red color, which shades off into yellowish and whitish tints, and, near the top and bottom of the series, show frequently reddish-gray bands. The deep brick-red color, however, is so persistent as to form one of the most clearly-defined geological horizons of the uplifted sedimentary beds.

The group reaches its greatest development to the southward in Colorado, between the Big Thompson and Cache la Poudre, while north of the railroad it appears much thinner, and, between Lodge Pole and Horse Creek, reaches its minimum. Still farther to the northward, in the region of the Chugwater, it again thickens, but scarcely attains the thickness in Colorado. A section at Chugwater shows between 500 and 600 feet of strata, and another at Box Elder Creek, 650 feet. Sandstones form by far the greater part of the entire series of strata. Even the conglomerates, shales, clays and earthy beds, which occur in-

---

[*] Geo. Expl. 40th Parallel.

terstratified, appear more or less arenaceous, and are really closely allied to true sandstones, only showing considerable diversity in texture and mechanical conditions. Deposits of gypsum are very common in the upper beds.

The Triassic is exposed along the Laramie river, exhibiting a series of nearly horizontal strata, 1,000 feet in thickness. In one place a deposit of pure solid gypsum, 22 feet in thickness, occurs, lying between two beds of hard red sandstone. In the North Park the thickness is estimated at 1000 feet.

S. F. Emmons * found the Triassic in the vicinity of Rawling's Peak, 600 feet in thickness. And in the Uinta Mountains, from 3,700 to 4,000 feet. At its base is a series of clayey beds, having a thickness of 1,200 to 1,500 feet, about equally divided by a thin but persistent bed of limestone. This is succeeded by the Red Bed Group in a thickness of about 2,500 feet, principally of sandstones.

In Henry's Fork Basin, which is a narrow valley, extending 15 miles in either direction, east and west from Green river, with a width of about 3 miles, and whose average level is about 300 feet below the center of the Bridger Basin proper, the Triassic sandstones in Flaming Gorge Ridge, near Green river, are exposed in perpendicular cliffs, about 1,200 feet in height, while having at their base an undetermined thickness of clay beds.

In Emigration, Parley and Weber Canons, in Utah, the Triassic is exposed from 800 to 1,000 feet in thickness. The Jurassic is also present, and in some places has an estimated thickness of 1,500 to 1,800 feet.

The Triassic, in the Desatoya and New Pass Mountains of Nevada, contains highly fossiliferous calcareous shales and limestones. In the Pah-Ute Range in the region of Dun Glen Pass, fossils indicating Jurassic and Triassic ages are found associated together.

The Triassic is represented in the West Humboldt Range, Nevada, in Cottonwood, Buena Vista, Coyote, Bloody and Star Canons. Single sections expose strata 1,500 feet or more in thickness.

Arnold Hague* estimated the thickness of the Jurassic on the outlying ridges and foot hills of the Colorado range at 250 feet, down to 50 feet and less. The rocks consist of loose friable sandstones, limestones, marls, and impure clays, presenting great variety in color and texture, and passing from one to the other by almost imperceptible

---

* Geo. Sur. 40th Parallel.

grades. The line separating this group from the Triassic is not clearly defined, and the separation therefore is somewhat arbitrary.

The group attains its greatest thickness in the region of Big Thompson Creek, in Colorado. In Wyoming, along Lodge Pole and Horse Creeks, it is represented by only about 75 feet of strata. Still farther to the northward it expands again to a thickness of 150 feet. On the Laramie Plains west of Antelope Creek the thickness is estimated at 200 feet. On Como Ridge, in the extreme northwestern corner of the Laramie Plains, just west of the 106th Meridian, the Jurassic rocks exhibit all the characteristic strata that have been observed in other localities, associated with organic remains, and possessing a thickness of from 175 to 200 feet. Its thickness in the North Park is estimated at from 200 to 250 feet.

S. F. Emmons* estimated the average thickness of the Jurassic in the Uinta Mountain Region at from 600 to 800 feet, in which the limestones are highly fossiliferous, and have a thickness of 200 or 300 feet, the remainder being made up of sandstones, shales and clay beds, remarkable, where well exposed, for their bright, variegated colors.

In Henry's Fork Basin, a thickness of 300 to 400 feet is observed in the cliffs overlooking Sheep Creek.

In the Montezuma Range, Nevada, the shales have a thickness of between 3,000 and 4,000 feet, and rest directly upon granite. North of Indian Pass, and at Antelope Peak, they reach a development of 4,000 feet.

F. B. Meek† described, from the Triassic at Buena Vista Canon, Nevada, *Sphæra whitneyi, Modiomorpha ovata, Modiomorpha lata, Gymnotoceras rotelliforme, Arcestes perplanus, A. gabbi, Acrochordiceras hyatti, Eutomoceras laubei, Eudiscoceras gabbi.* Hall and Whitfield, from the Trias of Pah-Ute Range, Nevada, *Spirifera alia, Edmondia myrina.*

Prof. E. D. Cope,‡ from the Trias at Phoenixville, Pa., *Palaeoctonus appalachianus*, a gigantic carnivorous dinosaurian, *P. aulacodus*, now *Suchoprion aulacodus, Clepsysaurus veatleianus, Suchoprion cyphodon, Thecodontosaurus gibbidens,§* and *Palacosaurus frazeranus*, from Texas, *Eryops megacephalus*; and from Painted Canon, in Southeastern Utah,‖ *Dystrophaeus viæmalæ.*

---

* Geo. Sur. 40th Parallel.
† U. S. Geo. Expl., 40th Parallel, vol. 4.
‡ Pal. Bull., No. 26.
§ Proc. Am. Phil. Soc.
‖ Wheeler's Sur. W. 100th Mer., vol 4.

Prof. C. A. White,* from the Jurassic south of Dirty Devil river, Utah, *Ostrea strigilecula;* from the North Fork of Virgin river, *Inoceramus crassalatus;* and from Camp Cottonwood, Old Mormon Road, Nevada, *Myophoria ambilineata.*

F. B. Meek,† from the Jurassic at New Pass, Desatoya Mountains, Nevada, *Lima erecta;* from the Weber Canon, Wasatch Range. *Pinna kingi, Cucullæa haguei, Myacites inconspicuus, Myacites weberensis,* and from Cottonwood Canon, *Belemnites nevadaensis.* Hall and Whitfield, from the Jurassic at Flaming Gorge, Uinta Range, Utah, *Rhynchonella myrina, Lima occidentalis,* from Chalk Creek, *Astarte arenosa,* from Shoshone Springs, Augusta Mountains, Nevada, *Terebratula angusta, Aviculopecten augustensis, Septocardia typica, S. carditoidea,* from Wyoming *Natica lelia, Camptonectes pertenuistriatus, Trigonia quadrangularis.*

Prof. E. D. Cope first suggested that the rocks at Canyon City, Colorado, supposed by Prof. Hayden to belong to the Dakota Group (and also those in the same horizon, 100 miles north, supposed by Prof. Marsh to be lower Cretaceous), are Jurassic, and described‡ *Camarasaurus supremus, Compsemys plicatulus.§ Caulodon diversidens, Tichosteus lucasanus, Amphicœlias altus,‖ A. latus, Symphyrophus musculosus,¶ Caulodon leptogamus, Lœlaps trihedrodon.***

And Prof. O. C. Marsh†† described, from the Upper Jurassic rocks on the eastern flank of the Rocky Mountains, *Stegosaurus armatus, Atlantosaurus montanus, Apatosaurus ajax, A. grandis, Allosaurus fragilis, Nanosaurus rex.*

In 1878, J. F. Whiteaves‡‡ pointed out the Jurassic Age of certain rocks exposed on Iltasyouco river, in British Columbia, and described *Pinna subcancellata, Grammatodon iltasyoucoensis* and *Trigonia dawsoni.*

Prof. E. D. Cope, from near Canyon City, Colorado,§§ *Hypsirophus discurus, Brachyrophus altarkansanus, Amphicotylus lucasi, Tichosteus æquifacies,* and *Ephanterias amplexus.‖‖*

Prof. O. C. Marsh¶¶ described, from the Upper Jurassic of Colorado, *Atlantosaurus immanis, Morosaurus impar, Allosaurus lucaris,*

---

* Wheeler's Sur. W. 100th Mer., vol. 4.
† U. S. Geo. Expl. 40th Parallel.
‡ Pal. Bul., No. 25.   § *Ibid.*, No. 26.   ‖ *Ibid.*, No. 27.   ¶ *Ibid.*, No. 28.
** Bull. U. S. Geo. Sur. Terr., No. 3.
†† Am. Jour. Sci. & Arts, 3d ser., vol. 14.
‡‡ Geo. Sur. Can.
§§ Bull U. S. Geo. Sur., vol. 4. No. 1.   ‖‖ Am. Nat., vol. 13.
¶¶ Am. Jour. Sci. & Arts, 3d ser., vol. 15 and 16.

Creosaurus atrox, *Laosaurus celer,* *L. gracilis,* *Dryolestes priscus,* *Pterodactylus montanus.*

In 1879, Geo. M. Dawson* found on Nicola Lake, in British Columbia, a great formation built up almost exclusively of volcanic products, which have frequently a characteristically green color, and hold toward the base beds of gray, subcrystalline limestone, intermingled in some places with volcanic material, and holding occasional beds of water-rounded detritus, which he regarded as of Triassic Age.

Dr. C. A. White described,† from the Jurassic of southeastern Idaho, *Terebratula semisimplex,* *Aviculopecten pealei,* *A. altus,* *Meekoceras aplanatum,* *M. gracilitatis,* *M. mushbachanum,* and *Arcestes cirratus.*

Prof. O. C. Marsh‡ described, from the Jurassic of the Rocky Mountains, *Stylacodon gracilis,* *Ctenacodon serratus,* *Dryolestes arcuatus,* *Tinodon robustus,* *T. lepidus,* *Brontosaurus excelsus,* *camptonotus amplus,* *C. dispar,* *Cœlurus fragilis,* and *Stegosaurus ungulatus.*

And Prof. E. D. Cope§ described, from the Jurassic of Colorado, *Camarasaurus leptodirus* and *Hypsirhophus seeleyanus.*

Jurassic strata were determined at Cook's Inlet, in Alaska, as early as 1848, and Grewingk described,‖ from this place, *Ammonites wosnessenski,* and identified *A. biplex,* *Belemnitella paxillosa,* and *Unio liasinus.* And in 1857, Jurassic strata were determined at Point Wilkie on Prince Patrick Land, far north of British America. It was from this place that Capt. McClintock collected the fossils described by Prof. Haughton¶ as *Ammonites macclintocki* and *Monotis* (*Avicula*) *septentrionalis.*

In taking a general view of the Triassic and Jurassic strata, we see them in the eastern part of the continent consisting of narrow belts, having an immense thickness. The thickness in the Connecticut Valley is but little short of four miles, while in New Jersey it exceeds five miles. Israel C. Russell has argued that the physical history of these beds, in New Jersey and Connecticut, tends strongly to show that the two areas are the borders of one great estuary deposit, the central portion of which was slowly upheaved, and then removed by denudation. That the trap sheets were derived from a reservoir beneath the estuary deposits, and represent in part the force that caused the upheaval. The outburst of trap must have been the closing event of the Triassic changes, and have occurred after the sedimentary beds had been up-

* Geo. Sur. Can. † Bull. U. S. Sur., Vol. 5, No. 1.
‡ Am. Jour. Sci. & Arts, 3d ser., vol. 18. § Am. Nat., vol. 13.
‖ Verhandluugen der Russisch-Kaiserlichen mineralogischen Gesellschaft zu St. Petersbourg.
¶ Jour. Roy. Dub. Soc., Ireland.

heaved and eroded. And that the detached areas, even to North Carolina, must have been part of the same estuary formation, now broken
up and separated through the agency of upheaval and denudation.

Much denudation has evidently taken place, which must be added
to the enormous thickness which still exists to ascertain the original
dimensions of the deposit. All this points to a great depth of the sea,
or the bays, as the case may have been, in which the deposits were
made.

But when we turn to the Triassic and the Jurassic of the West, we
observe them extending from Mexico far into British Columbia, and
covering hundreds of thousands of square miles. Over extended areas
the Triassic is more than a mile in thickness, and superimposed upon
it is a great thickness of the Jurassic ; and again the Jurassic is
found more than a mile in thickness resting upon the heavy-bedded
Triassic strata. The maximum thickness, therefore, of these formations over great tracts of country is more than two miles, and the questions very naturally arise, what age do they represent ? Could the
deposits have been rapidly made, and therefore represent only a brief
space of time, or were they extremely slow and indicative of the lapse
of millions of years? Were the deposits made in shallow water, or in
the depths of mid-ocean? Is there a deposit now taking place that
bears any resemblance to these, and if so, what light if any does it
throw upon the subject? And what does palæontology, the criterion
by which all rocks are to be judged, offer to enlighten us in regard to
the secrets of this vast accumulation of detrital material?

All deep-sea dredgings have shown, that at great depths in the Atlantic and Pacific Oceans, there is a deposit of red mud constantly
taking place. We think it bears some resemblance to the red sandstone of the Triassic and Jurassic periods, and in order that a comparison may the more readily be made, we quote from the most suc-
cessful of the many exploring and deep-sea dredging expeditions.

Sir C. Wyville Thomson says,* speaking of the first time that the
dredge brought up the mud from the bottom of the Atlantic at the
depth of 3,600 fathoms:

"This haul interested us greatly. It was the deepest by several
hundred fathoms which had yet been taken, and, at all events coinci-
dently with this great increase in depth, the material of the bottom was
totally different from what we had been in the habit of meeting with in
the depths of the Atlantic. For a few soundings past, the ooze had been
assuming a darker tint, and showed on analysis a continually lessening

---

* Voyage of the Challenger, vol. 1, 1878.

amount of calcareous matter, and, under the microscope, a smaller number of foraminifera. Now calcareous shells of foraminifera were entirely wanting, and the only organisms which could be detected, after washing over and sifting the whole of the mud with the greatest care, were three or four tests of foraminifera of the cristellarian series, made up apparently of particles of the same red mud. The shells and spines of surface animals were almost entirely wanting; and this is the more remarkable, as the clay-mud was excessively fine, remaining for days suspended in the water, looking in color and consistence exactly like chocolate, indicating therefore an almost total absence of movement in the water of the sea where it is being deposited. When at length it settles, it forms a perfectly smooth red-brown paste, without the least feeling of grittiness between the fingers, as if it had been levigated with extreme care for a process in some refined art. On analysis it is almost pure clay, a silicate of alumina and the sesquioxide of iron, with a small quantity of manganese."

After a great deal of experience in sea dredging, he says:

"According to our present experience, the globigerina ooze is limited in the open oceans—such as the Atlantic, the Southern sea, and the Pacific—to water of a certain depth, the extreme limit of the pure characteristic formation being placed at a depth of somewhere about 2,250 fathoms.

"Crossing from these shallower regions occupied by the ooze into deeper soundings, we find universally that the calcareous formation gradually passes into, and is finally replaced by an extremely fine pure clay, which occupies, speaking generally, all depths below 2,500 fathoms, and consists almost entirely of a silicate of the red oxide of iron and alumina. The clay is often mixed with other inorganic matter, particularly with particles, graduating up to the size of large nodules, of peroxide of manganese; and in volcanic regions, or in their neighborhood, with fragments of pumice. The transition is very slow, and extends over several hundred fathoms of increasing depth; the shells gradually lose their sharpness of outline, assume a kind of 'rotten' look and a brownish color, and become more and more mixed with a fine amorphous red-brown powder, which increases steadily in proportion until the lime has almost entirely disappeared. This brown matter is in the finest possible state of subdivision, so fine that when, after sifting it to separate any organisms it might contain, we put it into jars to settle it remained for days in suspension.

"We recognize the gray ooze as, in most cases, an intermediate stage between the globigerina ooze and the red clay; we find that on one side, as it were, of an ideal line, the red clay contains more and more of the material of the calcareous ooze, while on the other the ooze is mixed with an increasing proportion of red clay.

" From Teneriffe to Sombrero, the depth goes on increasing to a distance of 1,150 miles from Teneriffe, when it reaches 3,150 fathoms ; there the clay is pure and smooth, and contains scarcely a trace of lime. From this great depth the bottom gradually rises; and with decreasing depth the gray color and the calcareous composition of the ooze return. Three soundings in 2,050, 1,900, and 1,950 fathoms, on the ' Dolphin Rise,' gave highly characteristic examples of the *globigerina* formation. Passing from the middle plateau of the Atlantic into the western trough, with depths a little over 3,000 fathoms, the red clay returned in all its purity; and our last sounding, in 1,420 fathoms, before reaching Sombrero, restored the globigerina ooze with its peculiar associated fauna.

" The distance from Teneriffe to Sombrero is about 2,700 miles. Proceeding from east to west, we have about 80 miles of volcanic mud and sand ; 350 miles of globigerina ooze; 1,050 miles of red clay; and 330 miles of globigerina ooze; 850 miles of red clay; and 40 miles of globigerina ooze, giving a total of 1,900 miles of red clay to 720 miles of globigerina ooze.

" The nature and origin of this vast deposit of clay is a question of the very greatest interest; and although I think there can be no doubt that it is in the main solved, yet some matters of detail are still involved in difficulty. My first impression was, that it might be the most minutely divided material, the ultimate sediment, produced by the disintegration of the land by rivers, and by the action of the sea on exposed coasts, and held in suspension and distributed by ocean currents, and only making itself manifest in places unoccupied by the globigerina ooze. Several circumstances seemed, however, to negative this mode of origin. The formation seemed too uniform; whenever we met with it, it had the same character, and it only varied in composition in containing less or more carbonate of lime.

" Again, we were gradually becoming more and more convinced that all the important elements of the globigerina ooze lived on the surface; and it seemed evident that, so long as the conditions on the surface remained the same, no alteration of contour at the bottom could possibly prevent its accumulation; and the surface conditions in the Mid-Atlantic were very uniform, a moderate surface current of a very equal temperature passing continuously over elevations and depressions, and everywhere yielding to the tow-net the ooze-forming foraminifera in the same proportion. The Mid-Atlantic swarms with pelagic mollusca; and in moderate depths, the shells of these are constantly mixed with the globigerina ooze, sometimes in number sufficient to make up a considerable portion of its bulk. It is clear that these shells must fall in equal numbers upon the red clay; but scarcely a trace of one of them is ever brought up by the dredge on the red clay area. It might be possible to explain the absence of shell-secreting animals *living on the bottom* by the supposition that the nature of the deposit was injurious to them; but the idea of a current sufficiently strong to sweep them away, if falling from the surface, is negatived by the extreme fineness of the sediment which is being laid down. The absence of surface shells appears to be intelligible only on the supposition that they are in some way removed by chemical action.

" We conclude, therefore, that the red clay is not an additional substance introduced from without, and occupying certain depressed regions on account of some law regulating its deposition; but that it is produced by the removal, by some means or other, over these areas, of the carbonate of lime, which forms probably about 98 per cent. of the material of the globigerina ooze. We can trace, indeed, every successive stage in the removal of the carbonate of lime, in descending the slope of the ridge or plateau where the globigerina ooze is forming, to the region of the clay; we find, first, that the shells of pteropods and other mollusca, which are constantly falling on the bottom, are absent; or, if a few remain, they are brittle and yellow, and evidently decaying rapidly. These shells of mollusca decompose more easily, and disappear sooner than the smaller, and apparently more delicate shells of rhizopods. The smaller foraminifera now give way, and are found in lessening proportion to the larger; the coccoliths first lose their thin outer border and then disappear; and the clubs of the rhabdoliths get worn out of shape, and are last seen, under a high power, as minute cylinders scattered over the field. The larger foraminifera are attacked, and instead of being vividly white and delicately sculptured, they become brown and worn, and finally they break up, each according to its fashion: the chamber-walls of *Globigerina* fall into wedge-shaped pieces, which quickly disappear; and a thick rough crust breaks away from the surface of *Orbulina*, leaving a thin inner sphere, at first beautifully transparent, but soon becoming opaque and crumbling away.

" In the mean time, the proportion of the amorphous, red clay to the calcareous elements of all kinds increases, until the latter disappear, with the exception of a few scattered shells of the larger foraminifera, which are still found, even in the most characteristic samples of the red clay.

" There seems to be no room left for doubt that the red clay is essentially the insoluble residue, the ash, as it were, of the calcareous organisms which form the globigerina ooze after the calcareous matter has been by some means removed. An ordinary mixture of calcareous foraminifera with the shells of pteropods, forming a fair sample of globigerina ooze from near St. Thomas, was carefully washed, and subjected, by Mr. Buchanan, to the action of weak acid; and he found that there remained, after the carbonate of lime had been removed, about one per cent. of a reddish mud, consisting of silica, alumina, and the red oxide of iron. This experiment has been frequently repeated with different samples of globigerina ooze, and always with the result that a small proportion of a red sediment remains, which possesses all the characters of the red clay. I do not for a moment contend that the material of the red clay exists in the form of the silicate of alumina and the peroxide of iron in the shells of living foraminifera and pteropods, or in the hard parts of animals of other classes. That certain inorganic salts other than the salts of lime exist in all animal tissues, soft and hard, in a certain proportion, is undoubted; and I hazard the speculation that during the decomposition of these tissues in contact with sea water and the sundry matters which it holds in solution and suspension, these salts may pass into the more stable compound of which the red clay is composed.

" Shortly after the red clay has assumed its most characteristic form, by the total removal of the calcareous shells of the foraminifera, at a depth of say 3,000 fathoms, the deposit in the Pacific Ocean in many cases begins gradually to alter again, by the increasing proportion of the shells of Radiolarians, until, at such extreme depths as 4,575 fathoms, it has once more assumed the character of an almost purely organic formation—the shells of which it is chiefly composed being, however, in this case siliceous, while in the former they were calcareous. The radiolarian ooze, although consisting in great part of the tests of Radiolarians, contains even in its purest condition a very considerable proportion of red clay. While foraminifera are apparently confined to a comparatively superficial belt, Radiolarians exist at all depths in the water of the ocean.

" The distribution over the bed of the ocean may be broadly defined thus: the globigerina ooze covers the ridges and the elevated plateaus, and occupies a belt at depths down to 2,000 fathoms round the shores, outside the belt of shore deposits; and the red clay covers the floor of the deep depressions, the eastern, the northwestern, and the southwestern basins. An intermediate band of gray ooze occurs in the Atlantic at depths averaging perhaps from 2,100 to 2,300 fathoms.

"Over the red-clay area, as might have been expected from the mode of formation of the red clay, the pieces of pumice and the recognizable mineral fragments were found in greater abundance; for there deposition takes place much more slowly, and foreign bodies are less readily overwhelmed and masked; so abundant are such fragments in some places, that the fine amorphous matter, which may be regarded as the ultimate and universal basis of the deposit, appears to be present only in small proportion.

"The clay which covers, broadly speaking, the bottom of the sea at depths greater than 2,000 fathoms, Mr. Murray considers to be produced, as we know most other clays to be, by the decomposition of feldspathic minerals: and I now believe that he is in the main right. I can not, however, doubt that were pumice and other volcanic products entirely absent, there would still be an impalpable rain over the ocean-floor of the mineral matter, which we know must be set free, and must enter into more stable combinations, through the decomposition of the multitudes of organized beings which swarm in the successive layers of the sea; and I am still inclined to refer to this source a great part of the molecular matter which always forms a considerable part of a red-clay microscopic preparation."

It is quite clear, that it would require millions of years for an accumulation to take place two miles in thickness, at the progress now in operation in the Atlantic, at depths from three to five miles. And one can not help thinking that such deposits bear strong resemblance to the red sandstones of the Jurassic and Triassic strata, and concluding, unless there is some reason to be drawn from other sources, to infer a more rapid deposition in the formation of the latter strata than that which prevails at the present time, that one is the representative of the other as to the depth of the ocean and the material and method of the deposit. If this be so, the Triassic and Jurassic rocks represent an age of vastly greater duration than the combined Cretaceous, Tertiary and Post-pliocene periods.

Palæontology may not follow such a comparison all the way to the final conclusion, but it walks hand in hand so far that we are at a loss to imagine where the separation may be made. There are many classes and orders of animals that never find a tomb in the great depths of the Atlantic, there are others that start for that goal but reach it only in the shape of an impalpable powder, "the insoluble residue, the *ash* as it were of the calcareous organisms." And as to the rest they are sparsely distributed. In this respect the comparison with the Triassic and Jurassic is most favorable, as the rarity of fossils in the hands of the collectors very clearly testifies.

But when we examine the fossils that have been discovered, and note the evolution of forms, and compare these with the progress in other ages, we are most profoundly impressed with the immense lapse of time that must be ascribed to these periods. As not a single species that is found in rocks earlier than the Triassic, and not one that is found in rocks more recent than the Jurassic, has ever been found in either the Triassic or Jurassic strata, we are sent at once to the genera for comparison. Let us first turn to the Vegetable Kingdom.

It is represented in the Triassic and Jurassic of North America by 66 described species, distributed among 30 genera. Twelve of these genera are also of palæozoic age, viz.: *Calamites, Chondrites, Cyclopteris, Dadoxylon, Fucoides, Neuropteris, Noeggerathia, Odontopteris, Pecopteris, Sphenopteris, Tæniopteris,* and *Walchia;* and seven genera are found in the Cretaceous, or more recent strata, viz.: *Chondrites, Equisetum, Neuropteris, Pecopteris, Pterophyllum, Sphenopteris* and *Tæniopteris.* This shows that five genera only, or one sixth of all that are known, passed through this period, and that during this period 16 genera, or more than half of what are known, came into existence, and also became extinct. The change of forms, as thus indicated, is greater than that which has occurred to the Cretaceous flora during all the ages that have elapsed to the present time.

The evidence furnished by the invertebrate kingdom is no less striking.

Thus far no species belonging to the Annelida or Crustacea, has been described from these rocks, and the only articulated animal-found fossil, so far as I have ascertained, is the *Mormolucoides articulatus,* described by Prof. Hitchcock, in 1858,—a genus unknown in other rocks. The class Pteropoda and the Rudista are unknown. The class Polypi is not represented by a described species, and the *Cavea prisca* alone represents the Bryozoa—another genus unknown in other rocks.

The Echinodermata is represented by an *Asterias* and a *Pentacrinus,* genera unknown in the Palæozoic age, but one of them passed up into the Cretaceous, and the other into the Tertiary period.

The Brachiopoda are represented by eleven species belonging to the genera *Lingula, Rhynchonella, Spirifera* and *Terebratula.* All of these are Palæozoic genera, and all of them have continued an exis-

tence to recent times, except *Spirifera*, which, so far as known, terminated its career in the Jurassic age.

The Gasteropoda is represented by nine species belonging to eight
genera. Two of these genera, *Dentalium* and *Turbo*, had an existence
in the Palæozoic age, and continued to live until the Tertiary period.
One genus *Lioplacodes* commenced and terminated during the age in
question. The other five genera, *Neritella*, *Neritina*, *Planorbis*, *Valvata* and *Viviparus* are counted among the living Gasteropoda.

The Cephalopoda is represented by thirty species distributed among
seven genera. One, the *Nautilus*, is a palæozoic and living genus.
Two, *Goniatites* and *Orthoceras*, are palæozoic genera that closed their
existence in the Jurassic age. One, *Meekoceras*, is confined to rocks of
the age in question. The other three genera, *Ammonites*, *Belemnites*
and *Ceratites*, commenced their existence in the age in question, and
terminated their career in the Cretaceous period.

The Lamellibranchiata is represented by 125 species distributed
among 51 genera. Six genera, *Avicula*, *Cardium*, *Lima*, *Mytilus*, *Ostrea* and *Pinna*, are reckoned among the palæozoic and living. Of the
other 45 genera, eleven of them are palæozoic, but only 24 have yet
been found in the Cretaceous. 19 of these are Tertiary, and 7 are living,
all of which are marine except *Unio*, which is now a fresh-water genus.
Or looking at this most numerously represented class of the Inverte-
brata in another light, we observe that of the 51 genera represented in
the rocks in question, 13 genera, or more than 25 per cent., are still living. 21 genera had passed away before the Cretaceous period, leaving
30 genera only in the latter period; and consequently only 17 of these
genera have expired since the dawn of the Cretaceous.

The vast changes in the vertebrate kingdom during this period, and
the grand passage from the Batrachia to the Mammalia, evidences the
same great laspe of time that is indicated by other organic remain,
and inferred from the vast thickness and extensive distribution of the
strata.

The class Pisces is represented by fifteen species belonging to nine
genera. Two of these genera, *Amblypterus* and *Palæoniscus*, are also
of Palæozoic age. The other seven are not represented so far as
known in rocks of older or younger age.

The class Aves is represented only by *Palæonornis struthionoides*,
a bird named by Prof. Emmons, in 1857—a genus, however, not yet
clearly defined or understood.

The class Reptilia is represented by 41 genera, none of which are of
Palæozoic age, and only two, *Lælaps* and *Pterodactylus*, are said to

reach the Cretaceous era. Where tracks have been described new genera in all instances have been proposed.

The Mammalia are represented in the Triassic rocks by *Dromatherium silvestre*, described by Prof. Emmons, in 1857. Four genera have been named from the Jurassic, viz: *Ctenacodon, Dryolestes, Stylacodon* and *Tinodon*. These genera are not only confined to the rocks in question, but they are not referred to families found in other rocks.

Or taken as a whole, the vertebrate kingdom is represented by 57 genera, two of which only are referred to rocks of earlier date, and only two to a later period.

These calculations are based upon our present knowledge of the fauna and flora, but as new discoveries are being made almost daily, we can not tell how much they may be modified in future. It will be observed, however, that an increased number of species will not change the calculations, and that an increase of the genera is more likely by adding new ones, than by the discovery of either Palæozoic or Cretaceous genera in these rocks.

Amphicælias fragillimus was described by Prof. Cope, from near Canyon City, Colorado, in 1878. (See *Am. Nat.* for August.) *Hypsirhopus seeleyanus* should have been referred to the Jurassic of Wyoming, instead of Colorado; and *Palæoctonus appalachianus, Suchoprion aulacodus, Clepsysaurus reatleianus, Suchoprion cyphodon, Thecodontosaurus gibbidens*, and *Palæosaurus frazeranus* should have been referred to York County, Pennsylvania, instead of Phœnixville.

We will now pass the Triassic and Jurassic periods for the purpose of considering the Cretaceous or last period that is referred to Mesozoic age.

CRETACEOUS.

The existence of the Cretaceous formation, upon this continent, was first determined in the year 1827, when Dr. S. G. Morton and Lardner Vanuxem* compared the marl of New Jersey with the Cretaceous of Europe, called by the French *la craie inférieure ou ancienne*, and by the English the Green Sand formation or Ferruginous Sand-series.

In 1828, Dr. J. E. DeKay† described, from New Jersey, *Ammonites hippocrepis*, now *Scaphites hippocrepis*, and *A. placenta*, now *Placenticeras placenta*.

[To be Continued.]

---

* Am. Jour. Sci. & Arts, vol, 12.
† Ann. Lyc. Nat. Hist., N. Y., vol. 2.

In 1829, Dr. Morton illustrated a section of Cretaceous rocks, 27 8-12 feet in height, found in a bluff, on the margin of Crosswick's Creek, New Jersey, and separated the Cretaceous of New Jersey and Delaware into the lignite strata and the marl. He relied, in determining the Cretaceous age of the rocks, upon the genera *Terebratula, Gryphæa, Exogyra, Ammonites, Baculites* and *Belemnites.* He described,* from an excavation for the Delaware and Chesapeake canal, *Ostrea falcata,* and from other places, *Terebratula harlani, T. fragilis, T. sayi, Gryphæa mutabilis* and *G. romer.*

In 1830,† he published his Synopsis of the Organic Remains of the Ferruginous Sand Formation of the United States, with geological remarks. He treated of the distribution of the strata, and mentioned many localities in the eastern and southern States where they are exposed, and also discussed the mineralogical characters of the marls. He described *Belemnites americanus, B. ambiguus, Cucullæa vulgaris,* now *Idonearca vulgaris, Ammonites delawarensis, A. vanuxemi, Spatangus stella, Ananchytes cruciferus, A. cinctus, A. fimbriatus* and *Anthophyllum atlanticum,* now *Montivaltia atlantica.* He also determined that two species, figured in Sowerby's Mineral Conchology, under the names of *Chama heliotoidea* and *C. conica,* belong to Say's genus *Exogyra.* Sowerby soon after adopted his determination, which was the first instance in which the genus of an American author was adopted in Europe, where it required the separation of the species which had been referred to an older genus.

In 1833,‡ he published a Supplement to his Synopsis, in which he illustrated and described *Rostellaria arenaria,* now *Anchura arenaria, Tornitella bullata, Conus gyratus, Cytherea excavata,* now *Cyprimeria excavata, Cardita decisa, Claragella armata, Plagiostoma gregale,* now *Spondylus gregalis, P. pelagicum,* now *Lima pelagica, Pecten perplanus, P. venustus, Anomia argentaria, Gryphæa plicatella, Ostrea falcata,* var. *nasuta, O. mesenterica, O. tortuosa, O. urticosa,*

⁰ Jour. Acad. Nat. Sci., vol. 6, part 1.

† Am. Jour. Sci. & Arts., vols. 17 & 18.

‡ Am. Jour. Sci. & Arts., vols. 23 & 24.

Teredo tibialis, now *Polarthrus tibialis, Terebratula harlani,* var. *dis-coidea, T. harlani,* var. *rectilatera, T. lachryma,* now *Terebratulina lachryma, Pholas cithara,* now *Martesia cithara, Balanus peregrinus, Cidarites diatretum,* now *Cidaris diatretum, Clypeaster florealis, G. geometricus,* and *Spatangus ungula.* Some of the species which he described at this time, and referred to the Cretaceous, are now regarded as of Eocene age. Among these we may mention, *Nummulites mantelli,* which has been the subject of much discussion, and is now referred to D'Orbigny's genus *Orbitoides,* and classed with the Protista.

In 1834, his Synopsis appeared, illustrated with nineteen plates, and having an appendix, containing a tabular view of the Tertiary fossils hitherto discovered in North America. He said that he cast it, as "a grain of sand, on the mountain of geological knowledge, which has been heaped up by the genius and industry of the naturalists of both hemispheres." But the carefulness with which the work was prepared, and the sound discrimination and learning displayed upon every page, are so obvious that one is struck with astonishment, in comparing it with the puerile and hypothetical essays which emanated, at that time, from the colleges and professed teachers of geology. It was not only a valuable contribution to knowledge, prepared by a physician, during the constant interruptions of a professional life, but it was the best work which had appeared, at that time, upon American Geology, and one that will continue to be a standard of science for many decades to come.

He separated the Cretaceous into two parts, the lower, Ferruginous Sand, and the upper, Calcareous Strata. The mineralogical characters of the Ferruginous Sand are extremely variable, consisting, for the most part, however, of minute grains, collected into friable masses of a bluish or greenish or grayish color, the predominant constituents of which are silex and iron. Iron pyrites is found in profusion; succinite, lignite and spheroidal masses, of a dark green color, and compact, sandy structure are not uncommon. The calcareous strata consist of several varieties of carbonate of lime, the principal of which are as follows: an extremely friable mass, containing silex and iron, and about 37 per cent. of lime, composed almost entirely of disintegrated zoophytes ; a yellowish or straw-colored limestone, full of organic remains ; a granular or subcrystalline limestone, intermediate in structure between the former two; and a white, soft limestone, not harder than some coarse chalks and replete with fossils. All these varieties are occasionally infiltrated by silicious matter, and contain masses of chert, and also present some appearances of the green grains so characteristic of the adjacent marls.

The Cretaceous formation is unequivocally recognized in New Jersey, from whence it may be locally traced through Delaware, Maryland, Virginia, North and South Carolina, Alabama, Mississippi, Tennessee, Louisiana, Arkansas and Missouri, it is also, probably, traced to Long Island, and probably forms the substratum of the islands of Nantucket and Martha's Vineyard. "These various deposits" he says, "though seemingly insulated, are doubtless continuous, or nearly so, forming an irregular crescent, nearly 3,000 miles in extent; and there is not only a generic accordance between the fossil shells scattered through this vast tract, but in by far the greater number of comparisons I have hitherto been able to make, the same species of fossils are found throughout: thus, the *Ammonites placenta, Baculites ovatus, Gryphœa romeri, G. mutabilis,* and *Ostrea falcata,* are found without a shadow of difference from New Jersey to Louisiana; although some species have been found in the latter State that have not been noticed in the former, and *vice versa.*"

The calcareous strata appear to be much less extensively distributed than the friable marls, and present considerable difference in their organic characters, and always when observed form the overlying beds of this formation.

Two sections of the strata, as observed in Delaware, are furnished. Localities of exposure are mentioned in Maryland, Virginia, North and South Carolina, Georgia, Alabama, Mississippi, Tennessee, Louisiana, Arkansas, Missouri, and in the level country between the Missouri river and the Rocky Mountains.

He described: *Nautilus dekayi, Ammonites navicularis, A. petechialis, A. telifer, A. conradi,* now *Scaphites conradi, A. conradi,* var. *gulosus,* now *Scaphites conradi,* var. *gulosus, Scaphites reniformis, A. vespertinus,* now *Mortoniceras vespertinum, A. syrtalis,* now *Placenticeras syrtalis, Baculites asper, B. carinatus, B. columna, B. labyrinthicus, Hamites arculus, H. torquatus, H. trabeatus, Trochus leprosus,* now *Phorus leprosus, Delphinula lapidosa,* now *Angaria lapidosa, Turritella encrinoides, T. vertebroides, Scalaria sillimani, S. annulata, Rostellaria pennata, Natica abyssina,* now *Gyrodes abyssina, N. petrosa,* now *G. petrosus, Cirrus crotaloides, Patella tentorium, Ostrea cretacea, O. plumosa, Pecten cralicula, Placuna scabra, Inoceramus barabini, I. alveatus, Avicula laripes, Pectunculus australis,* now *Axinea australis, P. hamula,* now *A. hamula, Arca rostellata,* now *Cibota rostellata, Cucullœa antrosa,* now *Idonearca antrosa, C. vul-*

garis, now *I. vulgaris, Crassatella radosa, Pholadomya occidentalis, Trigonia thoracica, Venilia conradi*, now *Veniella conradi, Terebratula floridana, Serpula barbata, Hamulus onyx, Cassidulus æquoreus, Clypeaster geometricus, Flustra sagena*, now *Pliophlæa sagena, Eschara digitata, Alveolites capularis, Turbinolia inauris* and *Gryphæa pitcheri.* The latter species was collected by Dr. Z. Pitcher, on the Kiamechia, a stream which empties into the Red river, a few miles above Fort Towson, when on a tour with a small military force, marking out a road from Fort Smith to Fort Towson. Dr. Pitcher and M. Jules Marcou referred the rocks to the Jurassic, and Marcou claims that the species is distinct from that which abounds in the Cretaceous of Texas, and farther west which is now so universally referred to this species. The weight of authority, however, is in favor of the identity of the fossils, and the Cretaceous age of the specimens described by Dr. Morton.

In 1835, in an appendix to his Synopsis of Organic Remains, he separated the Cretaceous into upper, middle and lower divisions. In the upper division he placed the Cretaceous of South Carolina, and the *Nummulite*, or *Orbitoides* limestone of Alabama, which has since been regarded as of Eocene age. The middle division is partially seen at Wilmington, North Carolina, and to a considerable extent in New Jersey. The lower division embraces the vast Ferruginous strata of the Atlantic and Southern States. He enumerated the fossil species which he regarded as most characteristic of these divisions, and described *Plagiostoma echinatum*, now *Spondylus echinatus*.

In 1834, Dr. Harlan described[*] *Ichthyosaurus missuriensis*, now *Mosasaurus missuriensis*.

In 1836, Dr. Dekay described[†] *Geosaurus mitchelli*, now *Liodon mitchelli*.

In 1838, Prof. Bronn described[‡] from the greensand, *Mosasaurus dekayi*.

In 1840, Prof. Henry Rogers[§] divided the Cretaceous, which is exposed in the southern half of New Jersey, northwest of a gentle undulating line, drawn from Shark Inlet, on the Atlantic coast, to Salem, into five separate beds, in ascending order, as follows:

First.—A group of sands and clays, of several colors, and of somewhat variable constitution, but frequently of extreme whiteness and

---

[*] Trans. Am. Phil. Soc., vol. 4.
[†] Ann. Lyc. Nat. Hist. N. Y., vol. 3.
[‡] Lethæa Geognostica.
[§] Geo. of New Jersey.

remarkable purity. Among these occur beds of pure potter's clay. This division of the general series rests along its northwest margin, from the Raritan to the Assunpink, in an unconformable manner upon the middle secondary rocks, and from the Shipetaukin to the Delaware, upon the upturned strata of the primary belt. It contains, toward its upper beds, much of the dark blue sandy clay, which is also associated with the overlying greensand, from which it is not separated by any well-defined limit.

Second.—A somewhat mixed group, consisting of beds almost wholly composed of greensand, in a loose and granular condition, alternating with and occasionally replaced by layers of a blue, sandy, micaceous clay. This is the "greensand formation," properly so called. Having been used, however, for agricultural purposes, it has acquired the name of marl. It comprises, strictly speaking, several subordinate beds, all belonging, however, to two principal varieties. In the first of these, the green, granular mineral is the predominant and characteristic ingredient. The second consists, on the other hand, of a dark-blue clay, mingled with more or less silicious sand. This latter material constitutes the usual floor upon which the true greensand deposit rests ; and it occurs, in like manner, especially in the northern and eastern portions of Monmouth county, both above the uppermost visible greensand, and included between its beds in one or more alternations.

Third.—Immediately overlying the greensand formation near its southeastern border, there are several limited exposures of a yellowish granular limestone, of rather crystalline structure, and frequently silicious composition. This rock exists in rather irregular, thin, flaggy bands, usually from one to three inches thick. Between these there are often thin layers of loose, granular, calcareous sand, identical, or nearly so, with the matter of the rock, but destitute of cohesion. This formation contains a profusion of organic remains, many of which belong in like manner to the underlying greensand, though some occur in it alone. Resting usually in direct contact with the greensand stratum, it contains often a moderate proportion of the green granular mineral, sprinkled throughout its mass. It is useful as a source of lime in a district where there is no other calcareous stratum.

Fourth.—A yellow, very ferruginous, coarse sand, containing sometimes a small proportion of the green mineral. This stratum is in some places thirty feet thick. In the Nevesink Hills, and in one or two other localities, it occurs as a soft sandstone, containing hollow casts of fossil shells. Throughout much of the central portion of the greensand region, this bed is in the condition of a loose sand, but abounds in organic remains in the state of solid casts.

Fifth.—Resting upon the former, and constituting the highest ascertained member of the Cretaceous series in the State, there occurs a coarse, brown ferruginous sandstone, sometimes passing into a conglomerate. It is composed of translucent quartzose sand, small fragments of felspar, and pebbles of white quartz, cemented together by a dark brown paste of oxide of iron. The green mineral in detached grains is likewise a common ingredient. The position of this rock is usually upon the summits of the insulated outlying hills, which rise occasionally above the general plain of the marl region.

This division into beds is merely descriptive of the local appearance of the Cretaceous of New Jersey, and has never been regarded as of any service in the separation of the Cretaceous, in other States, into groups, nor has it been retained in New Jersey, since the geologists have been able to separate the strata by their organic remains.

In 1841, James C. Booth, in his Memoir of the Geological Survey of Delaware, divided the Cretaceous of that State, which is found superimposed upon the primary rocks, and extends from the lower limit of the primary nearly to the southern border of New Castle county, into red clay, and green and yellow sands. He estimated the thickness at not less than 330 feet.

In this year, Prof. J. W. Bailey* discovered that a large part of the calcareous green sand of New Jersey, the limestone from Claiborne, Alabama, and a light cream-colored marl from a mission station on the Upper Mississippi, called "Prairie Chalk," is composed of microscopic shells belonging to the foraminifera.

In 1842, Dr. Morton† described, from the Cretaceous of the upper Missouri river, *Ammonites mandanensis, A. abyssinus* and *A. nicolletti,* all of which are now referred to the genus *Scaphites,* and to the Fox Hills Group; *Hipponyx borealis,* now *Anisomyon borealis, Cytherea missuriana,* now *Dione missuriana* and *Tellina occidentalis,* now *Lucina occidentalis.* And from the Cretaceous group, of New Jersey, *Ammonceratites conradi,* now *Crioceras conradi, Hamites annulifer,* now *Ptychoceras annuliferum, Pinna rostriformis, Terebratula atlantica, Planularia cuneata, Cidarites armiger.* And *Ptycodus mortoni,* by Mantell, from the Cretaceous, at Prairie Bluff, Alabama. Dr. James E. Dekay, described,‡ from the Cretaceous

---

* Am. Jour. Sci. and Arts, vol. 41.
† Jour. Acad. Nat. Sci., vol. 8, part 2.
‡ Zool. of New York.

greensand, *Gavialis neocæsariensis*, now *Thoracosaurus neocæsariensis.*

In 1843, Prof. Mather* ascertained that beneath the drift, and above the New Red Sandstone, there exists a deposit of sand, clay, gravel and pebbles, on the Island of New York, Staten Island, Long Island and Gardener's, Plum, Shelter, Governor's and Bellow's Islands, which he referred to the Cretaceous. Sections furnished by the digging of wells indicated a thickness of 80 or 90 feet. He also regarded the exposure of trappean rocks in Rockland and Richmond counties, New York, as more recent than the New Red Sandstone.

In 1844, Dr. Morton† described, from New Jersey, *Crocodilus clavirostris.* And Dr. Robert W. Gibbes, from the greensand near the Santee canal, about 3 miles from Cooper river, in South Carolina, *Dorudon serratus*, now *Thoracosaurus neocæsariensis*

In 1845,‡ Lyell and Sowerby described, from Timber creek, New Jersey, *Ostrea subspatulata*, Lyell and Forbes described *Lima reticulata*, *Terebratula vanuxemi*, now *Terebratella vanuxemi*, *Bulla mortoni*, and William Lonsdale described *Idmonea contortilis*, *Tubulipora megæra*, now *Filifascigera megæra*, and *Cellepora tubulata.*

Goldfuss described§ *Mosasaurus maximiliani*, now *M. missuriensis.*

In 1846, Dr. Ferdinand Roemer‖ ascertained the character of the Cretaceous rocks of Texas, and compared them with the chalk of Europe, and greensand of New Jersey, and claimed that they represented the upper part of the Cretaceous formation. He mentioned their occurrence at New Braunfels, and ranging very far on both sides of the Guadaloupe, and everywhere parallel to the chain of high hills which separate the Indian country from the settled part of Texas. He followed them as far as Austin on the Colorado, and collected fossils in them at San Antonio, and on the Pedernales river. East of a line drawn through San Antonio, New Braunfels and Austin, the surface is covered with strata more recent than the Cretaceous; it is generally composed of a thick diluvium of loose materials, consisting either of a fertile vegetable mould, or of rounded pieces of hydrate of iron, or of sand and gravel.

In 1848,¶ he stated that an ideal line, drawn from Presidio de Rio Grande, on the Rio Grande, in a N. E. direction, and crossing the San

* Geo. Sur. N. Y. † Proc. Acad. Nat. Sci.
‡ Quar. Jour. Geo. Soc., vol. 1.
§ Act. Nov. Leop. Caes. Nat. Cur.
‖ Am. Jour. Sci. and Arts, 2d ser., vol. 1.
¶ Am. Jour. Sci. and Arts, 2d ser, vol. 6.

Antonio river, at the town of the same name, the Guadaloupe at New Braunfels, the Colorado at Austin, the Brazos at the falls of this river, the Trinity below its forks, and reaching from there to the Red river in the same N. E. direction, divides the Tertiary strata, and the diluvial and alluvial deposits (of the level and rolling part of the country) from the Cretaceous and older formations (of the hilly and mountainous sections) of Texas. The tract of level country which extends like a broad belt along almost the whole coast of Texas, is diluvial and partly alluvial in character. Its small elevation of a few feet only above the level of the sea, and its perfectly level surface, indicate, at once, the recent origin of the soil. The fossil remains found in many places in the deposits of clay and sand, prove their modern age still more conclusively. At the head of Galveston Bay, and near the town of Houston, he found, at the height of 12 to 20 feet above the general level of the Bay, large deposits of shells of Guathodon, a bivalve mollusc, which lives abundantly in the brackish waters along the coast of the Gulf of Mexico, and in the Bay of Galveston, and a few oyster shells of the common kind, but no shells different from those living in the Bay. Everything tending to show that there had been no material change in the climate, nor other circumstances since the period of these deposits along the coast of Texas, except in the relative change of the level of land and sea. To the diluvial period he referred the deposits of clay and sand which form the banks of the Brazos, and probably all the other large rivers of the country wherein he found the bones of the Mastodon, Megalonyx, Tapir and other mammals. To the same period he referred the deposits of gravel and sand, which form a broad belt of barren or poor land covered with pine and post oak timber, in the rolling or undulating portion of Texas, and extending from west to east across a considerable part of the country. Following up the Colorado from Columbus to Bastrop, or the Guadaloupe from Gonzales to Seguin, we pass directly across this belt. The gravel is mostly composed of pebbles of silex, evidently derived from decomposed Cretaceous strata. Within the limits of this gravel formation, fossil wood of dicotyledonous trees, in smaller or larger fragments, is found almost everywhere, and occasionally whole trunks of trees are met with. Near the town of Caldwell, on the Upper Brazos, he found alternating strata of brown ferruginous sandstone, and of dark-colored plastic clay, both teeming with fossils belonging to the older divisions of the Tertiary period

The Cretaceous strata which makes the most important part in the geological constitution of Texas, and chiefly her upper hilly part, is found north of the line above indicated, covering the whole area of country with the exception of small exposures of Silurian and Carboniferous strata and granitic rocks. The Cretaceous strata constitute, generally, compact and hard rocks, some of them equaling in compactness the hardest strata of more ancient secondary formations. Generally there is an alternation of compact silicious limestones, and less compact beds of either pure or marly limestone. The former contain the silex as well diffused through their whole mass, as in separate concretions or nodules. The silicious character of these rocks, excluding the decomposing action of the atmosphere, almost entirely produces the general dry and barren aspect of the country which they occupy. He pointed out the differences between the Cretaceous fauna of Texas, and that of New Jersey and other northern localities, and compared the fauna with that of Europe, from whence he concluded that there must have existed at the time of the Cretaceous period between the continents of Europe and America, such a relation that, in both, the same modifications in the zoological character distinguished the marine fauna of the north from that of the south. From thence he drew the interesting conclusion, that the same southern inflection of the isothermal lines, which is at present so remarkable in their course from the west side of the continent of Europe, toward the east side of the continent of America, already existed at a period of the globe as remote as that of the Cretaceous formation.

In 1849, Prof Owen* described, from the greensand of New Jersey, *Crocodilus basifissus, C. basitruncatus.* now *Holops basitruncatus, Macrosaurus lœvis* and *Hyposaurus rogersi.*

In 1850, T. A. Conrad† described, from Timber Creek, New Jersey, *Catopygus oviformis.*

In 1851, Dr. Gibbes‡ described, from South Carolina, *Mosasaurus acutidens, M. brumbyi, M. carolinensis, M. couperi,* and *M. minor.* And Dr. Leidy§ described *Discosaurus vetustus,* now *Cimoliasaurus vetustus,* and *Conosaurus bowmani.*

In 1852, Dr. D. D. Owen‖ described, from the Fox Hills of Nebraska, *Ammonites nebrascensis, A. cheyennensis,* now *Scaphites cheyennensis, A. opalus, A. moreauensis,* now *S. moreauensis, A. lenticularis,* now *Placenticeras lenticulare, Scaphites comprimus, S. nodosus, Ino-*

---

* Quar. Jour. Geo. Soc., vol. 5.
† Jour. Acad. Nat. Sci., 2d ser. vol. 2.
‡ Smithsonian Contributions, vol. 2.
§ Proc. Acad. Nat. Sci.
‖ Rep. Geo. Sur Wis., Iowa and Minn.

ceramus sagensis, *I. nebrascensis*, and *Cucullæa nebrascensis*, now *Idonearca nebrascensis*.

Dr. Joseph Leidy* described, from the greensand of New Jersey, *Crocodilus dekayi*.†

Dr. Ferd. Roemer‡ described, from the Cretaceous rocks of Texas, *Actæonella dolium*, *Ammonites dentato-carinatus*, *A. flaccidicosta*, *A. guadalupæ*, *Arcopagia texana*, *Astarte lineolata*, *Astrocænia guadalupæ*, *Avicula convexo-plana*, *A. pedernalis*, *A. planiuscula*, *Caprina crassifibra*, *C. guadalupæ*, *Caprotina texana*, *Cardium elegantulum*, now *Leiopistha elegantula*, *C. sanctisabæ*, *Chemnitzia gloriosa*, *Cyphosoma texanum*, *Cypricardia texana*, *Diadema texanum*, *Eulima texana*, *Exogyra arietina*, *E. laviuscula*, *E. ponderosa*, *E. texana*, *Fusus pedernalis*, *Globiconcha coniformis*, *G. planata*, *Hemiaster texanus*, *Hippurites texanus*, *Holectypus planatus*, *Homomya alta*, *Inoceramus confertim-annulatus*, *I. undulato-plicatus*, *Lamna texana*, *Lima crenulicosta*, *L. wacoensis*, *Modiola concentrico-costellata*, now *Volsella concentrico-costellata*, *M. granulato-cancellata*, now *Crenella granulato-cancellata*, *M. pedernalis*, now *Volsella pedernalis*, *Monopleura subtriquetra*, *M. texana*, *Mytilus semiplicatus*, *M. tenuitesta*, *Natica pedernalis*, now *Lunatia pedernalis*, *N. prægrandis*, *Nerinea acus*, *N. texana*, *Orbitulites texanus*, now *Tinoporus texanus*, *Ostrea anomiæformis*, *O. crenulimargo*, *O. aucella*, *Pecten duplicicosta*, now *Neithea duplicicosta*, *P. texana*, now *N. texana*, *Pholadomya pedernalis*, *Psammobia cancellato sculpta*, now *Gari cancellato-sculpta*, *Radiolites austinensis*, *Scalaria texana*, now *Anchura texana*, *Scaphites semicostatus*, *S. texanus*, *Solen irradians*, *Spondylus guadalupæ*, *Terebratula guadalupæ*, *T. wacoensis*, *Toxaster texanus*, *Turrilites brazoensis*, and *Turritella seriatim-granulata*.

In 1853, T. A. Conrad§ described, from San Felipe creek, near Rio Grande, Texas, *Exogyra caprina*; from New Jersey, *Avicula abrupta*, *A. petrosa*, *Solenomya planulata*, now *Legumen planulatus*, *Crassatella subplana*, *Arca uniopsis*, *Tellina densata*, *Lucina pinguis*, now *Tenea pinguis*, *Pecten quinquenaria*, now *Neithea quinquenaria*, *Cardium protextum*, *Veniila rhomboidea*, now *Veniella rhomboidea*, *Astarte parilis*, *Dentalium subarcuatum*, *Inoceramus peroralis*, *Requienia senseni* and *Pholas pectorosa*.

---

* Jour. Acad. Nat. Sci., 2d ser. vol. 2
† Smithsonian Contributions, vol. 2.
‡ Kreid. von Texas.
§ Jour. Acad. Nat. Sci., 2d ser. vol. 2.

In 1854,* the Cretaceous formation of Nebraska was subdivided by Hall and Meek, in ascending order, as follows:

1. Sandstone and clay, 90 feet.

2. Clay containing a few fossils, 80 feet.

3. Calcareous marl, containing *Ostrea congesta*, scales of fishes, etc., 100 to 150 feet.

4. Plastic clays, with calcareous concretions, containing numerous fossils, 250 feet. This is the principal fossiliferous bed of the Cretaceous formation on the Upper Missouri.

5. Arenaceous clays passing into argillo calcareous sandstones, 80 feet.

These subdivisions were referred to, by these numbers, until 1861, when Meek and Hayden, in accordance with the laws of nomenclature, gave them the following geographical names: No. 1, Dakota Group; No. 2, Fort Benton Group; No. 3, Niobrara Group; No. 4, Fort Pierre Group; and No. 5, Fox Hills Group.

They described from No. 5, at Fox Hills, *Pecten rigida*, now *Syncyclonema rigidum*, from the Bad Lands of Dakota, *Baculites grandis*; from No. 4, at the Great Bend of the Missouri, below Fort Pierre, *Avicula haydeni, Inoceramus convexus, I. tenuilineatus, I. sublævis, Nucula subnasuta*, now *Nuculana subnasuta, Buccinum vinculum*, now *Trachytriton vinculum, Ammonites complexus, Turrilites cochleatius*, now *Heteroceras cochleatum;* from Sage creek, *Nucula ventricosa*, now *Yoldia ventricosa, Crassatella evansi, Lucina subundata, Dentalium gracile, Actæon concinnus*, now *Cinulia concinna, Fusus tenuilineatus*, now *Closteriscus tenuilineatus, Natica concinna*, now *Lunatia concinna, Natica paludiniformis*, now *Amauropsis, paludiniformis, Fusus constrictus*, now *Odontobasis constricta;* from No. 2, near the mouth of Vermilion river, *Inoceramus fragilis*; from below the mouth of James river, *Cytherea orbiculata*, now *Callista orbiculata* and *C. tenuis;* from No. 1, at the mouth of Big Sioux river, on the Missouri, *Pectunculus siouxensis*, now *Trigonarca siouxensis.*

Dr. Geo. G. Shumard† found the Cretaceous rocks at Fort Washita, and extending from there uninterruptedly to the southwestern boundary of the Cross Timbers, in Texas. It usually consists of grayish yellow sandstone, with intercalations of blue, yellow and ash colored clays, and beds of white and bluish white limestone. The limestone reposes on the clays and sandstones, and in some places attains a

---

* Mem. Am. Acad. Arts & Sci., vol 5.
† Expl. of Red River, of Louisiana, by Marcy.

thickness of 100 feet. It is usually soft and friable, and liable to disintegrate rapidly when exposed to the action of the weather. At Fort Washita he found Ammonites several feet in diameter, and weighing between 400 and 500 pounds.

Dr. B. F. Shumard described *Ostrea suborata, Astarte Washitensis, Cardium multistriatum,* now *Protocardia multistriata, Panopœa texana, Terebratula choctawensis, Globiconcha elevata, G. tumida, Eulima subfusiformis, Ammonites acuticarinatus, A. marcianus, Hemiaster elegans,* now *Toxaster elegans,* and *Holaster simplex.*

Dr. Leidy* described, from near Greenville, Clark county, Arkansas, *Brimosaurus grandis,* now *Cimoliasaurus grandis.* And Evans & Shumard described, from Sage creek, Nebraska, *Avicula linguiformis, A. triangularis, Solarium flexistriatum,* now *Margaritella flexistriata, Pholadomya elegantula,* and *Rostellaria nebrascensis,* now *Anchura nebrascensis;* and from Fox Hills, *Mytilus galpingnus* now *Volsella galpinana.*

In 1855, M. Tuomey† described, from Alabama and Mississippi, *Nautilus orbiculatus, N. spillmani, N. angulatus, Ammonites angustus, A. binodosus, A. carinatus, A. magnificus, A. ramosissimus, Turrilites alternatus, Turritella fastigiata Phorus umbilicatus,* now *Endoptygma umbilicata, Voluta cancellata, V. fusiformis, V. jugosa, V. spillmani, Fusus eufalensis, F. turriculus, Pyrula richardsoni,* now *Pyropsis richardsoni, P. trochiformis,* now *Pyropsis trochiformis, Cerithium nodosum, Teredo calamus. Panopœa cretacea, Pholadomya tenua, Cardium hemicyclus, Cucullœa ungula, Inoceramus biformis, I. inflatus, I. proximus, I. salebrosus, I. triangularis, Radiolites ormondi, R. aimesi, R. undulatus, Ichthiosarcolites cornutus, I. loricatus, I. quadrangularis.*

T. A. Conrad‡ described, from Dallas county, Miss., *Baculites annulatus, Hamites larvatus, H. rotundatus;* from Arkansas, *Ancyloceras approximans,* and *Cardium arkansasense,* now *Protocardia arkansasensis;* from Alabama, *Caprina quadrata,* and from Texas, *Rostellites texanus, Turritella irrorata, Caprina occidentalis, C. planata, Neithea occidentalis. Mactra texana, Exogyra fimbriata,* and *E. fragosa.*

Dr. Joseph Leidy§ described, from the greensand near Pemberton, New Jersey, *Pristis curridens.*

* Proc. Acad. Nat. Sci., vol. 7.
† Proc. Acad. Nat. Sci., vol. 7.
‡ Proc. Acad. Nat. Sci., vol. 7.
§ Proc. Acad. Nat. Sci., vol. 7.

Dr. James Schiel* found the Cretaceous rocks west of Fort Atkinson, and described *Inoceramus pseudo-mytiloides*. W. P. Blake, from the fossils collected by Capt. John Pope, identified these rocks on the banks of the Red river, near Preston; Big Springs of the Colorado; Elm Fork of the Trinity river; and a point 20 miles east of the Sand Hills, on the Llano Estacado.

In 1856, William P. Blake† announced generally the Cretaceous age of the extensive table lands on the 35th Parallel from the 101st to the 110th Meridian, known as the Llano Estacado. Though Wislizenus as early as 1848 had described it in the bluffs of Gallinas Creek, and Dr. Schiel, Dr. Randall and Lieutenant Simpson as well as Jules Marcou had testified to its existence in various places in the exposed bluffs found upon these plains. The strata are nearly horizontal, and principally white or grey and highly calcareous, but sometimes intercalated with grey or blue marl or clay. Prof. James Hall described, from False Washita and other localities in the west, *Gryphæa pitcheri*, var. *navia*.

Meek & Hayden‡ described, from near Fort Union, Nebraska (later called Fort Union Group), *Cyclas formosa*, now *Sphærium formosum*, *C. subellipticus*, now *S. subellipticum*, *Bulimus teres*, now *Columna teres*, *B. vermiculus*, now *C. vermicula*, *Pupa helicoides*, *Limnea tenuicosta*, which is made the type of the genus *Pleurolimnea*, *Physa longiuscula*, now *Bulimus longiusculus*, *P. rhomboidea*, now *B. rhomboideus*, *P. nebrascensis*, *Velletia minuta*, now *Acroloxus minutus*, *Paludina leai*, now *Viviparus leai*, *P. retusa*, now *V. retusus*, *P. leidyi*, now *V. leidyi*, *P. trochiformis*, now *V. trochiformis*, *Valvata parvula*, *Melania minutula*, now *Micropyrgus minutulus*, *M. multistriata*, now *Campeloma multistriatum*, *M. nebrascensis*, now *Goniobasis nebrascensis;* from Moreau river, *Cyrena moreauensis*, now *Corbicula moreauensis*, *Cyrena intermedia*, now *Corbicula nebrascensis;* from the Bad Lands of the Judith river, *Cyrena occidentalis*, now *Corbicula occidentalis*, *Corbula subtrigonalis*, *C. perundata*, *C. mactriformis*, *Unio priscus*, *Physa subelongata*, now *Bulimus subelongatus*, *Planorbis subumbilicatus*, *Paludina vetula*, now *Campeloma vetulum*, *P. conradi*, now *Viviparus conradi*, *Melania convexa*, now *Goniobasis convexa;* from Fort Clark, *Bulimus limnæformis*, now *Thaumastus limnæformis*, *Paludina multilineata*, now *Campeloma multilineatum*, *P. peculiaris*, now *Viviparus peculiaris;* from Little Horn river, *Planor-*

* Expl. & Sur. R. R. Miss. River to Pacific Ocean, vol. 2.
† Pacific R. R. Sur. vol., 3.
‡ Proc. Acad. Nat. Sci., vol. 8.

bis convolutus; from the Yellow Stone, *Melania anthonyi,* now *Hydro-bia anthonyi;* and from near the head waters of the Little Missouri, *Cerithium nebrascense,* now *Cerithidea nebrascensis.*

From (the Fort Pierre Group)* No. 4, of the Cretaceous in Nebraska, *Actæon subellipticus, Turbo nebrascensis,* now *Margarita nebrascensis, Rostellaria biangulata,* now *Aporrhais biangulata, Helcion sexsulca-tus,* now *Anisomyon sexsulcatum. H. patelliformis,* now *A. patelli-forme, H. alveolus,* now *A. alveolus, H. subovatus,* now *A. subovatum, Bulla occidentalis,* now *Haminea occidentalis, Turritella convexa, Ammonites halli,* now *Phylloceras halli, Ancyloceras nebrascense,* now *Heteroceras nebrascense, A. cheyenense,* now *H. cheyenense, Aricula fibrosa,* now *Pseudoptera fibrosa;* from near the mouth of Milk river, *Bulla subcylindrica,* now *Haminea subcylindrica, Venus circularis,* now *Thetis circularis, Cytherea pellucida,* now *Callista pellucida; Cucullæa exigua,* now *Trigonarca exigua, Gervillia subtortuosa, Ino-ceramus incurens,* and *Ostrea patina;* from the Great Bend of the Mis-souri, *Nucula obsoletastriata.*

From (the Fox Hills Group)† No. 5, of the Cretaceous in Nebraska, *Scalaria cerithiformis,* now *Chemnitzia cerithiformis, Natica am-bigua,* now *Vanikoro ambigua, Natica occidentalis,* now *Lunatia occi-dentalis, Turbo tenuilineatus,* now *Spironema tenuilineatum, Fusus dakotensis, F. galpinanus,* now *Fasciolaria galpinana, F. contortus, F. flexuocostatus,* now *Fasciolaria flexuocostata, F. newberryi,* now *Pyrifusus newberryi, F. culbertsoni,* now *Fasciolaria culbertsoni, Py-rula bairdi,* now *Pyropsis bairdi, Fasciolaria cretacea, F. buccinoides, Buccinum nebrascensis,* now *Pseudo-buccinum nebrascense, Bulla vol-varia,* now *Cylichna volvaria, B. minor,* now *Haminea minor, Turri-tella moreauensis,* now *Cerithiopsis moreauensis, Belemnitella bulbosa.*

From the mouth of Judith river, on Cherry creek, and on Moreau river, *Pholadomya undata,* now *Cymella undata, Goniomya americana, Solen subplicatus,* now *Solenomya subplicata, Tellina gracilis,* now *Thracia gracilis, Tellina Cheyennensis, T. scitula, T. subelliptica,* now *Corbicula subelliptica, T. prouti,* now *Thracia prouti, Cytherea de-weyi,* now *Callista deweyi, Cytherea nebrascensis,* now *Callista nebras-censis, Corbula moreauensis,* now *Neaera moreauensis, C. ventricosa,* now *N. ventricosa, C. gregaria,* now *Corbulamella gregaria, Astarte gregaria,* now *Eryphyla gregaria, Nucula scitula,* now *Yoldia scitula, N. evansi,* now *Y. evansi, N. æquilateralis,* now *Nuculana æquilater-*

---

alis, N. subplana, N. cancellata, N. planimarginata, Pectunculina parvula, now *Limopsis parvula, Cucullæa cordata,* now *Idonearca cordata, C. shumardi,* now *I. shumardi, Mytilus attenuatus,* now *Volsella attenuata, Inoceramus pertenuis, Pecten nebrascensis, Natica subcrassa,* now *Lunatia subcrassa, Natica tuomeyana,* now *Vanikoropsis tuomeyana, Panopæa occidentalis,* now *Glycimeris occidentalis, Mactra formosa, M. warrenana, M. alta, Tellina subtortuosa,* now *Thracia subtortuosa, Cytherea owenana,* now *Callista owenana, Hettangia americana,* now *Tancredia americana, Cardium speciosum,* and *Mytilus subarcuatus.*

Professor L. Harper* described, from the bed of the Tuscaloosa, or Black Warrior river, near Erie, Greene county, Alabama, about twelve miles above the confluence of the Tombigbee and Black Warrior rivers, *Ceratites americanus.*

Dr. Joseph Leidy† described, from the greensand of Burlington county, New Jersey, *Chelonia ornata,* now *Peritresius ornatus Polygonodon vetus, Ischyrhiza mira, Edaphodon mirificus,* now *Ischyodus mirificus;* from Neuse river, North Carolina, *Ischyrhiza antiqua;* and from the Upper Missouri, *Cladocyclus occidentalis* and *Enchodus shumardi.* And from the Fort Union Group, at Long Lake, Nebraska, *Emys obscurus,* now *Compsemys obscurus, Compsemys victus,* and *Mylognathus priscus;* from the lowest lignitic of Grand river, Nebraska, *Thespesius occidentalis,* and from the Bad Lands of Judith River, *Palæoscincus costatus, Trachodon mirabilis, Troodon formosus, Trionyx foveatus, Deinodon horridus,* now *Amblysodon horridus, Crocodilus humilis,* now *Bottosaurus humilis, Lepidotus haydeni, L. occidentalis,* and *Ischyrotherium antiquum,* now *Ischyrosaurus antiquus.*

In 1857, Arthur Schott‡ described the Cretaceous basin of the Rio Bravo. The main portion, from Las Moras to the vicinity of Reynosa, forms a belt of 380 to 400 miles in width. The upper part of this belt commences in the vicinity of Las Moras, and terminates some few miles above Laredo, a distance of about 200 miles, whilst the lower part begins where the former ends, and reaches as far as the vicinity of Reynosa, showing a width of about 340 miles. Both of these parts are distinctly characterized by strata of greensand (chloritic chalk), which change, according to the amount of oxide of iron they contain, into

---

* Proc. Acad. Nat. Sci., vol. 9.
† Proc. Acad. Nat. Sci., vol. 8.
‡ U. S. & Mex. Bound. Sur., vol. 1.

variously tinted sandstone shoals. The solidity of the strata varies very much. They are sometimes formed into very solid rocks, well suited for mechanical or architectural operations; again, they consist of loose and coarsely grained sandstone slate, which rapidly crumbles on exposure to the air. The general characteristic of this belt and its subdivisions is the strict horizontality of its strata. It is only here and there that some slight local disturbance has taken place, as for instance, near Laredo, and again, some 40 or 50 miles above, where a dip of about 8° W., S. E. and E. is exposed.

From Las Moras to the vicinity of Arroyo Sombreretillo, which is about 10 miles above Laredo, lignite coal occurs quite frequently. On both sides of the mouth of Elm creek, near Eagle pass, particularly on the north bank of this water course, layers are exposed from 3 to 4 feet thick. On the slope of Lizard Hills, below the deserted Rancho Palafox, coal occurs from 4 to 5 feet thick.

Septariæ abound in the lower belt, especially below the mouth of Arroyo Sombreretillo; on the oyster-terraces, some 40 miles below Laredo, and near the Rancho San Ignacio; on the slope of Red Ridge, of Shady Bluffs, and Septariæ Hills. Their most common shape resembles very much a small flat loaf of bread. Both on the outside and inside large irregularly-shaped divisions, like a net work of veins occur, which are composed of crystals of gypsum, a mineral very abundant in these rocks.

C. C. Parry found the Cretaceous in the basin of the Rio Grande, where the Comanche trail crosses from Upper Texas into Mexico, near the Mexican settlement of San Carlos. The rock exposure exhibits a very variable dip, mostly inclined toward the west, occasionally at a very sharp angle. It rises at various points in the adjoining table-land, forming ochreous colored rocky bluffs, where at several points the gravelly table-land is seen to rest unconformably on the sharply-tilted strata. Further down the river, in an eastern direction, the Cretaceous assumes a nearly horizontal position, and a closer texture. It is here seen overlaid by a variable sheet of dark-colored lava rock. This sheet of igneous rock conforms closely to all the inequalities of the underlying limestone, exhibiting, in the walls of the Canon below, a distinct line of separation traceable for a long distance. The westerly dip of the Cretaceous underneath gradually thins out this upper igneous capping, which finally disappears, and solid limestone walls continue along the line of the river.

At one point on the line of the trail leading round the broken ranges of the mountain ledges, directly bordering the river, to reach its bed some eight miles below the Comanche Ford, the sides of a deep washed ravine bring to view the successive and relative thickness of the rocks. We here see the upper members of the Cretaceous rocks forming the tabled summits of the adjoining mountains, and marked by frequent Cretaceous fossils, resting on a bed of igneous trap-form rock 50 to 80 feet thick, this again overlaying the closer layers of the limestone strata below.

The gigantic canon of San Carlos, through which for ten miles the Rio Grande, pursuing a nearly due east course, makes its way, presents unbroken walls of Cretaceous limestone. The course of the river cutting the strata in a line directly opposed to the dip, there is a constantly increasing elevation of the canon walls. These walls commence with a height of between 200 and 300 feet; but the fall of the water, combined with the rise of the strata, develops, in the course of ten miles, a clear perpendicular height of at least 1,500 feet above the river level.

A faint conception only can be formed from these facts of the truly awful character of this chasm. Its course can be marked along the mountain slope in a regular zigzag line, terminating by an opening cleft, which rises high and clear above the surrounding mountain ranges. The surface of the ground adjoining the river bank, is a slightly broken slope, extending to the east, and showing a continuous development of the range to the north and south. The general surface presents no indication of a river course, and you are not aware of its presence till you stand suddenly on its abrupt brink; even here the running water is not always visible, unless advantage be taken of the projecting points, forming angles, along the general course of the river. From this dizzy height the stream below looks like a mere thread, passing in whirling eddies, or foaming over broken rapids; a stone hurled from above into this chasm passes completely out of sight behind the over-hanging ledges, and one can often count thirty before the last deadened splash announces that it has reached the river bed. From the point formed by its last projecting ledges the view is grand beyond all conception. You can here trace backward the line of the immense chasm, which marks the course of the river, till it emerges from its stupendous outlet.

The mountain range forming the San Vincente canon, lower down the Rio Grande, is exclusively Cretaceous. The eastern slope of the Sierra Carmel shows the Cretaceous limestone inclining eastward at an angle of about 20°; and the Cretaceous continues to be exposed almost uninterruptedly to the mouth of the Pecos river. From here to

Eagle Pass is an open country, occupied by low swells of Cretaceous limestone, thus merging into that character of country pertaining to the region of central Texas.

Prof. James Hall, to whom the minerals and fossils collected by the Boundary Commission were referred for geological examination, compared the Cretaceous of Texas and New Mexico with that of Nebraska and the eastern states. He furnished the following section of the successive beds comprising the Cretaceous formation of New Jersey, which had been communicated to him by Prof. Geo. H. Cook for comparison, to wit:

8. Greensand, Third or Upper Bed. [Probably of Eocene Age.]

This bed admits of a triple division, the central portion is nearly destitute of fossils, while those of the upper and lower divisions are mostly dissimilar.

7. Quartzose Sand, resembling Beach Sand.

This bed is (so far as known), quite destitute of fossils.

6. Greensand, Second Bed.

(*a*) Yellow limestone of Timber Creek. Characterized by *Eschara digitata, Montivaltia atlantica, Nucleolites crucifer, Ananchytes cinctus, A. fimbriatus,* Morton.

(*b*) A bed of nearly unchanged shells. Among the characteristic fossils of this bed are *Gryphæa vomer, G. convexa,* and *Terebratula harlani.*

(*c*) Greensand, etc.

Cucullæa vulgaris is the most characteristic fossil of the lower division.

5. Quartzose Sand highly Ferruginous throughout, and Argillaceous in its upper parts.

This rock is sometimes indurated or cemented by oxyd of iron. *Exogyra costata, Ostrea larva, Belemnitella mucronata, Pecten (Neithea) quinque-costatus;* and many other fossils mostly in the condition of casts of the interior, or impressions of the exterior.

4. Greensand, First or Lower Bed.

Several subdivisions may be recognized depending on the character of the marl, etc. *Exogyra costata, Ostrea larva, Belemnitella mucronata, Terebratula sayi, (Gryphæa convexa* and *G. mutabilis), Ostrea vesicularis.*

3. Dark Colored Clay, containing Greensand in Irregular Stripes and Spots.

Ammonites delawarensis, A. placenta, A. conradi, Baculites ovatus, casts of Cardium.

Equivalent to Nos. 4 and 5 of the Nebraska Section.

2. DARK COLORED CLAY. [Position of beds Nos. 2 and 3 of the Nebraska section.]

At the present time the evidence tends to show that No. 1 of the Nebraska section is represented by Nos. 1 and 2, and that Nos. 2 and 3 of the Nebraska section are wanting, and would find a place between Nos. 2 and 3 of this section if existing.

This bed contains large quantities of fossil wood (no animal remains are known to occur in it).

1. FIRE CLAY AND POTTER'S CLAY.

This bed contains fossil wood, and numerous impressions of leaves; but no animal remains.

In making the comparison of the strata he placed a large part of the Cretaceous fossils of the boundary survey in the same parallel with beds Nos. 2 and 3 of the Nebraska section, and below those beds in New Jersey and Alabama, which contain *Baculites ovatus, Nautilus dekayi* and *Ammonites placenta.* He described from Leon Springs, *Pyrina parryi.*

Prof. T. A. Conrad described, from between El Paso and Frontera, *Turbinolia texana, Cucullæa terminalis, Arca subelongata, Cardium mediale, Cytherea texana, Ostrea vellicata, Nodosaria texana;* from Leon Springs, *Trigonia texana. Protocardia filosa, Cardita eminula, Lima leonensis, Cytherea leonensis, Ammonites geniculatus, A. leonensis, Capsa texana, Terebratula leonensis, Turritella leonensis;* from Rio San Pedro, *Cardium congestum, Natica collina, N. texana, Rostellaria collina, Buccinopsis parryi;* from Dry creek, Mexico, *Ostrea cortex, O. multilirata;* from Turkey creek, Leon and Eagle Pass roads, *Pholadomya texana;* from Jacun, three miles below Laredo, *Ostrea robusta, Ammonites pleurisepta;* from other places, *Corbula occidentalis, Inoceramus texanus, Astarte texana, Plicatula incongrua, Ostrea bella, O. lugubris, Turritella planilateris, Nerinea schotti,* and *Cardita subtetrica.*

Evans and Shumard* described, from Nebraska, *Avicula nebrascana, Limopsis striato-punctata. Cardium subquadratum,* and *C. rarum,* now *Protocardia subquadrata,* and *P. rara, Arca sulcatina,* now *Nemodon sulcatinus, Leda fibrosa,* now *Neæra fibrosa, Mytilus meeki,* now *Volsella meeki, Ostrea subtrigonalis, Pleurotoma minor, Fusus nebrascensis, Turritella multilineata, Rostellaria americana,* now *Anchura americana,* and *Ammonites galpini.*

Meek and Hayden† described, from the Great Bend of the Missouri,

---

* Trans. St. Louis Acad. Sci., vol. 1.
† Proc. Acad. Nat. Sci., vol. 9.

and other places in Nebraska (Fort Pierre Group), *Ptychoceras mortoni, Fusus subturritus,* now *Pyrifusus subturritus, F. intertextus,* now *P. intertextus, Xylophaga elegantula,* now *Turnus elegantulus, X. stimpsoni,* now *T. stimpsoni;* from (Fox Hills Group), near the mouth of Heart river, *Fusus vaughni;* and from other parts of Nebraska, *Fusus* (?) *scarboroughi,* now *Fasciolaria scarboroughi, Pholadomya subventricosa, Cyprina cordata,* now *Sphæriola* (?) *cordata, C. compressa, C. sublumida, C. ovata, Pectunculus subimbricatus,* now *Axinæa subimbricata, Ostrea translucida, Hemiaster humphreysanus;* from the mouth of Judith river, *Vitrina obliqua, Planorbis amplexus, Helix occidentalis,* now *Hyalina occidentalis, H. vitrinoides,* now *H. vetusta, Melania omitta,* now *Goniobasis omitta, M. subtortuosa,* now *G. subtortuosa, M. sublævis,* now *G. sublævis, M. invenusta,* now *G. invenusta, Unio danai, U. deweyanus, U. subspatulatus, Ostræa glabra;* from the Fort Union Group, Lignite beds at Fort Berthold on the Missouri river, *Planorbis fragilis,* now *P. planoconvexus, Melania tenuicarinata,* now *Goniobasis tenuicarinata, M. warrenana,* now *Hydrobia warrenana;* from the Fort Benton Group, at the mouth of Vermilion river, *Serpula tenuicincta;* from the Dakota Group, near the mouth of Vermilion river, *Solen dakotensis,* now *Phorella dakotensis,* and *Cyprina arenaria,* now *Cyrena arenaria.*

The rocks* of the Lower Cretaceous, in Mississippi, consist principally of stratified sand, mixed with a large proportion of silicate of iron or glauconite, which imparts to it a greenish color of different hues, and has given origin to the very appropriate name of greensand. The indurated greensand is generally full of fossils. It is exposed in the western part of Tishamingo, eastern part of Tippah, northwestern part of Itawamba, northeastern part of Pontotoc, and northeastern part of Lowndes county.

The Upper Cretaceous has sometimes been called the rotten limestone, and occupies a larger area than the lower division. It occupies part of Kemper, Noxubee, Lowndes, Ocktibbeha, Chickasaw, Monroe, Pontotoc and Itawamba counties. This division is also highly fossiliferous where well exposed. The estimated thickness of the whole is placed at from 1,200 to 1,500 feet.

Dr Leidy† described, from Columbus, Miss., *Hadrodus priscus;* from Nebraska, *Phasganodus dirus,* and from the greensand of New Jersey, *Pycnodus robustus.*

---

* Geo. of Miss.
† Proc. Acad. Nat. Sci., vol. 9.

In 1858, Dr. Geo. G. Shumard* described the Cretaceous rocks near the mouth of Delaware Creek, on the Rio Pecos, in New Mexico, where he found a thickness of 960 feet. The first 100 feet consists usually of a hard limestone, of a light cream color and earthy texture, and contains numerous spheroidal cavities, from a fourth to a half an inch in diameter, which are sometimes partially filled with loose, ferruginous earth. In other places it is softer and lighter colored, resembling impure chalk. Beneath this limestone, deposits of gypsum, clay and sandstone occur. In some places the strata are much disturbed, and are found dipping in opposite directions, at angles of 40° or 50°. He also referred to the Cretaceous† certain strata in the bluffs of the Mississippi, above Commerce, Missouri, having a thickness of 158 feet, but no fossils were obtained.

The Cretaceous rocks‡ occupy a belt across the State of Alabama, from 50 to 100 miles in width. The counties, either in whole or in part, exposing these rocks on the surface, are Barbour, Russell, Pike, Macon, Montgomery, Butler, Lowndes, Autauga, Wilcox, Dallas, Perry, Marengo, Greene, Choctaw, Sumpter and Pickens.

T. A. Conrad§ described, from Tippah county, Mississippi, *Pholadomya tippana, Periploma applicata, Siliquaria biplicata,* now *Leptosolen biplicatus, Legumen ellipticus, L. appressus, Dosinia densata, Meretrix tippana,* now *Aphrodina tippana, Papyridea bella, Cardium ripleyense, C. spillmani, C. tippanum, Opis bella, O. bicarinata, Tellina ripleyana, Nucula percrassa, Cibota lintea, Cucullæa capax, C. tippana,* now *Idonearca tippana, Dreissena tippana, Pinna laquata, Gervillia ensiformis, Lima acutilineata, Inoceramus argenteus, I. costellatus, Ostrea confragosa, O. peculiaris, O. denticulifera, Exogyra interrupta, Pulvinites argentea, Anomia sellæformis, Strombus densatus,* now *Pugnellus densatus, Aporrhais decemlirata,* now *Anchura decemlirata, Harpago tippanus,* now *Pterocerella tippana, Rimella currilirata, Conus canalis, Drillia tippana, D. novemcostata, Turris ripleyana, Fusus novemliratus, F. bellaliratus; Pyrifusus subdensatus, Ficus octoliratus, Rapa supraplicata, Volutilithes cretacea,* now *Volutomorpha cretacea, Chemnitzia distans, C. interrupta, Trichotropis cancellaria, Turritella altilis, T. tippana, Lunatia rectilabrum, Solidulus linteus, Bulliopsis cretacea, Baculites spillmani, B. tippaensis, Scaphites iris,* and *Cytherina tippana.*

Meek and Hayden‖ described, from (now the Fox Hills Group) Long

---

* Trans. St. Louis Acad. Sci. vol. 1.
† Proc. Am. Ass. Ad. Sci.
‡ Geo. of Ala., 1858.
§ Jour, Acad. Nat. Sci., 2d ser. vol. 3.
‖ Proc. Acad. Nat. Sci.

Lake, Nebraska, *Corbula inornata*, *Pholas cuneata*, now *Martisia cuneata* and *Actæon attenuata*.

From (now the Fort Pierre Group) near Fort Clark, *Teredo globosa*, *Helicoceras tortum*, now *Heteroceras tortum*, *Turrilites cochleatus*, now *Heteroceras cochleatum*, *H. tenuicostatum*, *Turrilites umbilicatus*, now *H. umbilicatum*, and *Ancyloceras uncum*.

From (now the Fort Benton Group) Fort Benton, on the Upper Missouri, *Inoceramus umbonatus* now *Volviceramus umbonatus*, and from the Black Hills, *Scaphites larvæformis*.

Dr. Leidy described, from the marl of Haddonfield, Camden county, New Jersey, *Hadrosaurus foulki*.

F. B. Meek* described, from Vancouver's Island, *Nucula traskana*, *Arca æquilateralis*, *A. vancouverensis*, *Cardium scitulum*, *Pholadomya borealis*, *P. subelongata*, *Trigonia evansana*, *Thracia occidentalis*, *T. subtruncata*, *Dentalium nanaimoense*, and *Ammonites ramosus*.

Dr. B. F. Shumard† described, from the same island, *Inoceramus vancouverensis*, *Pinna calamitoides*, and *Pyrula glabra*.

Prof. E. Emmons‡ described, from the Greensand of North Carolina, *Sphenodus rectidens*, and *Belemnitella compressa*.

Prof. Oswald Heer,§ of Zurich, Switzerland, described, from Nebraska, *Liriodendron meeki*, *Sapotacites haydeni*, *Leguminosites marcouanus*, now *Bumelia marcouana*, *Populus cyclophylla*, now *Cissites cyclophylla*, *Phyllites obcordatus*, and *P. obtusilobatus*.

In 1859, Prof. Henry Y. Hind‖ found the Cretaceous rocks in position on the Little Souris River, in longitude 100° 30′ W., and on the South Branch of the Saskatchewan, in longitude 106° 35′ W., and between these widely separated points in many places on the Assiniboine, the Qu'Appelle, and their affluents. Fifteen miles from the mouth of the Little Souris, the rocks consist of a very fissile, dark blue argillaceous shale, holding numerous concretions, containing a large per cent. age of iron. This exposure is 70 feet thick, and the layers are perfectly horizontal. The whole is supposed to be of the age of the Fort Pierre Group. The Cretaceous of this latitude appears to repose directly upon the Devonian, as the former is found undisturbed and nearly horizontal at altitudes from 400 to 600 feet above exposures of Devonian age, recognized *in situ* 30 miles to the east.

Prof. F. B. Meek¶ described, from the Little Souris River, *Anomia flemingi*, *Inoceramus canadensis*, *Leda hindi*, now *Nucalana hindi*, and

---

* Trans. Alb. Inst., vol. 4. † Trans. St Louis Acad. Sci., vol. 1.
‡ Geo. Sur. N. Carolina. § Proc. Acad. Nat. Sci.
‖ Assiniboine and Saskatchewan Expl. Exped.
¶ Rep. on Assiniboine and Saskatchewan Expl. Exped.

from the valley of Mackenzie's River, *Ammonites, barnstoni;* and *A. billingsi.*

Prof. Leo. Lesquereux* described, from Vancouver's Island, and Bellingham Bay, *Populus rhomboidea, Salix islandicus, Quercus benzoin, Q. multinervis, Q. evansi, Q. gaudini, Q. platinervis, Planera dubia, Cinnamomum heeri,* now *Daphnogene heeri, Persoonia oviformis,* and *Diospyros lancifolia.*

In 1860, Dr. B. F. Shumard† subdivided the Cretaceous strata of Texas in descending order, as follows: 1. Caprina limestone, having a thickness of 60 feet, and consisting of a yellowish white limestone, usually massive, sometimes of a finely granular structure, and sometimes made up of rather coarse, subcrystalline grains, cemented with a chalky paste. It has an extended geographical range. 2d. The Comanche Peak Group, having a thickness of 300 to 400 feet, and made up of soft, yellowish and whitish chalky limestone, and buff and cream-colored limestones of greater or less compactness, being highly fossiliferous, and having a great geographical extension. 3. The Austin limestone and fish bed, having a thickness from 100 to 120 feet. The Austin limestone consists of cream-colored and bluish earthy limestones, and the fish beds of shaly layers of dark-bluish-gray calcareous sandstone. This is supposed to represent Nos. 4 and 5 of the Nebraska section, by Meek & Hayden. 4. Exogyra arietina marl, having a thickness of 60 feet, and consisting of an indurated blue and yellow marl with occasional bands of gray limestone, and thin seams of selenite interstratified. 5. Washita limestone, having a thickness from 100 to 120 feet, a wide geographical range, and consisting of white, yellow, gray and blue limestones, some of which are moderately hard, and others disintegrate rapidly. This is supposed to be parallel with the lower part of No. 3 of the Nebraska section, by Meek and Hayden. 6. Blue marl, having a thickness of 50 feet, and consisting of an indurated arenaceous marl, of a schistose structure, with small nodules of iron pyrites and irregular masses of lignite disseminated through it. It is not observed south of Grayson county, and is supposed to correspond with No 2 of the Nebraska section. 7. Caprotina limestone, having a thickness of 55 feet, and forming the basis of what is called the Upper Cretaceous Group. It is composed of light gray and yellowish gray earthy limestone, with intercalated bands of yellow marl and sometimes flint, and is exposed at the base of the hills near Comanche Peak, and underlying the Washita limestone near the Colo-

---

* Amer. Jour. Sci. and Arts, 2d Series, vol 27.
† Trans. St. Louis Acad. Sci., vol. 1.

rado, at the foot of Mt. Bonnell. 8. The Arenaceous Group and fish bed, having a thickness of 80 feet, and consisting of light yellow and blue sandstone, and beds of sandy clay, with crystals of selenite and some lignite. This is supposed to be the same as B, C and D of the Pyramid Mount Section of Prof. Marcou, and by him referred to the Jurassic period, and to be equivalent to No. 1 of the Nebraska Section. 9. Marly Clay or Red River Group, having a thickness of 150 feet, and supposed to represent the lower part of the Pyramid Mt. Section, which Prof. Marcou referred to the Trias.

He described *Nautilus texanus, Ammonites inæquiplicatus, A. swallovi, A. meekanus, A. graysonensis, A. brazoensis, Scaphites vermiculus, Ancyloceras annulatum, Baculites gracilis, Cerithium bosquense, Phasianella perorata, Avellana texana, Natica acutispira, Neritopsis biangulatus, Venus sublamellosus, Cardium choctawense,* now *Protocardia choctawensis, C. coloradoense, C. brazoense,* now *Protocardia brazoensis, Cytherea lamarensis,* now *Dione lamarensis, Tapes hilgardi, Arca proutana, Lucina sublenticularis, Nucula haydeni, N. serrata, Corbula graysonensis, C. tuomeyi, Pachymya austinensis, Panopæa newberryi, P. subparallela, Inoceramus capulus, Gervillia gregaria, Janira wrighti, Ostrea belliplicata, O. quadriplicata, Cidaris hemigranosus.*

Wm. M. Gabb* described, from Prairie Bluff, Alabama, *Chemnitzia meekana, Straparollus subplanus, Scousia alabamensis, Cancellaria alabamensis,* now *Turbinopsis alabamensis,* and *Bulla macrostoma;* from the marl of New Jersey, *Actæonia naticoides,* now *Cinulia naticoides, Phasianella punctata, Volutilithes biplicata,* now *Rostellites biplicatus, V. bella,* now *R. bellus, V. nasuta,* now *R. nasutus, V. conradi,* now *R. conradi, Fusus retifer, Rapa elevata, Morea naticella, Bulla recta, Mysia gibbosa, Dione delawarensis, Crassatella delawarensis, C. monmouthensis, Cardita subquadrata, Leda pinniformis,* now *Nuculana pinniformis, L. protexta,* now *N. protexta, Cultellus cretaceus, Pecten burlingtonensis;* from Tennessee, *Volutilithes saffordi,* and *Cardium abruptum;* from New Jersey, *Actæonina biplicata,* now *Solidula biplicata, Solarium abyssinus,* now *Margaritella, abyssinus, Volutilithes abbotti, Turbinella subconica, T. parva, Cancellaria septemlirata, Purpuroidea dubia, Fusus trivolvis, Rapa pyruloidea, Pleurotoma mullicaensis, Arca quindecemradiata, Cibola multiradiata,* and *Leda angulata,* now *Nuculana angulata, Desmatocium trilobatum,* and from Eufala, Alabama, *Cassidulus micrococcus.*

Wm. M. Gabb* described, from Eufaula, Alabama, *Fusus holmesan-us, Cancellaria eufaulensis, Dentalium ripleyanum, Venus meekana Astarte octolirata, Trigonia eufaulensis, Axinæa rotundata, Nucula distorta, N. eufaulensis, Arca eufaulensis,* now *Nemodon eufaulensis, Hamulus major;* from Hardeman county, Tennessee, *Neptunea impressa, Fasciolaria saffordi, Turritella hardemanensis, T. pumila, T. saffordi, T. tennesseensis, Venus ripleyana, Corbula subcompressa, C. crassiplicata, Modiola saffordi,* now *Volsella saffordi, Arca saffordi, Ostrea crenulimarginata;* from New Jersey, *Rostellaria rostrata,* now *Anchura rostrata, Cypræa mortoni, Lunatia halli, Pholas cretacea,* now *Martesia cretacea, Teredo irregularis,* now *Polarthrus irregularis, Gastrochæna americana,* now *Polarthrus americanus, Isocardia conradi,* now *Opis conradi, Modiola ovata,* now *Volsella ovata, Leda slackiana,* now *Nuculana slackana, Serpula habrogramma, Dentalina pulchra,* now *Phonemus pulcher;* from the Indian Territory, near the Choctaw Mission, *Chemnitzia occidentalis;* and from Oregon, *Discoidea occidentalis;* Gabb and Horn described, from Hardeman county, Tennessee, *Platytrochus speciosus;* from Prairie Bluff, Alabama, *Flabellum striatum;* from New Jersey, *Trochosmilia conoidea, Acerviclausa vermicularis, Heterocrisina abbotti,* now *Bicrisina abbotti, Hippothoa irregularis,† Cellepora carinata,* now *Reptoporina carinata, C. typica,* now *Escharifora typica, Reticulipora sagena,* and *Multicrescis parvicella.*

T. A. Conrad‡ described, from Barbour county, Alabama, and Tippah county, Mississippi, *Pholadomya anteradiata, P. papyria, P. postsulcata, Sanguinolaria cretacensis, Tellina eufaulensis, T. limatula, T. eborea, Dosina depressa,* now *Cyprimeria depressa, D. obliquata, Mysia parilis,* now *Tenea parilis, Cardium linteum,* now *Cymbophora lintea, Crassatella lintea, C. pteropsis, Linearia metastriata, Kellia cretacea, Sphærella concentrica, Crenella sericea, Cucullæa maconensis,* now *Trigonarca maconensis, Nucula cunei-*

* Jour. Acad. Nat. Sci., 2d ser. vol. 4
† Proc. Acad. Nat. Sci.
‡ Jour. Acad. Nat. Sci., 2d ser. vol. 4.

frons, *N. peræqualis*, *Leda longifrons*, now *Nuculana longifrons*, *Venilia trapezoidea*, now *Veniella trapezoidea*, *Cardium eufaulense*, *Dione eufaulensis*, *Astarte crenalirata*, *Corbula eufaulensis*, *Plicatula saffordi*, *P. tetrica*, *Pecten argillensis*, *P. mississippiensis*, *P. simplicius*, now *Syncyclonema simplicium*, *Turrilites spiniferus*. *Anchura abrupta*, *Turritella trilira*, *Daphnella eufaulensis*, *D. lintea*, *D. subfilosa*, *Drillia distans*, *Fusus tippanus*, *Strepsidura ripleyana*, *Volutilithes eufaulensis*, *Actæon modicellus*, *Chemnitzia corona*, *C. melanopsis*, *C. spillmani*, *C. laqueata*, *C. trigemmata*, *Pyropsis perlata*, *Neritella densata*, *Gyrodes alveatus*, *G. crenatus*, *Turbinopsis hilgardi*, *Tuba bella*, now *Spironema bellum*, *Morea cancellaria*, *Thylacus cretaceus*, *Placunanomia saffordi*, *Cassidulus abruptus*, and *C. subquadratus.*

Prof. E. W. Hilgard* subdivided the Cretaceous rocks of Mississippi into four groups as follows : 1. The lowest, the Eutaw Group, as characterized by Tuomey, near Eutaw, Alabama. 2. Tombigbee Sand Group. 3. Rotten Limestone Group. 4. Ripley Group of Conrad.

The Eutaw Group consists of bluish black, or reddish, laminated clays, often lignitic, alternating with, and usually overlaid by non-effervescent sands, mostly poor in mica, and of a gray or yellow tint. It contains beds of lignite, and rarely other fossils. It is displayed at a few places in Tishamingo, Itawamba, Monroe, and Lowndes counties.

The Tombigbee Group is usually a fine-grained micaceous sand, more or less calcareous, usually of a greenish tint, but not unfrequently gray, bluish, black, yellowish and sometimes even orange red. The greenish tint is imparted to these sands not by greensand grains, as is the case in the marls of the Ripley Group, but is caused by a greenish incrustation, covering thinly a portion of the quartz grains, but the presence of glauconite in the incrustation has not been detected. Clays and non-calcareous sands are subordinate to the greenish sand. This Group forms a narrow belt on the western exposure of the Eutaw Group, and extending from Lowndes county through Monroe, Itawamba and Tishamingo, to the southern border of Tennessee.

The Rotten Limestone Group possesses the same characteristics ascribed to it by Tuomey in Alabama, and appears as a soft, chalky rock, of a white or pale bluish tint, with very little sand ; consisting of variable proportions of fat, tenacious clay, and white carbonate of lime in crystals extremely minute, and with some shells of infusoria. It is generally highly fossiliferous, and irregular, rounded nodules of iron pyrites of a radiated structure called "sulphur balls" are common. It is of great thickness on its southwestern border in Chickasaw, Octibbeha, Noxubee and Kemper counties, where borings have been made in it from 700 to 1000 feet, but there is a gradual thinning out northward, through Pontotoc, Itawamba, Tippah and Tishamingo counties to the line of Tennessee. The surface area of this subdivision in Mississippi is greater than that of the other three combined.

The Ripley Group forms the border of the western exposure of the Cretaceous, from a point in Chickasaw through the central part of Pontotoc, the eastern part of Tippah and western part of Tishamingo to the south line of Tennessee. It consists of hard crystalline limestone, more or less sandy and glauconitic, which forms the highest strata; and bluish micaceous marls, more or less sandy, and often interstratified with subordinate ledges of sandy limestone, which latter become less and less frequent as we descend in the series toward the strata forming a transition into the Rotten limestone.

Meek and Hayden* described, from (Fort Benton Group) the mouth of Vermilion river, *Ammonites vermilionensis*, now *Mortoniceras vermilionense ;* from near the Black Hills, *Scaphites warreni ;* from Little Blue river, *Inoceramus ariculoides*, now *I. problematicus*, var. *ariculoides ;* from (Fox Hills Group), Moreau river, *Phylloteuthis subovatus*, *Dentalium pauperculum*, now *Entalis paupercula*, and *Cylichna scitula ;* from 20 miles below the mouth of Cannon Ball river, *Tellina formosa*, now *Linearia formosa ;* from the north branch of Cheyenne river, *Cyprina humilis*, now *Veniella humilis ;* and from Long Lake, *Aricula subgibbosa*, now *A. linguiformis*, var. *subgibbosa ;* from the mouth of Judith river (Judith river Group), *Helix evansi*, now *Hyalina evansi;* from the mouth of Grand river, *Sphærium planum*, *S. recticardinale*, *Cyrena cytheriformis*, now *Corbicula cytheriformis*, and *Inoceramus subcompressus*, now *I. cripsi*, var. *subcompressus ;* from (Fort Pierre Group) the head of the south branch of Cheyenne River, *Helicoceras angulatum*, now *Heteroceras angulatum*, *Ammonites placenta*, var. *intercalaris;* now *Placenticeras placenta*, var. *intercalaris*, from the Yellow Stone river, *Scaphites nodosus*, var. *plenus*, *Aporrhais parra*, now *Anchura parra*, *A. sublaevis*, and *Mactra gracilis;* from Fort Clark, *Teredo selliformis ;* from White river, *Inoceramus vanuxemi*, *I. balchi;* from Bijou Hill, *Anomia subtrigon-*

alis, and from the great bend of the Missouri river below Fort Pierre, *Ostrea inornata.*

From (Niobrara Group) near the mouth of the Niobrara river, *Anomia obliqua;* from (Dakota Group) near the mouth of the Big Sioux river, *Mactra Siouxensis.* F. B. Meek described, from near Bear river, on Sulphur Creek, *Anomia concentrica, Corbula concentrica, C. engelmanni, C. pyriformis,* and *Melania humerosa,* now *Pyrgulifera humerosa;* from the North Platte, *Inoceramus simpsoni;* from Ham's Fork, northeast of Fort Bridger, *Melampus priscus,* now *Rhytophorus priscus, Melania simpsoni,* now *Goniobasis simpsoni, M. arcta, M. nitidula,* now *Limnæa nitidula, L. similis, L. vetusta, Planorbis spectabilis, P. utahensis,* and from near Fort Bridger, *Unio haydeni.* Some of the latter species probably belong to the Lower Eocene.

In 1861, Meek and Hayden,[*] as before mentioned, separated the Cretaceous rocks of the Missouri region into five subdivisions, in ascending order, as follows :

1. Dakota Group, consisting of yellowish, reddish, and occasionally white sandstone, with, at places, alternations of various colored clays and beds and seams of impure lignite ; also silicified wood, and great numbers of leaves of the higher types of dicotyledonous trees, with casts of *Pharella dakotensis, Axinaea siouxensis,* and *Cyprina arenarea.* Found at the hills back of the town of Dakota; also extensively developed in the surrounding country in Dakota county, below the mouth of Big Sioux river, thence extending southward into northeastern Kansas and beyond. Estimated thickness, 400 feet.

2. Fort Benton Group, consisting of dark gray, laminated clays, sometimes alternating near the upper part with seams and layers of soft gray and light colored limestone, *Inoceramus problematicus, I. tenuirostratus, I. latus, I. fragilis, Ostrea congesta, Venilia mortoni, Pholadomya papyracea, Ammonites mullani, A. percarinatus, A. vespertinus, Scaphites warreni, S. larvaformis, S. ventricosus, S. vermiformis, Nautilus elegans,* etc. Extensively developed near Fort Benton, on the Upper Missouri; also along the latter from ten miles above James river to Big Sioux river, and along the eastern slope of the Rocky Mountains as well as at the Black Hills. Estimated thickness, 800 feet.

3. Niobrara Group, consisting of lead-gray calcareous marl, weathering to a yellowish or whitish chalky appearance above, containing

---

large scales and other remains of fishes, and numerous specimens of *Ostrea congesta*, attached to fragments of *Inoceramus*. Passing down into light yellowish and whitish limestone, containing great numbers of *Inoceramus problematicus*, *I. pseudomytiloides*, *I. ariculoides*, fish scales, etc. Found in the bluffs along the Missouri, below the Great Bend, to the vicinity of Big Sioux River; also below there on the tops of the hills. Estimated thickness, 200 feet.

4. Fort Pierre Group, consisting of dark beds of very fine unctuous clay, containing much carbonaceous matter, with veins and seams of gypsum, masses of sulphuret of iron, and numerous small scales, fishes local, filling depressions in the bed below. Lower fossiliferous zone, containing *Ammonites complexus*, *Baculites ovatus*, *B. compressus*, *Helicoceras mortoni*, *H. tortum*, *H. umbilicatum*, *H. cochleatum*. *Ptychoceras mortoni*, *Fusus vinculum*, *Anisomyon borealis*, *Amauropsis paludiniformis*, *Inoceramus sublaevis*, *I. tenuilineatus*, bones of *Mosasaurus missouriensis*, etc. Middle zone, nearly barren of fossils. Upper part consisting of dark gray and bluish plastic clays, containing, near the upper part, *Nautilus dekayi*, *Ammonites placenta*, *Baculites ovatus*, *B. compressus*, *Scaphites nodosus*, *Dentalium gracile*, *Crassatella, evansi*, *Cucullaea nebrascensis*, *Inoceramus sagensis*, *I. nebrascensis*, *I. vanuxemi*, bones of *Mosasaurus missouriensis*, etc. Found on Sage creek, Cheyenne river, White river above the Mauvaises Terres, Fort Pierre, and out to Bad Lands, down the Missouri on the high country, to Great Bend and near Bijou Hill on the Missouri. Estimated thickness, 700 feet.

5. Fox Hills Group, consisting of gray, ferruginous and yellowish sandstone and arenaceous clays, containing *Belemnitella bulbosa*, *Nautilus dekayi*, *Ammonites placenta*, *A. lobatus*, *Scaphites conradi*, *S. nicolletti*, *Baculites grandis*, *Busycon bairdi*, *Fusus culbertsoni*, *F. newberryi*, *Aporrhais americana*, *Pseudobuccinum nebrascense*, *Mactra warrenana*, *Cardium subquadratum*, and a great number of other molluscous fossils, together with bones of *Mosasaurus missouriensis*, etc. Found at Fox Hills, near Moreau river, near Long Lake, above Fort Pierre, along the base of Big Horn Mountains, and on North and South Platte rivers. Estimated thickness, 500 feet.

In Nebraska the sandstones of the Dakota Group rest directly upon rocks of the age of the Upper Coal Measures, or of Permian Age.

They described from the Fort Pierre and Fox Hills Groups, at Deer Creek, near the north branch of the Platte river, *Leda bisulcata*, now *Nuculana bisulcata*, *Gervillia recta*, *Crenella elegantula*, *Cardium pertenue*, now *Protocardia pertenuis*, *Tellina nitidula*, now *Mactra nitidula;* from the mouth of the Big Horn river, *Lingula nitida,* and from the head of Gros-Ventres river, *Ostrea gabbana,* and *Cardium curtum.* And from the Fort Union Group, on the Lower Fork of Powder river, *Viviparus reynoldsanus.*

F. B. Meek described, from Vancouver and Sucia Islands, *Dosinia tenuis, Inoceramus subundatus, Mactra gibbsana, Baculites inornatus, B. occidentalis, Ammonites complexus,* var. *suciensis, A. vancouverensis,* and *Nautilus campbelli.*

W. M. Gabb[*] described, from Crosswicks and other places in New Jersey, *Turritella granulicosta, Crassatella transversa,* now *Etea transversa, Axinæa subaustralis, Ctenoides squarrosa,* now *Lima squarrosa, Terebratulina hallana, Actæon cretaceus, A. ovoideus, Natica infracarinata, Lunatia altispira, Gyrodes obtusivolvus, G. abbotti, Turbinopsis depressa, Architectonica abbotti,* now *Margaritella abbotti, Fasciolaria slacki, Voluta delawarensis, V. kanei, V. mucronata;* from Comanche Peak, Texas, *Globiconcha curta;* from Mississippi, *Gyrodes spillmani, Ostrea pandiformis;* from Alabama, *Trochus mortoni, Gryphæa thirsæ;* from New Jersey, *Teredo contorta,* now *Turnus contortus, Anatina elliptica,* now *Periploma elliptica, Venilia trigona,* now *Veniella trigona, Arca altirostris, Cucullæa neglecta,* now *Idonearca neglecta, C. transversa,* now *I. transversa, Pecten tenuitesta, Eudea dichotoma;* and from Tennessee, *Ctenoides denticulicosta,* now *Lima denticulicosta.*

Isaac Lea[†] described, from Haddonfield, New Jersey, *Corbula foulkei, Dosinia haddonfieldensis,* and *Modiola juliæ,* now *Volsella juliæ.*

In 1862, Gabb and Horn[‡] described, from Timber Creek and Mullica Hill, New Jersey, *Cellepora prolifica, C. exserta, C. pumila, Reptocelleporaria aspera, Escharinella muralis, Reptescharellina prolifera, Escharipora distans, E. abbotti, E. immersa, Reptescharipora marginata, Biflustra torta, B. disjuncta, Membranipora abortiva, M. perampla, M. plebia, Flustrella capistrata, F. cylindrica, Reptoflustrella heteropora, Retelea ovalis, Fascipora americana, Spiropora calamus, Entalophora, quadrangularis, E. conradi, Diastopora lineata, Stomatopora regularis,* now *Alecto regularis, Reticulipora dichotoma, Crescis labiata,* and from near Yazoo, Mississippi, *Cellepora janewayi.*

Meek and Hayden[§] described, from the Fort Benton Group, at

* Proc. Acad. Nat. Sci.
† Proc. Acad. Nat. Sci.
‡ Jour. Acad. Nat. Sci., 2d ser., vol 5.
§ Proc. Acad. Nat. Sci.

Chippewa Point, below Fort Benton, on the Upper Missouri, *Scaphites ventricosus, S. vermiformis, Ammonites mullananus, Nautilus elegans,* var. *nebrascensis, Inoceramus undabundus, I. exogyroides, I. tenuirostratus, Venilia mortoni,* now *Veniella mortoni,* and *Pholadomya papyracea.*

In 1863, Dr. J. S. Newberry* described, from Orcas Island, *Asplenium elongatum, Tæniopteris gibbsi,* and *Ficus cuneatus;* from Vancouver's Island, *Aspidium kennerlyi,* and *Taxodium cuneatum.*

In 1864, W. M. Gabb† described, from Chico creek, California, *Callianassa, stimpsoni, Ammonites chicoensis, Turritella chicoensis, Meretrix lens, Dosinia inflata, Trochosmilia granulifera;* from Cottonwood creek, and other places in Shasta county, California, *Belemnites impressus, Ammonites breweri, A. haydeni, A. traski, A. hoffmanni, A. remondi, Ptychoceras æquicostatum,* now *Helicaucylus æquicostatus, Crioceras remondi,* now *Ancyloceras remondi, C. percostatum,* now *A. percostatum, Fusus kingi, Neptunea curvirostra, N. perforata, N. hoffmanni, Lunatia arellana, Ringicula varia, Nerinea dispar, Acteonina pupoides, Pugnellus manubriatus, Potamides diadema, Turritella infralineata, Nerita deformis, Anisomyon meeki, Actæon impressus,* now *Tornatellæa impressa, Turnus plenus, Panopæa concentrica,* now *Homomya concentrica, Meretrix arata, Eriphyla umbonata, Lithophagus oviformis, Arca brewerana,* and *Leda translucida,* now *Nuculana translucida ;* from Martinez and Clayton, and Canada de las Uva, *Aturia matthewsoni, Helicoceras vermiculare. Typhis antiquus, Fusus martinez, F. matthewsoni,* now *Surcula matthewsoni, F. aratus. F. flexuosus, Neptunea gracilis, Perissolax brevirostris, Buccinum iiratum,* now *Brachysphingus liratus, Nassa cretacea, N. antiquata, Pseudoliva lineata, Olivella matthewsoni;* from San Diego, *Ammonites cooperi, Hemifusus cooperi, Neptunea supraplicata, Tritonium diegoense,* now *Buccinofusus diegoensis, Ancillaria elongata, Fasciolaria læviuscula, F. sinuata, Margaritella crenulata;* from Vancouver Island, *Hamites vancouverensis, Solen diegoensis, Barbatia morsei;* from Martinez, *Mitra cretacea, Morio tuberculatus, Lunatia shumardana, Naticina obliqua,* now *Catinus obliquus, Cinulia matthewsoni;* from Pence's Ranch, Butte county, *Helicoceras breweri, H. declive, Ptychoceras quadratum, Neptunea ponderosa, Haydenia impressa, Gyrodes conradanus, G. expansus, Potamides tenuis, Martesia clausa, Anatina lata;* from Trinity river in Trinity county, *Crioceras latum;* from Fort Tejon, *Fasciolaria io,* now *Surcula io, Whitneya ficus, Lunatia horni.*

---

* Bost Jour. Nat Hist., vol. 7.
† Pal. of California, vol. 1.

L. nuciformis, Neverita secta, Architectonica horni, Conus horni, C. sinuatus, now *Surcula sinuata;* from Tuscan Springs, *Fusus averilli, Ficus cypræoides, Amauropsis oviformis, Cinulia obliqua ;* from Mount Diablo and other places, *Fusus diaboli, F. californicus, Hemifusus horni, H. remondi. Turris claytonensis. T. varicostata, Cordiera microptygma, Tritonium horni, T. paucivaricatum. T. whitneyi, Pseudoliva volutiformis ;* from Martinez, near Benicia, Tuscan Springs, Texas flat, in Placer county, Clayton, Fort Tejon, Alameda county, Pence's Ranch, Contra Costa county, Rag Canon and other parts of California, *Cinulia pinguis, Acteonina californica, Globiconcha remondi, Cylindrites brevis, Niso polita, Cerithiopsis alternata, Architectonica veatchi, A. cognata, A. inornata, Margaritella globosa, Discohelix leana, Straparollus paucivolvus, S. lens, Angaria ornatissima, Conus remondi, Rimella canalifera, R. simplex, Pugnellus hamulus, Tessarolax distorta. Aporrhais falciformis,* now *Anchura falciformis, A. angulata, A. californica, A. exilis, Cypræa bayerquei, Littorina compacta,* now *Ataphrus compactus, Turritella veatchi, T. robusta, Galerus excentricus,* now *Galeropsis excentricus, Spirocrypta pileum, Nerita cuneata, Lysis duplicosta, Dentalium pusillum, D. cooperi, D. stramineum, Emarginula radiata, Patella traski, Helcion circularis, H. dichotoma, Bulla horni, Cylichna costata, Megistoma striatum, Solen parallelus,* now *Plectosolen parallelus, Pharella alta, Corbula primorsa, C. traski, C. cultriformis, C. horni, C. parilis, Anatina tryonana, A. inæquilateralis, Pholadomya breweri, P. nasuta, Neæra dolabriformis, Mactra ashburneri,* now *Cymbophora ashburneri, Lutraria truncata, Asaphis undulata, Gari texta, Tellina longa, T. remondi, T. hoffmannana, T. monilifera, T. ovoides, T. matthewsoni, T. decurtata, T. quadrata, T. ashburneri, T. parilis, T. horni, T. californica, Venus varians, V. veatchi, V. lenticularis, V. tetrahedra, Meretrix horni, M. nitida, M. longa, M. ovalis,* now *Cyprinopsis ovalis, Dosinia elevata, D. gyrata,* now *Lucina gyrata, Tapes conradana, T. quadrata, Trapezium carinatum, Cyprinella tenuis, Cardium annulatum, C. remondianum, C. cooperi, C. breweri, C. placerensis, Cardita horni, Lucina nasuta, L. postradiata, L. subcircularis, L. cumulata, L. cretacea, Loripes dubia,* now *Clisocolus dubius, Mysia polita, Astarte conradana, A. matthewsoni, A. tuscana, Crassatella grandis, Anthonya cultriformis, Unio penultimus, Mytilus pauperculus, M. ascia, Modiola ornata,* now *Volsella ornata, M. cylindrica,* now *V. cylindrica, Septifer dichotomus, Crenella concentrica,* now *Stalagmium concentricum, Avicula pellucida, Inoceramus piochi, Pinna breweri, Trigonia tryonana. Meekia sella,*

M. radiata, M. navis, Area horni, A. gravida, A. decurtata, Cucullæa matthewsoni, C. truncata, Axinæa veatchi, A. sagittata, A. cor, Limopsis transversa, Pecten traski, P. operculiformis, P. californicus, Lima microtis, L. appressa, Plicatula variata, Ostrea breweri, Exogyra parasitica, Terebratella obesa, Flabellum remondianum, Trochosmilia striata, Astrocœnia petrosa, Ficus mammillatus, now *Ficopsis mammillatus, Natica uvasana, Scalaria matthewsoni, Turritella infra-granulata, Chione angulata,* now *Callista angulata, Tapes cretacea, Cardita veneriformis, Yoldia nasuta, Placunanomia inornata;* from Siskiyou Mountains of Oregon, *Siliqua oregonensis, Tellina whitneyi, Dosinia pertenuis, Modiola siskiyouensis,* now *Volsella siskiyouensis,* and *Ostrea malleiformis.*

In 1865, J. D. Whitney[*] described the Contra Costa Hills, which consist of a subordinate group of elevations, lying west of Martinez and the San Ramon and Livermore Valleys, and extend through Contra Costa county into Alameda and Santa Clara, and finally become merged in the Mount Hamilton division of the Monte Diablo Range. They are made up of Tertiary and Cretaceous strata, usually but little metamorphosed, although a belt extending along their western side is considerably altered from its original character.

Beginning at the northwest extremity of the group at Martinez, we have in the immediate vicinity of that place Cretaceous strata, well exposed in the bluffs along the Straits of Carquines. Here the rocks observed are sandstones, shales and argillaceous limestones, the latter forming bands and lenticular masses in the shales, generally but a few inches thick, although sometimes as much as three feet. Their strike is usually about N. 42° W, varying, however, from N. 39° W, to N. 44° W, and they dip southwest at an angle of from 35° to 60°. The rocks near Martinez have furnished a great many species of fossils.

In passing along the shore of the Straits of Carquines, west of Martinez, the Cretaceous strata occur for about seven miles, and are made up of shales and sandstones, the former containing frequent thin layers of hydraulic limestone. These rocks exhibit but few fossils. The dip and strike are variable, but generally about east and west magnetic, and the dip is also irregular, but almost always to the southwest, and at almost every angle from nearly horizontal to vertical ; the strike is nearly parallel with the line of the Straits. Near the upper limit of the Cretaceous, are sandstones very like those of Monte Diablo, which accompany the coal, and they contain a considerable quantity of carbonaceous matter, but no regular coal bed, so far as yet discovered.

---

[*] Geology of California, vol. 1.

Near these carbonaceous strata, and above them, is a narrow belt, partly altered and folded, and from 150 to 200 feet in width. The Rodes Valley marks the limit of the Cretaceous going west from Martinez, the Tertiary succeeding in that direction, and resting conformably on the strata beneath, and having the same general southwestern dip. South of Martinez, the Cretaceous strata have a higher dip, but in the same direction.

Near the "Walnut Creek House," a small patch of Cretaceous occurs, extending over a few acres, from which the overlying Tertiary, forming the crown of a low anticlinal, has been denuded.

Monte Diablo is one of the most conspicuous and best known landmarks of California. The central mass is made up of metamorphic rocks; it is about six miles long, and 1½ miles in width, and is surrounded on all sides by entirely unmetamorphosed strata. It is of an irregular crescent form, the concave side turned to the north northeast. It consists essentially of a central portion of very hard metamorphic sandstone, containing considerable epidote, flanked on both sides by jaspers, silicified shales and slates. The former constitutes the north peak, the latter the main peak, on Monte Diablo itself. Along the flanks of the ridge of which Eagle point is the culmination, one may observe the gradual passage of the argillaceous sandstone into the hard dioritic or trappean rock. The strata may be traced in all stages of passage, from the soft sandstone to the hardest and most crystalline rock. On the outside of the great central metamorphic mass, both on the north and south, are heavy accumulations of jaspery rock, one of the most peculiar features of the mountain, and the material of which the culminating point itself is made up. The jasper varies in color from a dull brick red to a brilliant vermillion hue, and may be traced in the ravines in which Bagley creek heads, passing into the unaltered shales of undoubted Cretaceous age, containing *Ammonites Inoceramus*, and other fossils. These jaspers are evidently the result of the alteration of the Cretaceous shales. Gold, copper and cinnabar have been found in these metamorphosed rocks.

The unaltered Cretaceous strata, consisting of shales, sandstones and argillaceous limestone, flank the sides of Monte Diablo, and run out into the plains of the San Joaquin in long, low, and almost parallel ridges. Coal has been found in the shales, but the extensive workable beds are included in the sandstones belonging to the upper part of the Cretaceous. The Arroyo del Puerto, Lone Tree Canon, and Hospital Canon, cut through sandstones and shales of Cretaceous age. The summit of Mount Oso is composed of jaspers, generally dull red, but often gray and green, with reticulations of quartz, like the rocks of Monte Diablo, and consists of metamorphosed Cretaceous rocks.

North of the mouth of San Luis creek the strata consist of conglomerates, sandstones and shales of Cretaceous age. The conglomerates generally form the crests of the ridges, and are very coarse, containing numerous boulders from one to two cubic feet in size. These consist of porphyry, granite, and various forms of metamorphic rock, entirely unlike the metamorphic Cretaceous of the center of the chain.

The jaspery beds of Chisnantuck are the exact counterpart of those of Monte Diablo, which we know to be Cretaceous, and those of Mine Hill, which contain the deposits of cinnabar, are evidently the continuation of those of Mount Chisnantuck. And as we trace them farther north, to the extremity of the peninsula, we find them still retaining the same lithological character, while we have there the evidence of fossils to prove them to belong to the Cretaceous epoch. Deposits of cinnabar have been found in rocks of Triassic and Tertiary age, but the large and valuable deposits are in the Cretaceous.

The larger portion of the rocks which make up the north end of the peninsula of San Francisco, are Cretaceous. The rocks in the vicinity of Clear Lake, when not of volcanic origin, are Cretaceous, and are the continuation of a great belt of strata of this age, which commences at Benicia, and stretches off to the northwest for an indefinite distance, apparently coming out to the ocean in the neighborhood of Cape Mendocino. The termination of the Coast Range at Benicia is of unaltered Cretaceous strata, much broken, and forming rounded hills, destitute of trees. Here as in Contra Costa county, the Cretaceous is well represented both by the bluish clay shales, with interstratified beds of argillaceous limestone, and by the overlying masses of blue and yellow sandstones, the latter in very heavy beds.

The Cretaceous formation, which is seen cropping out near the northern base of the twin sisters, is continuous from that place as far north as Capel Valley, at which point it becomes metamorphic and broken. The hills lying between the Sacramento and Suisun Valleys are of this age, and appear to form a line of foot hills along a high Cretaceous ridge, made up of unaltered shales and sandstones, running nearly northwest and southeast, and extending from Suisun Bay to Puta Creek. This range is about 3,500 feet high, and the ridge along the summit is formed by heavy bedded sandstones.

Cretaceous strata, in the San Emidio Canon, are seen resting on the granite and upturned edges of the mica and hornblende slates. At the Canada de los Alisos, which debouches into the plain four miles east of the Las Uvas Canon, the Cretaceous belt is over a mile wide, and forms hills about 1,000 feet above the plain.

The base of the foot hills of the Sierra Nevada is bordered, for a large part of the distance, between Tejon Pass and the head of the Sacramento Valley, by a series of beds of stratified Cretaceous materials-resting apparently undisturbed, and in a nearly horizontal position, on the upturned edges of the metamorphic slates and granitic rocks of which the foot-hills are formed. These materials, however, are not seen farther south than Folsom. Good exposures may be seen on Butte and Chico creeks. On the north side of the Cottonwood, beginning at Horsetown, and extending west to the Coast Ranges, Cretaceous strata lie at the base of the mountains, and form a table-land about 1,200 feet high, and have generally a southeasterly dip. Cottonwood creek runs nearly south from the summit of the Siskiyou Mountains to the Klamath river, a distance of 13 miles. The valley of which is about 10 miles long, and is excavated in the softer and unaltered Cretaceous strata, having on either side harder rocks, namely the auriferous slates on the west, and the modern volcanic on the east.

Dr. Joseph Leidy[*] described, from New Jersey, *Crocodilus tenebrosus, C. obscurus,* now *Holops obscurus, Cœlosaurus antiquus, Tomodon horrificus,* now *Diplotomodon horrificus, Chelone sopita,* now *Osteopygis sopitus, Emys firmus,* now *Agomphus firmus, E. beatus,* now *Adocus beatus,* and *Bothremys cooki;* from Maryland, *Astrodon johnstoni;* and from Minnesota, *Piratosaurus plicatus.*

T. A. Conrad,[†] from New Jersey, *Ostrea tuomeyi, Mortonia turgida,* and *Volutilithes lioderma,* now *Leioderma lioderma.*

In 1866, T. A. Conrad[‡] described, from Alabama, *Diploschiza cretacea,* and *Terebratulina filosa.*

Prof. E. D. Cope[§] described, from the greensand two miles south of Barnesboro, Gloucester county, New Jersey, *Lalaps aquilunguis,* and from Camden county, *Aturia paucifex.*

In 1867, Prof. F. V. Hayden[‖] referred the rocks at Yankton, the capital of Dakota Territory, located on the Missouri, about twelve miles above the mouth of the James, to the yellow, calcareous marl beds of No. 3, of the Niobrara division of the Cretaceous. The same rocks were found at Fort James, about twelve miles below the mouth of Firesteel creek, a branch of the James, and their thickness

* Cret. Reptiles, U. S., vol. 14, Smithsonian Contributions.
† Proc. Acad. Nat. Sci.
‡ Am. Jour. Conch., vol. 2.
§ Proc. Acad. Nat. Sci.
‖ Am. Jour. Sci. and Arts, 2d ser., vol. 43.

estimated at from 80 to 100 feet, underlaid by No. 2 of the Niobrara division. The entire surface of the country, from the latter place, in northeastern Dakota, to Fort Dakota, at Sioux•falls, on the Big Sioux river, is referred to the Cretaceous.

Prof. E. D. Cope* described, from Camden county, New Jersey, *Euclastes platyops*, and *Thoracosaurus brevispinus*, now *Holops brevispinus*.

The Cretaceous rocks occupy a belt or strip of country in New Jersey† which stretches obliquely across the State, from Raritan bay on the northeast, to the head of Delaware bay on the southwest. The extreme length of the formation, from the highlands of Navesink to the Delaware, above Salem, is about 100 miles. Its breadth at the northeast end, from Woodbridge to Deal, is 27 miles, and at the southwest end, from the mouth of Oldman's creek to Woodstown, it is 10¾ miles. The area included in this formation is about 1,500 square miles.

It is subdivided in ascending order, as follows:

1. Plastic clays, 210 feet.
2. Clay marls, 277 feet.
3. Lower marl bed, 30 feet.
4. Red sand, 100 feet.
5. Middle marl bed, 45 feet.
6. Yellow sand, 43 feet.

Total thickness, 705 feet.

The kaolin, which is dug so extensively, belongs to the plastic clay of the above subdivision. It is a very fine micaceous sand, with some fire-clay intermixed, and streaks of clay passing through it. It is of a bluish-white color, sandy in consistency when drained, but pasty when worked up in water.

Prof. E. D. Cope‡ described, from New Jersey, *Osteopygis emarginatus, Clidastes iguanavus, Nectoportheus validus, Emys petrosus*, now *Agomphus petrosus, Elasmosaurus orientalis*, and from the Niobrara Group, near the boundary line between Kansas and Colorado, a short distance north of the Smoky Hill fork of the Kansas river, *E. platyurus*.

Dr. Joseph Leidy described, from near Fort Hays, Kansas, *Ptychodus occidentalis*, and from the Judith River Group, *Aublysodon mirandus*, now *Aublysodon horridus*.

* Proc. Acad. Nat. Sci.
† Geo. of New Jersey. 1868.
‡ Proc. Acad. Nat. Sci.

Prof. Leo Lesquereux*described, from the Dakota Group, north of Fort Ellsworth, Nebraska, or its vicinity, *Pterophyllum haydeni, Glyptostrobus gracillimus, Sequoia formosa, Phyllocladus subintegrifolius. Arundo cretaceus, Liquidamber integrifolium, Populus lancastriensis, Populites elegans, P. flabellata, P. salisburyæfolia, P. ovata,* now *Ampelophyllum ovatum, P. quadrangularis,* now *Hamamelites quadrangularis, Salix proteæfolia, Betula beatriciana, Fagus polycladus, Quercus primordialis,* now *Dryophyllum primordiale, Q. hexagona, Q. ellsworthanus, Q. anceps,* now *Diospyros anceps, Q. semialatus,* now *Anisophyllum semialatum, Ficus (?) rhomboideus,* now *Phyllites rhomboideus, Ficus (?) fimbriatus,* now *Eremophyllum fimbriatum, Platanus aceroides,* var. *latior, P. obtusiloba, P. diminutivus, Credneria leconteana,* now *Protophyllum lecontcanum, Laurus macrocarpus. Sassafras mudgei, S. subintegrifolium, Lyriodendron giganteum, L. intermedium, Magnolia tenuifolia, Dombeyopsis obtusiloba,* now *Menispermites obtusilobus, Negundoides acutifolia, Paliurus membranaceus, Rhamnus tenax, Phyllites rhoifolius, Phyllites amorphus, P. umbonatus,* and *Prunus cretaceus.*

In Tennessee,† wherever the Cretaceous rocks are exposed, they lie upon Palæozoic strata. They are subdivided into, first, Coffee Sand Group; second. Green Sand Group, or the shell bed; and third, Ripley Group.

The Coffee Sand Group derives its name from the exposure at Coffee Landing, on the Tennessee river. It outcrops in Hardin and Decatur counties, and overlaps the Western beveled edge of the older rocks. Its outcrop occupies a belt of territory varying from two to eight miles in width, and running more than half way through the State. It has a maximum thickness of about 200 feet. It consists mostly of stratified sands, usually containing scales of mica. Thin leaves of dark clay are often interstratified with the sand, the clay leaves occasionally predominating. Sometimes beds of dark laminated or slaty clay of considerable thickness, from one to twenty feet or more, are met with in the series. It very generally contains woody fragments and leaves, converted more or less into lignite. Silicified trunks of trees are not uncommon. When it passes under Green Sand it becomes the reservoir which yields water when pierced by the well-borers. It is the equivalent of the Tombigbee Sand of Hilgard in Mississippi.

Fossil shells are so abundant in the Green Sand, at some points, that they are gathered by car loads and burned into lime. The maxi-

---

* Am. Jour. Sci. and Arts, 2d ser., vol. 46.
† Geo. of Tenn., 1869.

mum thickness ascertained from data, furnished by well-borers, is 350 feet. Its outcrop occupies a belt of the surface averaging about eight miles wide for at least half way through the State. This Group is the northern extension of the rotten, limestone of Mississippi and Alabama.

The Ripley Group occupies a belt of the surface along the Memphis and Charleston Railroad about fifteen miles wide, but having a less average width across the State. The high ridges dividing the waters of the Tennessee and Mississippi rivers lie mostly within its area. It has a thickness of 400 or 500 feet, and is mostly made up of stratified sands, though occasionally an interstratified bed of dark, slaty clay, 10 to 30 feet in thickness, occurs, or more frequently a sandy bed laminated with clayey leaves. The hills about Purdy, in McNairy, and about Lexington, in Henderson county, show these rocks well; but more interesting sections, on account of the fossils they contain, are found in Hardeman, near the M. & C. R. R.

In 1869, J. D. Whitney[*] divided the Cretaceous formation, which is found covering large areas on the west coast, from Vancouver and the adjacent islands of San Juan Archipelago on the north, through Washington Territory and Oregon to Southern California, as well as isolated patches in Eastern Oregon and in Mexico, into four groups, as follows:

1. The Tejon Group, the most modern member, is peculiar to California. It is found most extensively developed in the vicinity of Fort Tejon and about Martinez. From the latter locality it forms an almost continuous belt in the Coast Ranges to Marsh's, fifteen miles east of Mt. Diablo, where it sinks under the San Joaquin Plain. It is also found at various points on the eastern face of the same range, as far south as New Idrea, and in Mendocino county, near Round Valley, the latter locality being the most northern point at which it is yet known. It is the only coal-producing formation in California.

This group contains a large and highly characteristic series of fossils, the larger part peculiar to itself, while a considerable percentage is found extending below into the next group, and several species still further down into the Chico Group. Mr. Gabb considered it as the probable equivalent of the Maestricht beds of Europe.

2. The Martinez Group, which includes a series of beds, of small geographical extent, found at Martinez and on the northern flank of Monte Diablo.

---

[*] Pal. of Cal., vol. 2.

3. The Chico Group, one of the most extensive and important members of the Pacific coast Cretaceous. It is on the horizon of either the upper or lower chalk of Europe, and probably the equivalent of both. It is extensively represented in Shasta and Butte counties, and in the foot-hills of the Sierra Nevada, as far south as Folsom, occurring, also, on the eastern face of the coast ranges, bordering the Sacramento valley at Martinez, and again in Orestimba canon, in Stanislaus county. It includes all of the known Cretaceous of Oregon, and of the extreme northern portion of California, and is the coal-bearing formation of Vancouver's Island.

4. The Shasta Group, including all below the Chico Group. It contains fossils seemingly representing ages, from the Gault to the Neocomien, inclusive, and is found principally in the mountains west and northwest of the Sacramento valley. Two or three of its characteristic fossils have been found in the vicinity of Monte Diablo, and one of the same species has been collected in Washington Territory, east of Puget Sound. Few, or none, of its fossils are known to extend upward into the Chico Group.

W. M. Gabb* described, from Shasta county, from Martinez, Benicia, Colusa county, Tejon, and other places in California, *Philoteuthis foliatus, Belemnites impressus, Ammonites jugalis, A. stoliczkanus, A. fraternus, Ancyloceras lineatum, Diptychoceras laeve, Fusus tumidus, F. occidentalis, Neptunea cretacea, N. mucronata, Palaeatractus crassus, Surcula praeattenuata, S. inconspicua, Heteroterma trochoidea, Bela clathrata, Cordiera mitraeformis, Tritonium californicum, T. tejonense, T. fusiforme, Brachysphingus sinuatus, Bullia striata, Turbinella crassitesta, Urosyca caudata, Neverita globosa, Ampullina striata, Terebra californica, Cypraea mathewsoni, Anchura transversa, A. carinifera, Helicaulax bicarinata, H. costata, Loxotrema turrita, Alresius liratus, Turritella martinezensis, Nerita triangulata, Calliostoma radiatum, Ataphrus crassus, Margaritella angulata, Acmaea tejonensis, Aelaeonella oviformis, Liocium punctatum, Ringinella polita, Solen cuneatus, Anatina quadrata, Pholadomya oregonensis, Pleuromya papyracea, Arcomya undulata, Mactra tenuissima, Asaphis multicosta, Tellina undulifera, Donax latus, Venus acquilateralis, Meretrix~fragilis, Thetis elongata, Cardium translucidum, Crassatella compacta, Unio hubbardi, Modiola major, now Volsella major, Meleagrina antiqua, Inoceramus elliotti, I. whitneyi, Trigonia aequicostata, Nucula solitaria, Pecten martinezensis, P. complexicosta, P. interradiatus, Neithea grandicosta, Lima shastaensis,*

L. multiradiata, Anomia vancouverensis, Ostrea idriaensis,O. appressa. Rhynchonella whitneyi, Smilotrochus curtus.

And from the Sierra de las Conchas. near Arivechi, Sonora, Mexico. *Fusus mexicanus, Euspira tabulata, Chemnitzia zebra, Tylostoma mutabile, Anchura monilifera, Cerithium mexicanum, Angaria cingulata, Cinulia rectilabrum. Pholadomya sonorensis, Cardium sabulosum, C. granuliferum, Cardita alticosta, Remondia furcata, Cucullæa inermis, Gryphæa mucronata.*

Prof. E. D. Cope* described, from Raritan bay, *Ornithotarsus immanis;* from Western Kansas, *Macrosaurus proriger,* now *Liodon proriger;* from Sampson county, North Carolina, *Hypsibema crassicauda, Hadrosaurus tripos,* and *Polydectes biturgidus;* from New Jersey,† *Mosasaurus maximus,* and from Alabama, *Clidastes propython.*

Prof. O. C. Marsh‡ described, from the greensand marl, near Hornerstown, Monmouth county, New Jersey, *Mosasaurus copeanus, M. miersi, M. princeps, Halisaurus fraternus,* now *Baptosaurus fraternus,* and *H. platyspondylus,* now *B. platyspondylus.*

Prof. Leo Lesquereux§ described, from the Dakota Group, at Fort Ellsworth, Nebraska, *Populites microphyllus, Phyllites betulæfolius, Persea nebrascensis,* now *Laurus nebrascensis,* and *Sassafras leconteanum,* now *Persea leconteana.*

The Cretaceous is the lowest formation exposed in Louisiana,‖ and it comes to the surface only at the limestone hills of St. Landry and Winnfield. The borings that have been made for salt, however, show that it is more that 1,000 feet in thickness. The strata are saline, and pure beds of rock salt sometimes occur.

The Cretaceous rocks have been observed in Plymouth, Woodbury, Cass. Guthrie, Pottawattamie, Montgomery, Carroll and Greene counties,¶ Iowa. In all but the first two they appear as outliers. On account of the drift which covers the western half of this State, the area of the Cretaceous has not been determined. The exposure in Plymouth and Woodbury counties extends into Dakota, and belongs to the Dakota Group. The maximum thickness as far as observed is 350 feet. The rocks rest unconformably upon the coal measures beneath, and have a northwesterly dip, while the palæozoic strata dip southwesterly.

---

* Proc. Amer. Phil. Soc.
† Proc. Bost. Soc. Nat. Hist.
‡ Am. Jour. Sci. and Arts, 2d ser., vol. 45.
§ Trans. Am. Phil. Soc., vol. 13.
‖ Geo. of Lou., 1870.
¶ Geo. Sur. Iowa, 1870.

In 1870, Dr. Joseph Leidy described, from Middle Park, Colorado,* *Pœcilopleuron valens;* from Pickens county, Alabama, *Clidastes intermedius;* from Kansas, *Xiphactinus audax,* and from the Moreau river, *Nothosaurops occiduus.*

Prof. E. D. Cope† described, from the Green Sand of New Jersey, *Adocus agilis, A. pectoralis, A. syntheticus, Emys turgidus,* now *Agomphus turgidus, Bottosaurus tuberculatus, Catapleura repanda, Osteopygis chelydrina,* now *Catapleura chelydrina, Hyposaurus fraterculus,* now *Gavialis fraterculus, Holops cordatus, H. glyptodon, Lælaps macropus, Liodon congrops, L. perlatus, Lytoloma angusta, L. jeanesi, Mosasaurus depressus, M. fulciatus, M. maximus, M. oarthrus, Osteopygis platylomus, Peritresius, Platecarpus tympaniticus, Pneumatarthrus peloreus, Taphrosphys lesleyanus, T. longinuchus, T. molops, T. nodosus,* and *T. strenuus.*

He described, from the Niobrara Group, at a point six miles south of Sheridan, Kansas, *Liodon mudgei,* now *Platecarpus mudgei, Clidastes cineriarum, Saurocephalus prognathus, Ichthyodectes ctenodon;* from the north bank of Smoky Hill river, thirty miles east of Fort Wallace, Kansas, *Liodon ictericus,* now *Platecarpus ictericus;* from twenty miles east of Fort Wallace, *Saurocephalus phlebotamus;* from near Fort McRae, in New Mexico, *Liodon dyspelor,* from the bank of Solomon's river, in Kansas, 160 miles from its junction with the Kansas river, *Saurocephalus thaumas,* now *Portheus thaumas.*

Prof. O. C. Marsh,‡ from Green Sand near Barnsboro, New Jersey, *Hadrosaurus minor, Mosasaurus crassidens;* from Hornerstown, *Liodon laticaudus;* from Birmingham, *Laornis edwardsanus;* and also from the Green Sand of New Jersey,§ *Palæotringa littoralis, P. vetus, Telmatornis affinis, T. priscus.*

T. A. Conrad‖ described, from Crosswicks, New Jersey, *Inoceramus peculiaris, Crassatella prora,* now *Etea prora, Trigonarca passa, Goniosoma inflata, Axinea mortoni, Cyprimeria spissa, Dentalium falcatum;* and from Haddonfield, *Nucularia papyria, Scambula perplana,* now *Anthonya perplana, Gouldia decemnaria, G. declivis, Nemoarca cretacea, Trigonarca cuneiformis, Perrisonota protexta, Camptonectes bellisculptus, Liroscapha squamosa, Cancellaria subalta, Eulima cretacea, Gadus obrutus, Donax fordi;* from Mississippi,

Proc. Acad. Nat. Sci.
† Trans. Am. Phil. Soc. Ext. Batr. Reptilia N. Am.
‡ Proc. Acad. Nat. Sci.
§ Am. Jour. Sci. and Arts, 2d ser., vol. 49.
‖ Am. Jour. Conch., vol. 5.

Gemma cretacea; and from Haddonfield, New Jersey,* *Æora cretacea, Tenea parilis, Ænona papyria, Venilia elevata,* now *Veniella elevata. Cardium dumosum,* and *Solyma lineolatus.*

In 1871, Prof. F. B. Meek† said, the oldest beds of the Bear river country of Utah and Wyoming, properly belonging to the Tertiary, (they are now regarded as Cretaceous), and so intimately related to the latest Cretaceous, contain species of *Corbula, Cyrena (Corbicula)*, perhaps *Ostrea,* and a univalve related to *Melampus,* directly associated with several species of *Goniobasis,* two of *Unio,* one or two of *Melantho,* several species of *Viviparus,* one of *Tiara,* etc., showing clearly that these strata were deposited in brackish waters. These shells also exist in great numbers, and are preserved in a condition, showing that they could not have been transported far by currents, but that they must have lived and died, at least, near where we now find them.

All palæontologists are aware of the fact, that the remains of fresh and brackish water shells do not generally present such well marked peculiarities of form, ornamentation, etc., in beds of different ages, as we see in marine types, so that they can not always be relied upon, with the same degree of confidence in identifying strata, that we place in marine forms; some of those from oldest Cretaceous being, for instance, very similar to existing species. So far as I have been able to compare the species from this formation with described forms from other parts of the world, they generally agree most nearly with Lower Eocene types; the *Corbicula* and *Tiara* being very similar to forms found in the lower lignites of the Paris basin, and at the mouth of the Rhone in France. At the same time it is worthy of note, that most of these shells are quite unlike any of the known existing North American species, and one of them (*Tiara humerosa*) belongs even to a genus entirely unknown among the existing *Melania* of the American continent, though found inhabiting the streams of Madagascar, the Fejee Islands, etc. One of the *Uniones* (*U. belliplicatus*) resembles in its ornamentation some of the South American species, and the genus *Castalia,* much more nearly than it does any of the recent North American species, although having the form and hinge of a true *Unio:* and another abundant bivalve, found in the same association, *Corbula (Anisothyris) pyriformis,* seems to be allied in some respects to a peculiar group recently described from a Pliocene or Miocene formation, on the Upper Amazon of South America, by Mr. Gabb, under the

* Am. Jour. Conch., vol. 6.
† Advance pamphlet from Hayden's U. S. Geo. Sur. of Wyoming, etc.

name *Pachydon*, and afterward renamed *Anisothyris* by Mr. Conrad, because the name *Pachyodon* had been previously used for another genus.

Of course, comparisons of the shells, from this formation with those of the Tertiary beds of the Atlantic and Pacific slopes, afford no aid whatever in fixing its precise position in the series, because the species from the latter are, almost without exception, marine types. There is less difficulty, however, in drawing parallels between it and the Tertiary deposits of the Upper Missouri country, by a comparison of fossils, although the species are mainly different, so far as yet known, in these two districts. At least two of the known forms, however, from the Utah and Wyoming beds under consideration, are believed to be specifically identical, with species found in the oldest beds, referred to the Tertiary at the mouth of the Judith river, on the upper Missouri, under the name of the Fort Union Group. These are *Unio priscus*, and *Viviparus conradi*. In addition to this, the fossils at these two localities are in precisely the same state of preservation, and have a more ancient appearance than those of the later deposits of both districts, while they also agree exactly in their mixture of brackish and fresh water characters. Again, at both localities, these deposits are intimately associated, as already stated, with what appears to be the latest of the Cretaceous series; while in both districts they contain lignite, and are succeeded by later Tertiary beds of strictly fresh water origin.

He described,[*] from the Fort Pierre Group, near the great bend in the Upper Missouri,[*] *Isocardia hodgei*, now *Procardia hodgei*.

Prof. O. C. Marsh[†] named, from the Niobrara Group, on the North Fork of the Smoky river in Kansas, *Edestosaurus dispar*, now *Clidastes dispar*, *E. velox*, now *C. velox*, *Clidastes pumilis*, *C. wymani*, and *Pterodactylus oweni*.[‡]

Prof. E. D. Cope described,[§] from the Niobrara Group, near Fossil Spring canon. *Edestosaurus stenops*, now *Clidastes stenops*, *E. tortor*, now *C. tortor*, *Holcodus coryphæus*, now *Platecarpus coryphæus*, *Liodon curtirostris*, now *P. curtirostris*, *L. glandiferus*, now *P. glandiferus*, *Portheus molossus*, *P. angulatus*, now *Erisichthe angulatus*; from Butte creek. *Holcodus tectulus*, now *Platecarpus tectulus*, *Protostega gigas*; and from one mile southwest of Sheridan, near the Gypsum Buttes, *Liodon latispinus*, now *Platecarpus latispinus*.

---

* Proc. Acad. Nat. Sci.
† Am. Jour. Sci. and Arts, 3d series, vol. 1.
‡ Proc. Acad. Nat. Sci.   § Proc. Am Phil Soc

In 1872, Prof. E. W. Hilgard[*] showed that the Cretaceous of Alabama and Mississippi has a dip sensibly at right angles to the trend (*i.e.*, between W. and S.) at the rate of 20 to 25 feet per mile. That the lower division, called the Coffee Group, or the Eutaw Group is from 300 to 400 feet thick, and consists of noncalcareous sands, and blue or reddish laminated clays, with occasional beds of lignite, and rarely marine fossils, silicified, as at Finch's Ferry in Alabama. This group corresponds with Hayden's Dakota Group, and in its upper part, as at Finch's Ferry, probably with the Fort Benton Group.

The Middle or Rotten Limestone Group is not less than 1,200 feet in maximum thickness, consisting of soft, mostly somewhat clayey, whitish, micro-crystalline limestones, and calcareous clays; very uniform on the whole, with the exception of the locally important feature of the "Tombigbee Sand." The Cretaceous area of Arkansas, according to Owen's description, seems to fall within this group, as does also the greater part of the Cretaceous area of middle and northern Texas.

The Ripley Group consists of crystalline, sandy limestones, alternating with dark-colored glauconitic marls, containing finely preserved fossils, and has a thickness of 300 to 350 feet. It is the equivalent of the highest Cretaceous beds of New Jersey, and of the Fox Hills Group of the West. The series of isolated Cretaceous outliers, which traverse Louisiana, from the head of Lake Bistencau, in a S. S. E. direction to the great salt mass at Petite Anse, belong to this Group.

Prof. F. V. Hayden[†] said, that in Nebraska, the sandstones of the Dakota Group rest directly upon rocks of the age of the Coal Measures. Although they do not appear in full force until we reach a point near De Soto and beyond, yet remnants of the sandstones make their appearance within five or ten miles of Omaha, at any point north of the Platte river. It is quite probable that they once extended all over Nebraska, passing across into Iowa, and further eastward. The Coal-measure limestones are thus exposed, in northeastern Nebraska, by the erosion of the Cretaceous rocks.

Near the entrance of the Big Sioux river, into the Missouri, the Dakota Group disappears beneath the water-level, and is succeeded by a series of black, plastic, laminated clays, with lighter-colored arenaceous partings and thin layers of sandstone. Near the mouth of the Vermilion river the upper portion becomes more calcareous, and gradually passes up into the next group, called the Fort Benton Group. It is often immensely thickened, in the vicinity of the mountains, from the north

---

[*] Proc. Am. Ass. Ad. Sci.
[†] Hayden's U. S. Geo. Sur. of Wyoming.

line to New Mexico, but on the Lower Missouri, where it was first observed by geologists, it never reaches a thickness of more than 150 or 200 feet. In New Mexico it occurs as the most conspicuous of the Cretaceous divisions, and along the line of the Kansas Pacific Railway, in Kansas, it has yielded large quantities of the most remarkable reptilian remains.

The Niobrara Group is found, in some form, wherever the Cretaceous beds occur, from the north line to New Mexico, and probably much farther. As it is developed on the Lower Missouri, and southward through Nebraska, Kansas, into Texas and the Indian Territory, it contains thick, massive beds of chalky limestone. On the Kansas Pacific Railway, at Forts Hays and Wallace, this limestone is sawed into blocks of any desirable size, with a common saw, and used for building purposes; but along the flanks of the mountains, or in the far west, it never reveals its chalky character. It is found in thin, slaty, calcareous layers, but universally characterized by the presence of the oyster *Ostrea congesta*, and also some form of *Inoceramus*, or a few fish remains, but the little oyster is ubiquitous. In these three divisions there seems to be no well-marked line of separation, and the more we study them the more intimately do they seem to be blended together.

The Fort Pierre Group begins to overlap the Niobrara Group below the mouth of the Niobrara, and above that point, although the river cuts deep down into the chalk limestone, and long lines of cone-like bluffs extend up nearly to the Great Bend, yet the distant hills on either side of the river show plainly the dark shaly clays of this group. This group covers a vast area of country, perhaps 5,000 square miles or more, and wherever it prevails, it gives to the surface the aspect of desolation. The entire thickness of the group is filled with the alkaline material, which is so well known in the west, and wherever the water accumulates in little depressions and evaporates, the surface is covered with a deposit of the salt varying from an inch to several inches in thickness. The water that flows through these clays is usually impregnated with these salts and thus rendered unfit for use. Although these clays seem to be so sterile, and in the dry season are typical of extreme aridity, yet they are by no means destitute of vegetation. The various species of *chenopodiaceous* shrubs and herbs, that are peculiar to the west, find their natural habitat in these clays, and grow most luxuriantly. The *Sarcobatus* reaches its highest growth in this region. The somber appearance given to the country by the black clays is unfavorable to it. At the Great Bend there is a large thickness of the strata filled with concretions that are made up mostly of an aggregate of fossils, as *Ammonites, Baculites*, etc. Near Chain de Roche creek these concretions have been swept down into the Missouri by the swift current, during the spring floods, and in the low water of autumn they present a picturesque appearance.

Although the rivers cut deep channels through the different formations, we do not meet with the Fox Hills Group along the Missouri, until we reach nearly up to the mouth of Cannon Ball river, yet fifty miles or more before reaching that point it has overlapped the Fort Pierre Group. In traveling across the plain country westward from Fort Pierre, we find it occupying the entire area. Very soon after passing west of the Big Cheyenne river the traveler will readily recognize its presence by the more cheerful appearance that it gives to the surface, as well as by the greatly increased growth of vegetation. The water is pure and good, and springs become quite common in the hills.

An important feature in the geology of the West is the great lake basins, which seem to set in the older formations and in each other like dishes. The principal one is the Fort Union, or Great Lignitic Group, which forms the transition group, from the strictly marine condition of the Cretaceous period, to the epoch of the numerous freshwater lakes, which were scattered all over the country west of the Mississippi. This group was called the Fort Union or Lignitic Group in 1861,* and supposed to be of Eocene age.

It was described as consisting of beds of clay and sand, with round ferruginous concretions, and numerous beds, seams and local deposits of lignite, great numbers of dicotyledonous leaves, stems, etc., of the genera *Platanus, Acer, Ulmus, Populus*, etc., with very large leaves of true fan palms; also *Helix, Melania, Vivipara, Corbicula, Unio, Ostrea, Corbula* and scales of *Lepidotus*, with bones of *Trionyx, Emys, Compsemys, Crocodilus*, etc; as occupying the whole country around Fort Union, extending north into the British possessions to unknown distances, southward to Fort Clark, under the White River Group on North Platte river above Fort Laramie, and on the west side of Wind River mountains; and as having a thickness of 2,000 feet or more. The passage from the brackish to the fresh water beds seems not to be marked by any material alteration, in the nature of the sediments; nor have we any reason for believing, that any climatic or other important physical changes beyond the slow rising of the land, and the conse-

---

* Proc. Acad. Nat. Sci.

quent recession of the salt and brackish water, took place during the
deposition of this group.

Prof. Hayden proposed to call the strata found in the Judith basin
near the sources of the Missouri river, consisting of ancient lake deposits, and not differing materially from those of the Fort Union Group,
the Judith Group. It contains impure beds of lignite, fresh-water
mollusca, a few leaves of deciduous trees and a great number and
variety of reptilian remains.

There is no real physical break in the deposition of the sediments
between the well-marked Cretaceous and Tertiary strata. In some
localities the continuity is clear and beautiful in the highest degree.
On Green river, and in the Bitter Creek Valley, one can trace the
continuity step by-step, so far as the strata are concerned, from the
Cretaceous through the greatest thickness of clays, sands, and sandstones of the Lower Tertiary to the purely fresh water beds of Green
river shales, Washakie, or Bridger Groups. In these localities the
influence of the elevation of the mountain ranges has been such as to
expose the outcroping edges of all the strata, from the Cretaceous to the
sands of the most recent Tertiary, like the leaves of a book. In the
clays interspersed among the coal beds, in the Bitter Creek valley,
several species of oyster shells occur in seams. At Bear river, we
have well defined Cretaceous strata and from these we ascend, through
a series of sandstones and clays, with an abundance of shells of the
genus *Ostrea* and a few other marine forms, resembling Tertiary types
as much as Cretaceous. Soon we come to the coal-beds, which at
this locality are nearly vertical. Above them we find seams of oyster
shells, but no other marine forms. And finally, high up in the upper
beds of the coal group, we find the greatest profusion of brackish and
fresh water life. The coal group in Weber Valley, and at Coalville,
is referred to the Cretaceous.

Prof. F. B. Meek* said that some of the specimens from near
Bear river, and at Coalville, Utah, from a light-colored sandstone
containing beds of a good quality of brown coal, appear to belong to
a member of the Cretaceous series not corresponding to any of
those named in the Upper Missouri country; though it is, as he
believed, represented by a similar sandstone under the oldest estuary
Tertiary beds at the mouth of the Judith river, on the Upper Missouri. In 1860, Colonel Simpson brought from this rock, on Sulphur
Creek, a small tributary of Bear river, in Utah, some casts of *In-*

---

* Hayden's U. S. Geo Sur. of Wyoming, etc.

oceramus, and other fossils, and in some remarks on this collection* he referred this formation to the Cretaceous. The collections that have since been brought in from it, in Utah, by Mr. King's and Dr. Hayden's surveys, confirm the conclusion that it belongs to the Cretaceous, as they contain, among other things, species of *Inoceramus*, *Anchura* and *Gyrodes*—genera that seem not to have survived the close of the Cretaceous period. In addition to this, there is among Dr. Hayden's collections from this rock, at Coalville, a *Turritella* that he could not distinguish from *T. martinezensis*, and a *Modiola* which appeared to be specifically identical with *M. pedernalis*. Dr. Hayden also had, from a little above the coal beds at Coalville, specimens of oyster that seemed much like *O. idriaensis* and *O. breweri*, of Gabb, from the upper beds of the California Cretaceous. From the affinities of some of these fossils to forms found in the latest of the beds referred in California to the Cretaceous, and the intimate relations of these marine coal bearing strata of Utah to the oldest Tertiary of the same region, and the apparent occurrence of equivalent beds bearing the same relations to the oldest brackish-water Tertiary beds at the mouth of Judith river on the Upper Missouri, he was inclined to believe that these Coalville beds occupy a higher horizon in the Cretaceous than even the Fox Hills beds of the Upper Missouri Cretaceous series; or, in other words, that they belong to the closing or latest member of the Cretaceous.

All of the explorers of this region concur, in the statement, that the above mentioned Cretaceous beds are intimately related to the succeeding brackish water deposits that appear to belong to the oldest Tertiary; the two formations, wherever they occur together, being conformable and without any intermediate beds, so that the one seems to shade into the other, without any abrupt or sharply-defined line of separation; the change being mainly indicated by a gradual transition from beds containing Cretaceous types of only marine origin, to those with brackish and fresh water types, apparently most nearly allied to early Eocene species of the old world.

So far as yet known, there would appear to be no strictly marine Tertiary deposits in all this interior region of the continent; even the lower parts of the surface here having been apparently gradually elevated above the sea level, at, or very near, the close of the Cretaceous period. For the same reason all of the beds hitherto referred with confidence to the Cretaceous are of undoubted marine origin, as they contain only marine types.

---

* Proc. Acad. Nat. Sci.

These Cretaceous gulfs or seas, however, evidently did not occupy the whole country here, as we know from the absence of Cretaceous deposits throughout what were doubtless wide areas, or possibly, in some cases, smaller islands of dry land at that time. As the whole surface was gradually elevated, however, even the lowest portions rose finally to near the tide level, thus probably leaving large inlets and estuaries of brackish waters, that subsequently became so far isolated, by the continued elevation, and from sedimentary deposits, as to prevent the influx of the tides and form fresh-water lakes, in which the later fresh-water and terrestrial types of fossils only were deposited.

That this change from marine to brackish-water conditions was exactly contemporaneous with the close of the Cretaceous epoch, and the introduction of the Tertiary in Europe, is not certain; nor is it necessary that this should have been the case to constitute the older rock Cretaceous, and the later Tertiary, because in the use of these terms we have reference rather to the order of succession of certain great physical changes, affecting life in distantly separated parts of the earth, than to the exact time of the occurrence of these changes.

He described from Bear river, near Sulphur creek (now Laramie Group), *Goniobasis chrysalis.*

From the Dakota Group, twelve miles southwest of Salina, Kansas, *Crassatellina oblonga, Arca parallela, Yoldia microdonta, Corbicula nucalis, C. subtrigonalis, Cardium salinaense, C. kansasense, Arcopagella mactroides, Tellina subscitula, Leptosolen conradi, Turritella kansasensis,* and *Turbo mudgeanus.* From opposite Sioux City, in Dakota county, Nebraska, *Unio nebrascensis.*

From the Fort Pierre Group, near Medicine Bow Station, Union Pacific Railroad, *Inoceramus altus.* From the Fox Hills Group, at the mouth of Deer creek, on North Platte, in Wyoming, *Tapes wyomingensis.* From Box Elder and Colorado City, Colorado, *Anisomyon centrale.*

From the Fort Benton Group, at Oil Springs, twenty miles west of Fort Bridger, Wyoming, *Cardium pauperculum;* from Point of Rocks,[*] Wyoming, *Anomia gryphhorhynchus.*

From Salt Lake, Utah, *Pachymya truncata*; and from Canon City, *Mactra canonensis.*

Wm. M. Gabb[†] described, from Chihuahua, Mexico, *Lima kimballi.*

Alfred R. C. Selwyn[‡] described the Jackass mountain Conglomerate

---

[*] Hayden's 5th Rep. U. S. Geo. Sur. Terr.
[†] Proc. Acad. Nat. Sci.
[‡] Geo Sur. Canada.

Group of British Columbia. It consists of hard, close-grained and thick-bedded, greenish sandstones or quartzites, green and black shales, and above these, massive thick-bedded pebble conglomerates, dipping generally at low angles in various directions; some of the inclosed pebbles are of rocks belonging to the Cache creek series. At Jackass mountain the road is built round, or excavated out of vertical cliffs of these conglomerates, at from 800 to 900 feet above the river, into which you can almost drop a stone from the parapet of the road; and at a short distance back they rise into hills, not less than 3,000 feet above the valley, which they occupy to within about five miles from Lytton. This group belongs to the Upper Cretaceous, and is above what he called the Upper Cache Creek Group.

The road to Cariboo, between Clinton and Lillooet, runs through a valley transverse to the strike of the rocks, from one to two miles wide, on either side of which hills rise abruptly from 1,000 to 2,000 feet. The Upper Cache Group was first observed here by Mr. James Richardson, the base of which he supposed to be about two miles west of Clinton. The beds have generally a high westerly dip. They consist of a great volume of bluish, dove-colored, and white limestones, often a good marble, interstratified with brown dolomitic limestone, red and green shale, and epidotic and chloritic rocks, with others which closely resemble rocks of the Quebec Group, in the eastern townships of Canada. These rocks occupy the country westward for about six miles. On their strike to the northward, they can be easily traced by the eye, from the almost snowy appearance of the limestones for 20 or 30 miles; and in the opposite direction they can be traced, by the same characters, for 10 to 12 miles, to another transverse narrow valley called Marble canon. A narrow, deep lake, of clear water, occupies the bottom of this canon, the white cliffs of limestone rising on either side of the lake to heights of from 2,000 to 3,000 feet above the water. About half-way up, on the north side, the limestone beds stand up in masses, which look like detached columns of a diameter of from 50 to 100 feet, and from 300 to 400 feet high, due to the unequal weathering of the almost vertical strata. The limestones are succeeded by a considerable thickness of black shales, sometimes soft and calcareous, but often hard and flinty.

Mr. James Richardson* described numerous sections, in the Cretaceous rocks of Vancouver and adjacent islands, showing the coal seams: one of which occurs about five miles from the shore on the southwest

---

* Geo. Sur. Can.

side of Comox harbor, on a small tributary of the Puntledge river. A descending section is as follows:

1. Brownish or drab-colored, slightly calcareous sandstone, the grains of which are composed of quartz, feldspar and mica, with some of a black substance supposed to be peroxide of manganese, the beds being from one to five feet thick, 45 feet.

2. Coal, black and shining, clean and free from shale, 4 feet 6 inches.

3. Brownish-black argillaceous shale, and greenish-brown sandstone, interstratified with one another in thinish layers, the shale predominating, and both holding thin irregularly-distributed, lenticular patches of coal, which may constitute about one tenth of the mass, no indications of roots penetrating the upper part of the bed were observed, 15 feet.

4. Coal, apparently of good quality, 5 feet 4 inches.

5. Brownish-gray or light drab sandstone, in beds of from one foot to eighteen inches, 10 feet.

6. Coal, apparently clean and of good quality, 6 feet.

7. Brownish-gray or light drab sandstone interstratified with thin layers of black, soft, argillaceous shale, 3 feet.

8. Coal, without observed impurities, 10 feet.

Total, 98 feet 10 inches.

A section near Departure bay shows a thickness of 1,538 feet.

Prof. E. D. Cope* described the Cretaceous along the line of the Kansas Pacific Railroad, where it consists of the Dakota, Benton and Niobrara Groups. The Dakota Group constitutes the bluffs at Salina, one hundred and eighty-five miles west of the State line of Missouri, and continues as far as Fort Harker, thirty-three miles farther west. . They are a coarse, brown sandstone, containing irregular concretions of oxide of iron and numerous mollusks of marine origin. The Benton Group appears at this point, containing large quantities of dicotyledonous leaves and other forms of land vegetation. It appears also at Brookville, eighteen miles east, and at Bunker Hill, thirty-four miles west of this Fort. The Niobrara Group forms the bluffs at Fort Hays, seventy miles west of Fort Harker, and from this place to Fort Wallace, one hundred and thirty-four miles beyond. This group consists of two parts—a lower, of dark bluish calcareo-argillaceous character, often thin-bedded; and an upper, of yellow and whitish chalk, much more heavily bedded. Near Fort Hays the best section may be seen at a point eighteen miles north on the Saline river. Here the

---

* Hayden's 5th Rep. U. S. Geo. Sur. Terr.

bluffs rise to a height of 200 feet, the yellow strata constituting the upper half. Half way between this point and the fort, *Haploscapha grandis,* and *H. eccentrica* occur. Some of them are twenty-seven inches in diameter. Fragments of *Anogmius* occur in the yellow bed and *Inoceramus problematicus* in the blue.

Along the Smoky Hill river, 30 miles east of Fort Wallace, the strata have a gentle dip to the northwest. The yellow and the blue strata are about equally fossiliferous and pass into each other by gradations and by slight laminar alternations at their line of junction. *Cimolichthys semianceps, Liodon glandiferus,* and *L. dyspelor* occur in both classes of strata. The yellow strata are remarkably uniform in mineral con- tents, but the blue contain numerous concretions and great abundance of thin layers of gypsum and crystals of the same. Near Sheridan, concretions and septaria are abundant. In some places the latter are of great size, and being imbedded in the strata have suffered denuda- tion of their contents, and the septa standing out form a huge honey- comb. This region, and the neighborhood of Eagle Tail, Colorado, are noted for the beauty of their gypsum crystals. These are hex- agonal-radiate, each division being a pinnate or feather-shaped lamina of twin rows of crystals. The clearness of the mineral and the regular leaf and feather forms of the crystals give them much beauty. The yellow bed disappears to the southwest, west, and northwest of Fort Wallace, beneath a sandy conglomerate of Tertiary age.

He described, from the Fort Benton Group, at Bunker Hill station, Kansas, *Apsopelix sauriformis.*

He described, from the Niobrara Group, near Eagle Tail in Colorado, *Liodon crassartus,* now *Platecarpus crassartus;* from Kansas,[*] *Ornithochirus harpyia, O. umbrosus;* from near Butte creek, *Cynocer- cus incisus;* from Sheridan, *Plesiosaurus gulo.*[†]

He determined[‡] the Upper Cretaceous age of the Lignitic strata of the Bitter Creek Basin of Wyoming, and described, from near Black Buttes station, on the U. P. R. R., 52 miles east of Green river, and near the Hallville coal mines, *Agathaumas sylvestris.* This dinosaurian was discovered between the thinner or lower strata of the Bitter creek series of coal, which at this point occupy a position of elevation and crop out high on the bluffs. Two strata appear above the sandstone in which the bones occur, and one below it.

---

[*] Proc. Am. Phil. Soc.
[†] Proc. Acad. Nat. Sci.
[‡] Pal. Bull. No. 4 and No. 10, and Proc. Am. Phil. Soc.

Prof. O. C. Marsh* described, from Smoky Hill river, in Western Kansas, *Pterodactylus occidentalis, P. velox, P. ingens, Graculavus anceps, Hesperornis regalis, Lestosaurus simus,* now *Platecarpus simus, L. felix,* now *P. felix, L. latifrons,* now *P. latifrons, L. gracilis,* now *P. gracilis, Rhinosaurus micromus,* now *Liodon micromus, Edestosaurus rex,* now *Clidastes rex, Ichthyornis dispar, Colonosaurus mudgei,* now *Ichthyornis dispar;* and from the greensand at Hornerstown, New Jersey, *Graculavus velox, Graculavus pumilus,* and *Palæotringa vagans.*

Dr. Joseph Leidy† described, from Texas, *Otodus divaricatus;* from Kansas, *Oxyrhina extenta;* from New Jersey, *Acrodus humilis;* and from Mississippi, *Pycnodus faba.*

F. B. Meek and J. H. Kloos‡ found the Benton Group underlying the drift gravel and clay in the Sauk valley, in Minnesota.

T. A Conrad§ described, from the Yellow Chalk, near the Saline river, Kansas, *Haploscapha grandis,* and *H. excentrica.*

And Prof. Leo. Lesquereux described, from the hard ferruginous sandstone of the Dakota Group, in Kansas, *Pterospermites quadratus,* now *Pterophyllum quadratum, Pterospermites multinervis,* now *Pterophyllum multinerve, Pterospermites haydeni,* now *Pterophyllum haydeni, Magnolia ensifolia,* now *Celastrophyllum ensifolium, Quercus mudgei,* now *Protophyllum mudgei, Aralia quinquepartita, Platanus heeri,* and *Sassafras obtusus,* now *Cissites obtusus.* From the reddish, ferruginous, hard shale of the Laramie Group, below the Coal at Evanston, Utah, *Quercus negundoides, Betula stevensoni, Rhus evansi, Juglans rhamnoides;* from a grayish, fine-grained, hard shale on the divide between the source of Snake river and the southern shore of Yellowstone lake, *Gymnogramma haydeni;* and from six miles above Spring canon, and top of hills between Fort Ellis and Botteler's ranch, Colorado, *Myrica ambigua, Quercus ellisana,* and *Q. pealei.*

In 1873, Prof. Leo Lesquereux‖ described the Lignitic Group, from the Raton mountains, northward to Denver and Cheyenne, and then along the Union Pacific railroad to Evanston. In passing obliquely from the town of Trinidad to the Raton valley, in a northwest direction, the stage-road gently ascends about 150 feet to a plateau which, at the surface, consists of the black shale of the Fort Pierre Group, and

---

* Am. Jour. Sci. and Arts, 3d ser., vol. 3 and 4.
† Proc. Acad. Nat. Sci.
‡ Am. Jour Sci. and Arts, 3d ser., vol. 3.
§ 5th Rep. Hayden's U. S. Geo. Sur. Terr.
‖ Hayden's 6th Rep. U. S. Geo. Sur. Terr

contains well preserved, large, characteristic shells in ferruginous concretions. But soon the plain appears cut by undulations which already, one mile from Trinidad, have their tops strewn with large broken flags of sandstone, over which no other trace of fossil remains, but marine plants or fucoids are seen. A little farther from the town, the same sandstone is in place, immediately and conformably overlying the black shale; and in entering the small valley of the Raton, the road curves around steep hills, whose base rests upon the fucoidal sandstone, and whose sides, exposed by denudation, are blackened by outcrops of coal at different altitudes. A section, along a small branch, in whose banks the lignite-beds appear in succession down to Raton creek, and then down this creek to Purgatory river, where the Fort Pierre Group is exposed, shows the lignitic 300½ feet, succeeded by 178 feet of sandstone. The characters of the sandstone are as follows:

First.—Its general color is whitish-gray, so white indeed, sometimes, that the lower strata, seen from a distance, appear like banks of limestone.

Second.—Though generally hard, it weathers by exfoliation under atmospheric influences, and its banks are thus molded in round undulations; and as it is locally hardened by ferruginous infiltrations, it is often too concretionary or grooved in cavities, so diversified in size and forms, that sometimes the face of the cliffs shows like the details of a complicated architecture.

Third.—It is entirely barren of remains of animals.

Fourth.—From its lowest stratum to its upper part, it abounds in well-preserved remains of marine plants or fucoids, which, at some localities, are seen even in the sandstone over lignite-beds.

Fifth.—In its upper part, the sandstone or the shales of this group are mixed with broken *debris* of land-vegetation, with which also fucoidal remains are found more and more abundant in descending.

In passing from the black shale of the Fort Pierre Group to this group of sandstone beds overlaying it, the difference in the characters is striking, not only in considering their compounds, but in the class of fossil remains which they contain, the traces of deep marine animal-life predominating in the black shale, while here they have totally disappeared. In the sandstone, marine life still marks its activity only by the abundant remains of fucoids, indicating, by their growth, a comparatively shallow water. They point out, therefore, a slow upheaval of the bottom of the sea in which they appear to have lived; for their stems penetrate the sandstone in every direction. And this indication is still more manifest in the great abundance of *debris* of land-plants,

which seem as if ground by the waves, thrown upon the shore and mixed
in the sand with fucoidal remains. That this sandstone forms all
over and around the Raton mountains, the base of what is called the
Lignitic Group, and that it overlies the black shale of the Fort Pierre
Group, has been remarked by all the geologists who have explored the
country. Dr. Leconte, considering the strata as Cretaceous, mentions
them in his report as continuing southward of the Raton, along the
base of the Rocky mountains, forming an immense terrace, which
extends as far south as the valley of the Tonejo, and perhaps even to
the north bank of the Cimarron. From this place northward to the
base of the Spanish Peak, these sandstone beds, always with the same
characters and superimposed upon the Fort Pierre Group, form an immense terrace, perpendicularly cut, like a wall facing east, high above
the plain. They support the lignitic beds which still tower above
them, either ascending in steep declivities from the top of the perpendicular sandstone, or receding at some distance, where they have been
more deeply sapped by erosion. This abrupt front, says Dr. Hayden,
seems to form a sort of shore-line of a wonderful basin, as if a body
of water had swept along and washed against the high bluffs, as along
some large river. The stage-road from Trinidad to Pueblo follows
the base of these cliffs for thirty-two miles. South of Trinidad, the
lignitic measures have been followed nearly without interruption to
the Maxwell estate, about fifty miles. The area which they cover, at
and around the Raton mountains, may be estimated at 600 to 800
square miles. The same formation is reported farther south, near and
around Santa Fe; in the Gallisteo valley; along the mountains to
Albuquerque, and in the valley of the Rio Grande, as far south as
Fort Craig. Everywhere, with a single exception, these Lignitic
measures have exposed, by their relative position, by the absence of
animal remains in the thick beds of sandstone, which indicate their
base and constitute their foundation, by the homology of their marine
and land flora, as recognized in the remains of fossil-plants which
they contain in abundance, all the characters authorizing the separation of this group from the Cretaceous formation.

From Pueblo to Canon City, forty-five miles, the stage-road follows
a broad valley, closed on the south side by the Greenhorn mountains,
on the north side by the Rim Range of the Colorado mountains, over
which towers Pike's Peak, whose summit is visible all the time. The
whole valley is essentially Cretaceous; all the eminences, either near
the borders or in the middle, are hills of this formation, molded by the
erosions of the Arkansas river, which has dug numerous beds in this

soft material. The borders of its present bed, like those of its old ones, where the road sometimes meanders, as in a labyrinth, are picturesquely marked by rocks of diversified forms, resembling monuments built by the hand of man, towers, columns, ruins, etc., often strewn around in confusion. On the south side of the river, however, about fifteen miles before reaching Canon City, the aspect of the country is modified by the appearance of a group of hills of the Lignitic, filling the space from the base of the Greenhorn mountains to the borders of the river, three to four miles in width. The whole area covered here by the Lignitic is about 33 square miles. The lower strata, overlying the sandstone, rise abruptly about 50 feet above the Arkansas river, forming a kind of narrow plateau, over which the hills of the upper Lignitic rise up to about 500 feet. The whole thickness of the lignite bearing strata is estimated at about 600 feet. The lower sandrock, about 200 feet thick, is the equivalent of the lower fucoidal sandstone of the Lignitic of the Raton mountains, and it graduates into the Lignitic above. Indeed, in some places the lower sandstone includes in its divisions beds of lignite to its base.

From Pueblo northward no trace of the Lignitic is seen along the mountains till near the southern base of a range of hills, the Colorado pinery, which, in its eastern course, at right angles from the primitive mountains, forms the divide of the waters between the Arkansas and the Platte rivers.

The succession of the Cretaceous strata is clearly marked on the banks of Monument creek. In following it up from Colorado Springs, the formation can be studied to the top of the black shale of the Fort Pierre Group, and above this to a bed of brownish sandstone, separated from the black shale by thin layers of *Tuten clay* and soapstone, where the last remains of Cretaceous animals, especially fragments of *Baculites,* are still abundant. Over this is the sandstone, barren of any kind of remains, overlaid in the banks of the creek, by a bed of fire-clay, or very soft chocolate-colored shale, which marks the base of the following section at low-water level of the creek:

1. Brown, laminated fire-clay, or chocolate-colored soft shale, a compound of remains of rootlets, and leaves and branches of undeterminable conifers, 2 feet.

2. Coal, soft, disaggregating under atmospheric influence, 2 feet.

3. Chocolate-colored clay shale, like No. 1, with a still greater proportion of vegetable *debris,* 6 feet.

4. Soft, yellowish, coarse sandstone in bank, 8 feet.

5. Clay, shale and shaly sandstone covered slope, 130 feet.

6. Soft, laminated clay, interlaid by bands of limonite iron, thin lignite seams, and fossil-wood, 88 feet.

7. Lignitic black clay, in banks, 32 feet.

8. Fine-grained conglomerate, 112 feet.

9. Fine-grained sandstone, 4 feet.

10. Coarse conglomerate, 7 feet.

11. Sandstone, 3 feet.

12. Ferruginous hard conglomerate, 32 feet.

Total, 426 feet.

The soft chocolate-colored, laminated clay, Nos. 1 and 3 of this section, has the same composition, color, and characters as the clay under and above the coal-beds of the Raton mountains and of the Arkansas valley. It is the same, more or less darkly colored by bitumen, which prevails over the whole area of the Lignitic. This clay takes the place of the fire-clay so generally underlying the coal-beds of the carboniferous measures, where, as in the Lignitic, it forms, beside the floor, some bands, clay partings, separating coal strata, and soft shale overlying them. The dicotyledonous leaves, specifically identical with those found at Raton mountain and in the Arkansas valley, leave no doubt about the cotemporaneity of these Lignitic measures.

By far the most interesting member of this section is the conglomerate at the top. This is a compound of small grains or pebbles, mostly of white quartz, and of silex of various colors, varying in size, at least for the largest proportion, from that of a pea to that of the head of a pin. Pebbles as large as a walnut are abundant. This formation, 150 feet thick, at least, is conformable to the strata overlying the coal of the base of the section, and here, as it will be seen at other places, it overlies immediately thick banks of soft, laminated, bituminous, black clay. The materials forming this conglomerate are cemented together by a thin coating of carbonate of lime, which easily disaggregates under atmospheric influence, except in the upper stratum, where the cement has been hardened by ferruginous infiltration. Its greater resistance has then locally preserved the whole mass from destruction. These conglomerate cliffs, which, from the hotel of Colorado springs, arrest the view to the west, appearing like high bluffs of white sandstone, are evidently the mere vestiges of an extensive formation, originally covering the base of the mountains from the Arkansas river, extending far inland to the east. For hundreds of miles the ground of Colorado is formed by its *debris*. They have given to the soil, that apparent sterility of surface, which is so remarkably changed into fertility, by the culture of the substratum composed of softer-grained materials and lime. Nearer to and along the base of the Colorado pinery, whose lignitic hills have escaped destruction, by the upheaval of the ridge, these conglomerates, still detached from the common mass, and molded into the most diversified forms by disintegration, have scattered columns, pinnacles, round towers, and cupolas over a wide area, the far-famed Monument Park.

From the mouth of Bear creek into the Platte, a few miles west of Denver, the Lignitic formation abutting against the Cretaceous and diversely thrown up by the upheaval of the primitive mountains, follows the base of these mountains in a nearly continuous belt to Cheyenne. Though generally covered by detritus, the basin is deeply cut by all the creeks descending to the plain—Clear, Ralston, Coal, Erie, Boulder, Thompson creek and others, and the strata thus exposed. Golden is on the banks of Clear creek, at its outlet from a deep cañon, and in the middle of a narrow valley, shut up, on the west, by the slopes of the primitive rocks, and on the east by a high wall, a trapdike, which here follows the same trend as that of the mountain at a distance of one to one and one-half miles. As it is generally the case along the eastern base of the Rocky mountains, the more recent formations have been thrown up and forward, and their edges upraised to a certain degree nearest to the uplift, and thus succeeding each other by hog-backs facing the mountains, they pass toward the plains in diminished degrees of dip and soon take their horizontal position.

At Golden, the lignitic strata, compressed, as they are, between two walls of eruptive rocks, have been forced up on the western side, in a nearly perpendicular position, while on the other they were thrown up, at the same time, by the basaltic dike, and thus folded or doubled against their faces, in the same way, as the measures of the anthracite basin of Pennsylvania have been so often compressed in multiple folds between the chains of the Allegheny mountains. In that way the lowest strata of the Lignitic, which are nearly perpendicular, overlie the upper Cretaceous strata, which, following the slope of the mountain's plunge, incline in a less degree. The line of superposition of both formations is seen along a ditch opened for a canal of irrigation, about two hundred feet from the tunnels, made in a bank of clay which underlies the lower lignite bed, and which is worked for pottery. These upper Cretaceous strata are seen in the same position, and exactly of the same nature as at Gehrung's; thin beds of soapstone or laminated clay, with Cretaceous fossils, and above them the same kind of *Tuten-clay*, a few inches thick, under the lower sandstone of the Lignitic, which is there covered. The surface of the ridge formed by the upthrow is pierced by the edge of the perpendicular strata, especially of the hard sandstone, and there the characters of the lowest beds are recognized at many places as the same as those of the fucoidal sandstone of the Raton mountains. At the cut made across the measures by Clear creek, the lower sandstone appears proportionally thin, 10 to 20 feet. It is a white, soft-grained sandstone, hardened by metamorphism, containing, beside remains of dicotyledonous leaves, some species of finely preserved fucoids. In following the same sandstone to the south it is seen increasing in thickness, and near and under the Roe coal, five miles from Golden, it forms a high, isolated ridge, at least 200 feet thick, barren of any kind of remains, except some fucoids.

By its compound, the alternance of its coarse-grained and soft-grained strata, these being often mere clay or mud beds, its characters appear the same as those of the lower lignitic sandstone of the Raton mountains. It has, too, broken, undeterminable fragments of wood, Cyperaceæ, etc. Beside the species of fossil dicotyledonous leaves found in the white sandstone of Golden, most of them homologous, or even identical with some species of the Raton and other localities, it has one of those very rare land plants, which has been described and recognized in Europe as pertaining, as yet, exclusively to the Eocene.

The finest and best preserved specimens of fossil leaves that have ever been found in this country, with the exception, perhaps, of those of Black Butte, have been found at and around Golden, in the hard metamorphosed white sandstone, under and interlying the beds of coal, and in the beds of white clay upheaved against the sides of the basaltic dike; a clay, hard as silex from metamorphism, having mostly remains of palm leaves; and from three miles south of Golden, from a sandstone still upheaved, near the tail of the dike, but scarcely changed by heat, and easily cut in large pieces. The continuity of the Lignitic formation is traced north toward Cheyenne, where the conglomerate sandstone covers the face of the country, and all the facts discovered, tend to confirm the statement made by Dr. Hayden in 1868, that all the lignite Tertiary beds of the West are but fragments of one great basin, interrupted here and there by upheaval of mountain chains, or concealed by the deposition of newer formations.

At Medicine Bow, the line of connection with the underlying Cretaceous is, perhaps, more difficult to fix than at other localities, the fucoidal sandstone here being mostly barren of remains of marine plants. But from its base to its top, in a thickness of, perhaps, 200 feet, it is barren, too, of any remains of animals, while here and there branches of fucoids appear, as thrown by the waves, being generally mixed with fragments of wood and stems of dicotyledonous plants. From the cut of the railroad west of Medicine Bow, where this sandstone is seen overlying the Cretaceous, and where two fine mineral springs come out from its base, it is continuous to Carbon, in repeated and deeper undulations, forming basins, which at this place and around contain the upper Lignitic formation, with remarkably thick beds of combustible mineral. The coal is mined at Carbon Station by a shaft descending through the following strata:

1. Shale, clay, and sandstone at top, 35 feet.
2. Ferruginous shale, with dicotyledonous leaves, 3 feet.
3. Clay, shale, and sandstone, with plants at top, 18 feet.
4. Coal (main), 9 feet.
5. Fire-clay and shale, with dicotyledonous plants, 20 feet.
6. Coal, 4 feet.
7. Fire-clay and shale, 8 feet.
8. Coal, 4 feet.
Total, 101 feet.

In following the railroad from Black Butte westward, the Lignitic formation, already seen at the surface of the country from below Bitter Creek Station, forms an irregularly broken ridge, whose general dip toward the east is varied by low undulations. In that way the measures slowly ascend to Point of Rocks, where they overlie the black shale of the Fort Pierre Group, there constituting the axis of an anticlinal, which is cut below Point of Rocks, by the meanders of Bitter creek. The counterface of the axis appears westward, in corresponding strata, after passing Saltwell valley, and hence the dip to the west brings to the surface the upper strata of the Lignitic at Rock Spring. The section of the measures is perfectly clear and exposed in its whole length. At Point of Rocks, and near the highest part of the anticlinal axis, the Cretaceous strata are exposed 80 feet in thickness, immediately and conformably overlaid by 185 feet of the Lignitic sandstone, which, from its base, bears fucoidal remains. It has, moreover, the composition and mode of disintegration of the same formation at Raton, east of the station, 25 feet above the base of this sandstone, there is a bed of coal 8 feet thick. Farther east, at Hallville, a Lignitic bed, overlaid by shales where are imbedded a quantity of fossil shells, is worked near the level of the valley at a short distance from the railroad. At Black Buttes a bed of lignite is worked, too, above the sandstone. At Rock Spring, in boring for an artesian

well, 16 beds of coal making 48 feet in thickness were passed at   28
feet, and at 1,180 feet the sandstone of the Lower Lignitic had not
been pierced.   He found a remarkable analogy, not to say identity,
between the 547 feet of measures above the lignite beds at Evanston
and the conglomerate which tops the Lignitic at Colorado Springs and
other places.

The masterly review of the Lignitic Group, by Prof. Lesquereux,
lead him to the conclusion that it is of Eocene age.   He said that the
Upper Cretaceous is positively characterized as a deep marine formation.   Immediately over it, the sandstone shows, in its remains, the
result of the upheaval of a wide surface exposed to shallow marine
action, as indicated by fucoidal life.   The upheaval continuing, this
area is brought out of marine influence to be exposed to that of the
atmosphere.   It is a new land, cut in basins of various sizes, where
fresh water is by and by substituted to brine, where vegetable life
of another character appears, where swamps are filling with clay and
floating plants, where peat-bogs in their growth form deposits of combustible matter, etc.   To suppose that the marine action is totally
banished from such a land would demand the absurd admission of an
absolutely flat surface.   Of course estuaries penetrate into it at many
places;   their waters feeding marine species, brackish shells;   their
bayous inhabited by Saurians, and their remains mixed with leaves
of the trees growing on the borders and preserved together in a fossil
state, without impairing the true character of the formation by what
palæontology considers as types of different ages.   The surface of the
Eocene sandstone, before its separation from marine influence, was, of
course, uneven.   This sandstone has, therefore, the general characters
of the Eocene, while in some troughs, Cretaceous species, still living
in deep water, may have left their remains in the sand.   Even if these
remains were numerous, their presence does not change the age of the
formation.   But on this subject, and in comparing our Eocene sandstone to the other groups established by geology, we find, in its abrupt
and permanent separation from the Cretaceous, its lithological compounds, its total barrenness from animal remains, at least generally,
and the homogeneity of its flora, reliable and constant characters
better defined than in any geological division admitted by science.
This sandstone formation is inexplicable.   It can be compared to
nothing but the millstone-grit of the Carboniferous epoch.   How
to explain why, at once, animal life seems to disappear from the
bottom of the sea, to be superseded by marine vegetation?   May this
change have been caused, perhaps, by a rapid increase of temperature

of the water brought up by the force acting to the upraising of the bottom into land, and afterward into chains of mountains. Though it may be, this change is evident and proves the geological discrimination of the Eocene sandstone from the Cretaceous, a separation the more remarkable, that from numerous observations this sandstone is reported constantly conformable to the Upper Cretaceous beds. As Dr. Hayden remarks in his description of the Lignitic 'Group of Nebraska, when we bear in mind the fact that wherever this formation has been seen in contact with the latest Cretaceous beds, the two have been found to be conformable, however great the upheavals and distortions may be, while at the junction there seems to be a complete mingling of sediments, one is strongly impressed with the probability that no important member of either system is wanting between them.

After contrasting the distribution and character of the plants with those known from the Tertiary of other parts of the world, Prof. Lesquereux thought himself authorized to deduce the conclusion: That the great Lignitic Group must be considered as a whole and well characterized formation, limited at its base by the fucoidal sandstone, at its top by the conglomerate beds; that, independent from the Cretaceous under it, and from the Miocene above it, our Lignitic formations represent the American Eocene.

He described, from South Park, near Costello's ranch, *Ophioglossum alleni*, now *Salvinia alleni, Planera longifolia*: from Elko station, Nevada, *Sequoia angustifolia, Thuja garmani, Abies nevadensis;* from the Raton mountains, *Sphæria lapidea, Chondrites subsimplex, C. bulbosus, Halymenites major, H. striatus, Delesseria incrassata,* now *Caulerpites incrassatus, Delesseria lingulata;* from Gehrung's coal-bed, near Colorado Springs, *Dombeyopsis obtusa;* from Golden City, Colorado, *Sclerotium rubellum, Delesseria fulva, Pteris anceps, Carex berthoudi, Sabal goldana, Quercus stramineus, Ulmus irregularis,* now *Ficus irregularis, Ficus auriculata, F. spectabilis, Cissus lævigatus, Dombeyopsis trivialis, D. occidentalis,* now *Ficus occidentalis, Sapindus caudatus, Ceanothus fibrillosus,* now *Zizyphus fibrillosus, Rhamnus cleburni, R. goldanus, R. goldanus,* var. *latior;* from Erie Mines, Boulder Valley, *Caulinites fecundus, Cercis eocenica;* from Carbon station, Wyoming, *Populus decipiens, Ficus oblanceolata, Coccoloba lævigata, Asimina eocenica, Zizyphus meeki;* from Black Butte station, *Sphaeria myricæ, Opegrapha antiqua, Caulinites sparganioides, Myrica torreyi, Ficus planicostata, F. planicostata,* var. *latifolia, F. clintoni, F. corylifolius, F. haydeni, Vibernum marginatum, V. contortum, Cissus lobato-crenatus, Aleurites eocenica, Paliurus zizyphoides, Carpolithes falcatus;* from the Black Butte saurian bed, *Viburnum dichotomum;* from the Black Butte red baked shale, *Quercus wyominganus;* from Evanston, *Calycites hexaphylla, Carpolithes arachioides,* now *Leguminosites arachioides;* from Elk creek, near Yellowstone river, *Carpolithes osseus;* from six miles above Spring canon, near Fort Ellis, *Abies setigera* and *Nyssa Lanceolata.*

He described, from the Dakota Group, six miles south of Fort Harker, Kansas, *Hymenophyllum cretaceum, Caulinites spinosus, Populites fagifolia, Ficus sternbergi,* now *Persea sternbergi, Sassafras mirabile, S. recurvatum,* now *Platanus recurvata, S. harkeranum,* now *Cissites harkeranus, Laurophyllum reticulatum, Pterospermites sternbergi,* now *Protophyllum sternbergi;* from nine miles above Salina in the Saline Valley, Kansas, *Populites salinæ,* now *Menispermites salinensis, P. affinis,* now *Cissites affinis,* and *Pterospermites rugosus,* now *Protophyllum rugosum.*

Prof Meek* said that the coal-bearing rocks at Coalville, Utah, are undoubtedly of Cretaceous age, as he had from the first maintained, and he quoted in support of this view his remarks in Dr. Hayden's Report of 1870, page 299. He prepared a section running from the principal coal-bed, near Coalville, in a northwesterly direction, to Echo canon, a distance, by a right line a little obliquely across the strike of the rocks, of about three and a half miles. This section commences 393 feet below the heavy bed of coal, and furnishes a thickness of 3,980 feet below the conglomerate, or including the conglomerate, which is here 700 feet in thickness, 4,680 feet of strata. Several parts of this section contain marine Cretaceous fossils, the highest of which is gray, soft, sandstone, 30 feet in thickness, and 1,431 feet below the conglomerate. It contains many large *Inoceramus, Ostrea* and *Cardium.*

The conglomerate not only composes the towering walls of Echo canon at places forming perpendicular, or even overhanging escarpments, 500 to 800 feet in height, but also rises into mountain masses on the west side of Weber river, near the mouth of the canon. It probably attains a thickness in places of 2,000 feet. This he referred to Tertiary age because of its position above the Cretaceous, its non-conformability with the rocks below it, and its remarkably coarse material.

---

*6th Rep. Hayden's U. S. Geo. Sur. Terr.

Recurring to the Rocks at Coalville, he says: As I have, however, mentioned faults and lateral displacements of the strata here, it may be thought, by some, who are yet incredulous in regard to the Cretaceous age of these coals, that these disturbances of the strata may have given origin to erroneous conclusions respecting the position of the beds containing the Cretaceous types with relation to the coals. This, however, is simply impossible, because these fossils occur both above and below the coal-beds, even in local exposures, where all the strata, and included coal-beds can be clearly seen conformable and in their natural positions with relation to each other. We found both above and below the main coal-bed, *Inoceramus problematius*, a widely distributed species, that is very characteristic of the Niobrara and Benton Groups of the Upper Missouri, which there occupy positions below the middle of the series. Again, far above this, numerous specimens of a larger *Inoceramus,* which, if not really identical with one of these forms, is scarcely distinguishable from *I. sagensis* and *I. nebrascensis,* which occur in the later members of the Upper Missouri series. From these facts, it is more probable that we have here, at and near Coalville, representatives of the whole Upper Missouri series, with possibly even lower members, farther up Weber river, than any of the known Upper Missouri subdivisions of the Cretaceous. If this is so— and there seems to be but little reason to doubt it—the marked difference observed between almost the whole group of fossils found here, and those of the Upper Missouri Cretaceous, would seem to indicate, that there was no *direct* communication between the Cretaceous seas or gulfs of that region and those in which these Utah beds were deposited. Differences of physical conditions, however, probably also played an important part in the production of this diversity of life, since it is evident from the great predominance of clays and other fine materials in the Cretaceous beds of the Upper Missouri, that they were deposited in comparatively deeper and more quiet waters than those in Utah, in which coarse sandstones, with occasional pebbly beds, predominate.

The strata including the beds of coal exposed on Sulphur creek, near Bear river, in Western Wyoming, he regarded as of the same age as the Coalville series. His section here is 3,542 feet thick. The lower 1,213 feet he regarded as certainly Cretaceous, the next 2,049 feet he thought is probably Cretaceous, and the upper 280 feet he regarded as of Tertiary age.

The Bitter creek series, which is found along Bitter creek (a small tributary of Green river, in Wyoming), from Black Butte northwestward to Salt Wells Station, on the Union Pacific Railroad, and at Rock Spring, and some other points west of Salt Wells, consisting of a vast succession of rather soft, light-yellowish, lead-grey, and whitish sandstones, with seams and beds of various colored clays, shale, and good coal, the whole attaining an aggregate thickness of more than 4,000 feet, present a mingling of fresh, brackish, and salt water types of invertebrate fossils, such as *Goniobasis, Viviparus, Corbicula, Corbula, Ostrea, Anomia,* and *Modiola.* This is the Lignitic Group which Prof. Lesquereux determined from the character of the plants to be of Eocene age, and Prof. Cope, from the Dinosaurian remains, to be of Cretaceous age. Prof. Meek thought the Judith river brackish-water beds are of the same age, and that the invertebrate fossils alone left the question of the age of the series in doubt. He stated the information as to its age in the following summary:

1. That it is conformable to an extensive fresh-water Tertiary formation above, from which it does not differ materially in lithological characters, excepting in containing numerous beds and seams of coal.

2. That it seems also to be conformable to a somewhat differently composed group of strata (1,000 feet or possibly much more in thickness) below, apparently containing little if any coal, and believed to be of Cretaceous age.

3. That it shows no essential difference of lithological characters from the Cretaceous coal-bearing rocks at Bear river and Coalville.

4. That its entire group of vegetable remains (as determined by Prof. Lesquereux) presents exclusively and decidedly Tertiary affinities, excepting one peculiar marine plant (*Halymenites*), which also occurs thousands of feet beneath undoubted Cretaceous fossils, at Coalville, Utah.

5. That all of its animal remains, yet known, are specifically different from any of those hitherto found in any of the other formations of this region, or, with perhaps two, or possibly three exceptions, elsewhere.

6. That all of its known invertebrate remains are mollusks, consisting of about thirteen species and varieties of marine, brackish and

fresh water types, none of which belong to genera peculiar to the
Cretaceous or any older rocks, but all to such as are alike common to
the Cretaceous, Tertiary and present epochs, with possibly the exception of *Goniobasis*, which is not yet certainly known from the
Cretaceous.

7. That, on the one hand, two or three of its species belong to
sections or subgenera (*Leptesthes* and *Veloritina*) apparently characteristic of the Eocene Tertiary of Europe, and are even very closely
allied to species of that age found in the Paris basin; while, on the
other hand, one species seems to be conspecific with, and two congen-
eric with (and closely related specifically to) forms found in brackish-water beds on the Upper Missouri, containing vertebrate remains
most nearly allied to types hitherto deemed characteristic of the
Cretaceous.

8. That one species of *Anomia* found in it is very similar to a Texas
Cretaceous shell, and perhaps specifically identical with it; while a
Viviparus, found in one of the upper beds, is almost certainly identi-
cal with the *V. trochiformis* of the fresh-water lignite formation of the
Upper Missouri; a formation that has always, and by all authorities,
been considered Tertiary.

9. That the only vertebrate remains yet found in it are those of a
large reptilian (occurring in direct association with the *Viviparus*
mentioned above) which, according to Prof. Cope, is a decidedly Cretaceous type, being, as he states, a huge Dinosaurian.

He described from Coalville, Utah, *Ostrea soleniscus, Avicula
propleura, A. gastrodes, Modiola multilinigera*, now *Volsella multilinigera, Cyrena carletoni, Neritina bellatula, N. patelliformis, N.
carditoides, N. bannisteri, N. pisum, N. pisiformis, Admete rhom-
boides, A. subfusiformis, Turritella coalvillensis, T. micronema, T.
spironema, Fusus gabbi, F. utahensis, Turbonilla coalvillensis, Eulima chrysalis, E. inconspicua, Melampus antiquus, Valvata nana,
Physa carletoni*; from the Missouri river below Gallatin City, Montana,
Ostrea anomioides, Corbicula inflexa, Pharella pealei; from Bear
river city, on Sulphur creek, Wyoming, *Trapezium micronema, Corbicula aequilateralis, C. securis*, from near Cedar City, Southern Utah,
Corbula nematophora; from the Bitter creek series, at Point of
Rocks, Wyoming, *Ostrea wyomingensis, Corbula tropidophora*; from
Black Butte Station, *Corbicula bannisteri, Melania wyomingensis;*
and from Rock Spring Station, Central Pacific Railroad, Wyoming, *Corbula undifera* and *Goniobasis insculpta*.

Prof. E. D. Cope* described, from Solomon river, Kansas, *Portheus gladius*, now *Pelecopterus gladius*, and *Portheus lestrio.*

Prof. O. C. Marsh† described, from Kansas, *Apatornis celer.*

Dr. Joseph Leidy‡ described, from Smoky Hill river, Kansas, *Clidastes affinis;* and from Columbus, Mississippi, *Eumylodus laqueatus.*

Mr. James Richardson§ separated the Cretaceous rocks of Vancouver Island into seven divisions in ascending order as follows:

A. Productive coal measures.

B. Lower shales,

C. Lower conglomerate.

D. Middle shales.

E. Middle conglomerate.

F. Upper shales.

G. Upper conglomerate.

A section of division A., on Brown's river, is shown to be $739\frac{1}{2}$ feet thick. Division B., on Sable river and Denman Island, 1,000 feet. Division C., on Denman and Hornby Islands, between 900 and 1,000 feet. Division D, on Hornby Island, 70 feet. Division E., on Hornby Island, from 1,100 to 1,200 feet. Division F., near Tribune Bay, $776\frac{1}{2}$ feet. And Division G., on Tribune Bay, 320 feet. Making a total thickness of 5,000 feet.

Dr. Dawson‖ described, from the Lower Cretaceous of Queen Charlotte Islands, *Cycadeocarpus columbianus.*

In 1874, Dr. F. V. Hayden¶ said, that to one who has carefully studied the divisions along the Missouri river, the Cretaceous beds in Colorado and New Mexico, may be separated into five groups, without much difficulty. The Dakota group is well shown and is always characteristic, though seldom containing any organic remains. The Niobrara group is represented by a thin bed of impure gray limestone, and thin calcareous shale, with *Ostrea congesta* and a species of *Inoceramus.* The fossils are about the same as those occurring on the Missouri, but the rocks have little of the chalky texture, as observed in the northwest and in Kansas. The Fort Benton and Fort Pierre groups are black shaly clays, and do not differ materially from the same groups occurr-

---

* Proc. Acad. Nat. Sci.

† Am. Jour. Sci. and Arts, 3d series, vol. 5.

‡ Cont. to Extinct Vert. Fauna. W. Terr.

§ Geo. Sur. Can.

‖ Geo. Sur. Can.

¶ Ann. Rep. U. S. Geo. Sur. Terr.

ing in other localities to the northward. The Fox Hills group contains a great abundance of well marked Cretaceous fossils, many of the species identical with those found on the Missouri river. This group passes up into the lignite strata, apparently, without any marked unconformability. In passing upward in the Fox Hills Group, one by one the mollusca of purely marine character disappear until only some varieties of oysters remain, with the plants peculiar to the Lignitic Group.

The relation of the well-defined Cretaceous with the Lignitic Group forms one of the most important problems in Western geology, and the area for the solution of the question probably lies in the Laramie plains and westward toward Salt Lake, where the aggregate thickness is from 10,000 to 20,000 feet. So far, the evidence from the vegetable remains is wholly in favor of the Tertiary age of the coal group. The vertebrate remains, according to Prof. Cope, place the coal group with the Cretaceous, while the proof from the invertebrate fossils is not strong in any direction, although, perhaps, leaning toward the Tertiary. We must admit, however, that the lower coal-beds are of Cretaceous age so far as the evidence goes. For instance, the Coalville and Bear river beds are most probably Cretaceous, inasmuch as many undoubted Cretaceous types are found in strata above the coal, and further south, in New Mexico, Arizona, and Utah, there are coal-beds of undoubted Cretaceous age.

A. R. Marvine[*] described the Dakota Group between the Big Thompson and South Platte. It can be traced from one point to the other, though it is somewhat obscured near Golden City ; this is due to the fact, that its hardness is greater than the beds either above or below, and it forms a more persistent hog-back ridge than any other group. Between the cross-cutting streams for all this distance and beyond, it rises in its long characteristic ridge, capping the soft Jurassic beds below, and whether the dip be high or low, usually reaching to about the same level. The sandstones are usually clean, gritty, even-grained and silicious in texture, varying from a silicious conglomerate, on the one hand, to a hard quartzite on the other, and only occasionally becoming soft. Their color is usually light yellow or light gray, or even white, varying to rusty yellow, and only occasionally red in the softer portions. These are the hard and massive portions which characterize the group, and which are separated by thin, shaly layers, which may be quite argillaceous or even carbonaceous in character, with many broken remains of fossil plants. A section at Bear Canon

---

[*] Hayden's U. S. Geo. Sur. Terr.

shows a thickness of 240 feet, and another near the South Platte river, 385 feet.

The Fort Benton Group consists of a series of shaly beds, which may be either highly argillaceous or quite arenaceous in character, there being associated with them, in either case, a few thin, brown sandstones; the thickness from Big Thompson to South Platte varying from 100 to 400 feet. A section at Little Thompson creek shows a thickness of 400 feet, and one at Bear Canon 120 feet.

The Niobrara Group is decidedly calcareous, and usually contains numerous fossils. A section at Bear Canon shows 105 feet, and one at Little Thompson creek 150 feet. The Fort Pierre Group, at Bear creek is about 300 feet in thickness.

The total thickness of the Dakota, Fort Benton, Niobrara, Fort Pierre, and Fox Hills Group, at the Middle Park, is estimated at from 3,500 to 4,500 feet. A section of the Lignitic Group at Golden City shows a thickness of 3,360 feet, and the estimated thickness at Middle Park is 5,500 feet.

Prof. Leo Lesquereux* described, from the Lignitic Group at Golden, Colorado, *Woodwardia latiloba*, *Pteris erosa*, *P. subsimplex*, *P. affinis*, now *Osmunda affinis*, *Aspidium goldianum*, now *Lastrea goldana*, *Sphenopteris membranacea*, *Selaginella berthoudi*, *Hymenophyllum confusum*, *Flabellaria fructifera*, now *Sabalites fructiferus*, *Quercus goldanus*, *Ficus planicostata*, var. *goldana*, *F. zizyphoides*, *Platanus rhomboidea*, *Viburnum lakesi*, *Nelumbium lakesanum*, *Zizyphus distortus*, *Rhamnus inæqualis*; from Black Butte, Wyoming, *Woodwardia latiloba*, var. *minor*, *Sphenopteris nigricans*, *Quercus cleburni*, *Pisonia racemosa*; from the roof of coal mines at Sand creek, Colorado, *Pteris gardneri*, now *Gymnogramma gardneri*, *Equisetum lævigatum*, *Eriocaulon porosum*, *Nelumbium tenuifolium*; from Coal creek, Colorado, *Cornus holmesi*; from Evanston, *Laurus sessiliflora*, now *Tetranthera sessiliflora*; and from Mount Brosse, or Troublesome creek, *Persea brossana*, now *Laurus brossana*, and *Cornus impressa*.

Prof. E. D. Cope, from the evidence of the vertebrates, and especially from the evidence afforded by the remains of the *Dinosauria*, referred the Fort Union or Lignitic Group, the Judith River Group, the Bitter Creek Group, and the Bear River Group to the upper Cretaceous. And he described, from the Fort Union Cretaceous, of Colorado, *Cionodon arctatus*, *Polyonax mortuarius*, *Bottosaurus perrugosus*, *Trionyx vagans*, *Plastomenus punctulatus*, *P. insignis*, and *Adocus lineolatus*.

---

* Hayden's 7th Rep. U. S. Geo. Sur. Terr.

T. A. Conrad described, from Trout creek, near Fairplay, *Ptycho-ceras aratum* and *Meekia bullata;* from seven miles south-southeast of Fairplay, *Helicoceras vespertinum, Anchura bella ;* and from near Denver, *Haploscapha capax.*

Prof. Leo Lesquereux * described, from the Dakota Group at Fort Harker, Kansas, *Lygodium trichomanoides, Greviopsis haydeni ;* from Kansas, *Todea saportanea, Dioscorea cretacea, Flabellaria min-ima, Alnus kansasana,* now *Hamamelites kansasanus, Myrica obtusa, Quercus poranoides, Sassafras acutilobum, Oreodaphne cretacea, Em-bothrium daphneoides, Diospyros rotundifolia;* from Minnesota, *Ficus hallana;* from Decatur, Nebraska, *Hedera ovalis, Protophyllum ne-brascense;* from the bluffs of Salina River, *Protophyllum minum;* from Warner's quarry eight miles from Winnebago village, bluffs of the Missouri river, *Ptenostrobus nebrascensis.*

The Cretaceous is visible, in North Carolina,† only in the bluffs in the southeastern part of the State, from the Neuse and its tribu-tary Contentnea, southward. It is best exposed, in the bluffs, along the Cape Fear between Fayetteville and Wilmington. The rocks for 50 to 60 miles below Fayetteville consist of sandstones, clay slates and shales, 30 to 40 feet thick, in many places dark to black and very lig-nitic, with projecting trunks and limbs of trees, and at a few points full of marine shells. For 40 to 50 miles above Wilmington, and in all the other river sections, the rock is a uniform, dark, greenish-gray, slightly argillaceous sandstone, massive, and showing scarcely any marks of bedding. This sandstone everywhere contains a small per-centage of glauconite, and is the representative of the true greensand.

The Ripley Group was so named by Conrad from the town of Ripley, Mississippi,‡ in 1858, and some of the species of shells at that place are identical with species from North Carolina, Georgia, Eufaula, Alabama, and Haddonfield, New Jersey. The mineral character of the beds and state of preservation of the fossils are the same, proving not only a simultaneous deposit, but a similar depth of water, not in an estuary but in a marine basin. This group constitutes the great bulk of the Cretaceous strata east of the Mississippi, and, as Conrad supposed, corresponds most nearly in age with the Senonian stage of D'Orbigny, or that part of the Cretaceous which underlies and most nearly approaches in age the chalk.

---

* Cret. Flora, Hayden's U. S. Geo. Sur. Terr., vol. 6.
† Geo. of N. Carolina, 1875.
‡ Jour. Acad. Nat. Sci. 2d Ser. vol. 3.

In 1875, he described,[*] from the Ripley Group at Snow Hill, Greene county,North Carolina, *Anomia linifera,Radula oxypleura, Trigonarca triquetra, T. umbonata, T. perovalis, T. carolinensis, T. congesta, Nemodon brevifrons, Barbatia carolinensis, B. lintea, Arcoperna carolinensis, Inoperna carolinensis, Mytilus condecoratus, M. nasutus, Etea carolinensis, Brachymeris alta, Crassatella carolinensis, C. pteropsis, Arene carolinensis, Lucina glebula, Cardium carolinense, Protocardia carolinensis,Aphrodina regia,Cyclothyris alta, C. carolinensis, Baroda carolinensis, Oene plana, Linearia carolinensis, Valeda lintea, Cyprimeria depressa, Hercodon ellipticus, Cymella bella, Corbula carolinensis, C. bisulcata,C. perbrevis,C. subgibbosa, Diploconcha cretacea, Callonema carolinense, Leioderma thoracica, Lunatia carolinensis,* and from Cape Fear river, *Corbula oxynema,* and *Anomia lintea.*

The Cretaceous rocks,[†] corresponding in age with the great chalk formations of Europe, though very different from them in mineral character, are spread over a great extent of surface in the western part of British America. Except in a few localities, and those chiefly in proximity to the Rocky Mountain region of uplift, they are still almost as perfectly horizontal as when first deposited. The eastern edge overlaps Silurian and Devonian beds, and runs nearly parallel with the base of the Laurentian range for a distance of about 130 miles, from the 53d to the 55th parallel of latitude. Southward it trends to the East, and probably crosses the 49th parallel east of Red river; while in southwestern Minnesota it reposes in some places directly on granites which are no doubt Laurentian. The general course of the eastern outcrop is consequently about north-northeast; and it is marked, broadly, by a series of escarpments and elevations, including —from south to north—Pembina, Duck, Porcupine and Basquia Mountains. All these appear to be composed, for the most part, if not entirely, of Cretaceous rocks, though the extreme edge of the formation may often stretch beyond them. These mountains are, more correctly speaking, the salient points of the edge of the second plateau, and the generally horizontal position of the beds thus suddenly cut off to the east, attests the immense denudation which must have taken place in modern times. North of the Basquia Mountain the edge of the Cretaceous would appear to run westward and cross the Saskatchewan near Fort a la Corne, where, at Cole's Falls, a dark-colored shale has been referred to the lowest member of the series. The

* Geo. of N. Carolina, 1875.
† Dawson's Rep. Geo., 49th Parallel, 1875.

western border of the Cretaceous seems, in some places, to follow closely along the base of the Rocky Mountains, but many circumstances arise to complicate it in that region.

The Lignitic Group north of the 49th parallel is not bounded by any great physical features of the country, but adheres closely to the upper members of the Cretaceous. Though, no doubt, originally deposited in extensive basin-like depressions, it is now generally found forming slightly elevated plateaus. Denudation must have acted on these rocks on a vast scale, but they still cover an immense area, and contain the greatest stores of mineral fuel known to occur in the vicinity of the 49th parallel. The line of their eastern edge crosses the parallel near the 102d meridian, and thence appears to pursue a north-westward course, remaining for some distance nearly parallel with the edge of the third plateau. Beyond the elbow of the South Saskatchewan, though the same physical feature continues to the north, it is not known what relation it may bear to the outcrop of this formation, nor has its northern limit been ascertained.

On leaving the Lake of the Woods, and proceeding westward, the face of the country is found to be thickly covered with drift and alluvial deposits. The Silurian limestones, which probably exist at no very great depth, are not observed, and the first rocks seen are those of the Cretaceous along the base of Pembina mountain, which bounds the Red river valley on the west. From this point westward to the base of the Rocky Mountains no rocks are found older than the Cretaceous. About 25 miles north of the Line, where the Boyne river cuts through the Pembina escarpment the Niobrara Group is found exposed. The rock is a cream-colored or nearly white limestone, breaking easily along horizontal planes, parallel to the surfaces of the shells of *Ostrea congesta*, and *Inoceramus*, of which it is in great part composed. The rock also abounds with more or less perfect remains of Foraminifera, Coccoliths, and allied microscopic organisms. Prof. G. M. Dawson here proposes the name of Pembina Mountain Group for what he supposes may be the equivalent of the Fort Pierre Group. It is exposed in the valley, by which the Commission Trail ascends Pembina mountain, about ten miles north of the 49th parallel, and where the 49th parallel cuts the base of the Pembina escarpment rocks, and at various other places for about 40 miles west of the foot of Pembina mountain. In some places the exposures vary from 100 to 240 feet. From this point for 350 miles west no exposures of the Cretaceous occur on account of the drift deposits which cover the surface. When the rocks underlying the drift are again seen, near La Roche Percee, they belong to the Lignitic Group.

The Lignitic Group appears, in the valley of the Souris river, 250 miles west of Red river and affords numerous sections. The mollusca as well as the characters of the strata show that it is the equivalent of the Fort Union Group. A bed of lignite, 7 feet 3 inches in thickness, occurs in the Souris valley about a mile north of the position occupied by the Wood End depot. The strata appear to be nearly horizontal. West of Wood End, the Souris valley runs north-westward along the base of the Coteau, diverging rapidly from the boundary line. It loses, at the same time, its abrupt character, and no exposures of the rocks occur for a long distance. In following the 49th parallel, the escarpment of the third great prairie level is overcome, and it is not till after having passed through the broken Coteau belt, and reached the Great Valley, that exposures of the underlying rocks occur. This valley is the most eastern great channel of erosion which crosses the Line southward, toward the Missouri, and in it the beds of the Lignitic Group are exhibited on a grand scale. On the boundary-line, thus a space of 82 miles, from the 263 to the 345 mile point, is completely shrouded by drift. There is every reason to believe, however, that the Lignitic Group stretches uninterruptedly between the two localities, and an exposure some distance north of the line sustains this view.

In the Great Valley, the beds exposed are at an elevation of about 700 feet greater than those near Wood End, on the Souris river. They consist of shales, clays, and sandstones, with beds of lignite. The next stream crosses the line at the 351 mile point, called Pyramid creek, where the lignite beds are again exposed. They reappear on Porcupine creek, 35 miles farther west, and near the 393 mile point, on the line, an 18 feet bed of lignite occurs. The fossil plants here are nearly identical with those of the Fort Union Group. In the neighborhood of Wood Mountain, hard, grayish sandstones, belonging to this group, are exposed, in the sides of the hills and banks of the valleys. At 19 miles from Wood Mountain the edge of the plateau is reached, and a few miles further on, the junction of the lignite with the marine Cretaceous is crossed. Twenty miles south of the Wood Mountain settlement, on the 49th parallel, near the 425 mile point from Red river, the Lignitic Group is found superimposed upon the marine Cretaceous. The exposures are numerous, and are produced by the streams flowing from the southern escarpment of the water-shed plateau, which has been gashed by their action into most rugged *Bad Lands.*

This term has attached to it, in the western regions of America, a peculiar significance, and is applied to the rugged and desolate country formed where the soft clayey formations are undergoing rapid waste. Steep irregular hills of clay, on which scarcely a trace of vegetation exists, are found, separated by deep, nearly perpendicular-sided, and often well nigh impassable valleys; or when denudation has advanced to a further stage—and especially when some more resisting stratum forms a natural base to the clayey beds—an arid flat, paved with the washed-down clays, almost as hard as stone when dry, is produced, and supports irregular cones and buttes of clay, the remnants of a former high-level plateau. Denudation, in these regions, proceeds with extreme rapidity during the short period of each year, in which the soil is saturated with water. The term, first and typically applied to the newer White river Tertiaries of Nebraska, has been extended to cover country of similar nature in the lignite regions of the Upper Missouri and other areas of the West. In the Bad Lands, south of Wood Mountain, the hills assume the form of broken plateaus; degenerating gradually into conical peaks, when a harder layer of sandstone, or material indurated by the combustion of lignite beds, forms a resistant capping. Where no such protection is afforded, rounded mud-lumps are produced from the homogeneous, arenaceous clays. Waste proceeds entirely by the power of falling rain, and the sliding down of the half-liquid clays, in the period of the melting snow in spring. The clay hills are consequently furrowed, from top to base, by innumerable runnels, converging into larger furrows below. The small streams, rapidly cutting back among these hills, have formed many narrow, steep-walled gullies, while the larger brooks have produced wide, flat-bottomed valleys at a lower level, in which the streams pursue a very serpentine course. Denudation is even here, however, still going on as from the frequent change in the channel of the stream, it is constantly encroaching on the banks of the main valley, undercutting them and causing landslips.

The general section at this place, in descending order, is as follows:

1. Yellowish sand and arenaceous clay, sometimes indurated in certain layers and forming a soft sandstone. It forms the flat plateau—like tops of the highest hills seen. About 50 feet.

2. Clays and arenaceous clays, with a general purplish-gray color when viewed from a distance. It contains a lignite-bearing zone and beds, rich in the remains of plants, and in the lower part, the remains of vertebrate animals. About 150 feet.

3. Yellowish and rusty sands, in some places approaching arenaceous clays, often nodular. About 80 feet.

4. Grayish-black clays, rather hard and very homogeneous, breaking into small angular fragments on weathering, and forming earthy banks. This division belongs to the upper part of the Fox Hills Group, and only about 40 feet of it is exposed at this place.

The sombre clays of the Fox Hills Group may be traced almost continuously for a distance of about ten miles west, on the 49th parallel, where lower beds are exposed. Near the crossing of the 49th parallel and trail to Fort N. J. Turney, where the Wood Mountain Astronomical Station was established, good exposures of the Fort Pierre Group occur in the banks of the valley of a large brook. Taking into consideration the difference of level between this locality and that of the section above, it appears that the Fort Pierre Group must be at least 200 feet below the Fort Union or Lignitic Group.

Westward from these sections the continuity of the Cretaceous clays in the vicinity of the boundary line is indicated by occasional small exposures, and at a distance of 13 miles a tolerably good exhibition of the Fort Pierre Group occurs. Where the boundary line crosses White Mud river, or Frenchman's creek, numerous and very fine exposures occur. The stream flows in the bottom of a great trough, cut out of the soft Cretaceous strata, over 300 feet deep, and in some places fully three miles wide. The tops of the banks, on both sides of the valley, are formed of yellowish ferruginous sands referable to the base of the Fort Union Group. Below this the sombre clays of the Fox Hills and Fort Pierre Groups have a thickness of 273 feet to the water level of the river. A similar section occurs on the main trail going west from Wood Mountain in the Valley of the White Mud river, 16 miles north of the 49th parallel, and 23 miles northwest of the last described exposure.

On the western side of White Mud river, hilly ground occurs, and at about the 505 mile point from Red river, the prairie makes a very definite rise and forms a plateau, which extends along the 49th parallel to the 534 mile point. The plateau is composed of the Fort Union Group. On coming to the western edge of this plateau, a great area of barren and arid prairie, at a lower level, and based on the Fort Pierre Group is seen stretching westward toward Milk river. An interesting section of the Fort Pierre Group and lower strata occurs, in a deep valley, about six miles west of the East Fork of Milk river, on and near the 49th parallel. The thickness exposed is 893 feet. The Valley of the Milk river offers continuous and magnificent sections of the Fort Union Group. The country, on both sides of it, is seamed with tributary ravines and gorges, the banks of which are often nearly perpendicular, and which ramify in all directions. The banks of Milk river rise abruptly nearly 300 feet above the level of the stream, and are more than a mile apart. Sections of the Fort Union Group were obtained near the 49th parallel 284 feet in thickness. In the conlees and gorges which intersect the prairie on the west side of the Milk river, exposures of the same group continue to occur for many miles. Near the 620 mile point, west of Red river, a very interesting and highly fossiliferous section of the brackish water deposits of the Fort Union Group is exposed. In the valleys which seam the flanks of the hills, and furrow the surface of the prairie around East Butte numerous more or less extensive exposures of this group occur. But on the west side of West Butte, where a considerable brook issues from the central valley, a section of the Fort Pierre Group is exposed, 800 feet in thickness.

The exposures of the Fort Union Group continue to occur as we go west until the base of the Rocky Mountains is reached. They occur on the branches of Milk river, St. Mary river and the Belly river.

Prof. G. M. Dawson found the Lignitic or Fort Union Group everywhere conformable with the Fox Hills Group below. He referred it to Tertiary age, and estimated the thickness, assuming the horizontality of the beds and the rise in the general surface of the country, at not less than 1,000 feet.

Dr. J. W. Dawson* described, from the Fort Union Group, south of Woody Mountain, *Lemna scutata,* and from west of Woody Mountain, *Esculus antiqua.*

Prof. E. D. Cope† described, from the Fort Union Group, six miles west of First Branch of Milk river, near latitude 49°, *Cionodon stenopsis,* *Compsemys ogmius,* and from the Bad Lands of South Woody Mountain, *Plastomenus coalescens,* and *P. costatus.*

Speaking of the age of the Fort Union or Lignitic Group, the Bitter creek series and the Bear River Group he says‡ that Prof. Lesquereux, as is well known, pronounced this whole series of formations to be of Tertiary age. The material (fossil plants) on which this determination is based is abundant, and it must be accepted as demonstrated beyond all doubt. But that he regarded the evidence derived from the mollusks in the lower beds and the vertebrates in the higher as equally conclusive that the beds are of Cretaceous

* Rep. Geo., 49th parallel.
† Geo. Rep., 49th parallel.
‡ Vert. Cret. Form. of the West.—Hayden's-U. S. Geo. Sur. Terr., vol. 2.

age.   There is, then, no alternative but to accept the result that a
Tertiary flora was contemporaneous with a Cretaceous fauna, estab-
lishing an uninterrupted succession of life across what is generally
regarded as one of the greatest breaks in geologic time.

He described, from the Niobrara Group, of Colorado, *Syl-
læmus latifrons;* from the Fort Benton Group, two miles west
of Sibley, Kansas, *Pelycorapis varius;* from the Niobrara Group
or yellow chalk, near the Solomon river, Kansas, *Portheus arcuatus,
P. mudgei,* and *Pachyrhizodus leptopsis;* from Ellis county, Kansas,
Lamna macrorhiza, L. mudgei, and *Empo merrilli;* from Trego
county, *Empo contracta* and *Empo semianceps;* from the neighborhood of Fort Wallace, *Phasganodus carinatus, P. gladiolus, P. anceps;*
from Phillips county, *Tetheodus pephredo;* from Kansas, *Enchodus
dolichus, E. petrosus, Pelecopterus chirurgus, Toxochelys serrifer;*
from Stockton, Kansas, *Ptychodus janewayi;* from Spring creek, in
Rooks county, *Pelecopterus perniciosus;* from the Greensand of
New Jersey, *Osteopygis erosus, Enchodus oxytomus, E. tetraecus, Lep-
tomylus forfex, Diphrissa latidens, Bryactinus amorphus, Ischyodus
stenobryus, I. tripartitus, I. longirostris, I. incrassatus, I. gaskilli, I.
fecundus, Isotænia neocæsariensis.*

And he furnished a section of the Cretaceous rocks of the region
west of the Sierra Madre range of New Mexico as follows :*

Dakota Group, 500 feet.

Fort Benton Group, 2,000 feet.

Niobrara Group, 400 feet.

Fort Pierre Group, 1,500 feet.

Uncertain (concealed in the Sage plain), 500 feet.

G. K. Gilbert† found a section of the Cretaceous exposed by the
north fork of the Virgin river, from the vicinity of Mountain Lakelet
to Rockville, Southern Utah, 1,800 feet in thickness, and another on
the west fork of Paria creek, 935 feet.

Prof. G. F. Credner‡ described, from the Cretaceous of Texas, *Salenia
texana.*

J. J. Stephenson§ found the Cretaceous out-crop practically
unbroken from Golden, Colorado, to Mexico.   On the west side of the
front or eastern range, there is a narrow area, of which only isolated
portions remain in Huerfano, Wet mountain, Current creek, and South

---

* Proc. Acad. Nat. Sci.
† Geo. Sur. W. 100th Meridian, vol. 3.
‡ Zeitschrift fur d. gesammten Naturwiss.
§ Geo. Sur. W. 100th Meridian, vol. 3.

Parks. In the area of the San Juan they are the only rocks exposed between Macomb's trail and the New Mexico line, excepting the small patch of Triassic, on the Rio Florida, and Rio de las Animas. The rocks differ in detail, but as a whole, the series is made up of three divis.ons. The lower is a mass of sandstone 200 to 500 feet thick; the middle is composed of shales and limestones, with, in the eastern localities, marls and sandstones 1,000 to 1,500 feet; and the upper, chiefly sandstones, with intercalated shales and lignites 500 to 700 feet.

He referred the whole lignite bearing series exposed at Canon City, and at other localities along the eastern base of the Rocky mountains, to the upper Cretaceous.

In 1876, Prof. J. W. Powell* separated the Cretaceous rocks of the Plateau Province of the west, in ascending order, as follows:

1. Henry's Fork Group, 500 feet.
2. Sulphur Creek Group, 2,050 feet.
3. Salt Wells Group, 2,000 feet.
4. Point of Rocks Group, 2,000 feet.

The Henry's Fork Group consists of sandstones, bad land rocks, conglomerates and shales, with carbonaceous shales and lignitic coal. It has an out-crop parallel and approximately co-extensive with the Triassic and Jurassic; that is, like those groups, it was brought up by the great Uinta upheaval, and the elevation of the Yampa Plateau. The conglomerates have a much more extensive development on the south than on the north side of the Uinta mountains. On the south side of the Yampa plateau, where the Fox creek and Cliff creek flexures unite, they stand on edge, with a dip of about 85° to the southeast, and are firmly cemented, and stand as high walls, separated by a long, narrow valley, strewn with fragments of the conglomerate which have tumbled down from either side.

The Sulphur Creek Group consists of black shales, occasionally friable sandstones with carbonaceous shales and lignitic coal. It is well exposed near Hilliard station, on the Union Pacific railroad, in the hills cut by Sulphur creek; there are many fine exposures on the north and south sides of the Uinta mountains; on Henry's Fork; between the head of Dry Lake valley and Vermilion creek; in the Escalante valley, Paria valley, Kanab valley, and many other localities.

The Salt Wells Group consists of sandstones or arenaceous shales; often very friable, producing bad lands, with carbonaceous shales and

---

lignitic coal. The rocks are well exposed on Green river, about two and a half miles above Flaming Gorge; along the northern flanks of the Uinta mountains; in the Pink cliffs; at Gunnison's Butte, on Green river south of Gray canon, but especially in the cliffs and escarped hills of the Salt Wells basin, east of the debouchure of the Point of Rocks canon.

The Point of Rocks Group consists of sandstones, usually indurated, sometimes ferruginous, with many beds of carbonaceous shales and lignitic coal, and is divided into the Golden Wall Sandstone, the Middle Hogback Sandstone, and the Upper Hogback Sandstone. The rocks are well exposed at Point of Rocks Station on the Union Pacific Railroad, in the escarpments facing Bitter Creek, at Rock Springs, on Green river, 2 miles above Flaming Gorge, at the foot of Desolation Canon, and Gray Canon on Green river, in the Wahsatch Cliffs at the head of the Escalante river, and in the hills at the foot of the Pink Cliffs in Southern Utah.

Prof. C. A. White* described, from the Point of Rocks Group, near Point of Rocks, Wyoming, *Ostrea insecura, Odontobasis buccinoidea ;* from Upper Kanab, Utah, *Unio gonionotus, Planorbis kanabensis, Physa kanabensis, Helix kanabensis ;* and from Bear River Valley, near Mellis Station, Wyoming, *Rhytophorus meeki, Goniobasis cleburni, G. chrysaloidea, Viviparus panguitchensis;* from the Salt Wells Group, near Coalville, Utah, *Ostrea sannionis, Arca coalvillensis, Lunatia utahensis;* from Last Chance creek, Southern Utah, *Inoceramus gilberti;* and from Upper Kanab, Utah, and Hilliard Station, Wyoming, *Cyrena erecta.*

He described, from the Sulphur Creek Group at Upper Kanab, Utah, *Turnus sphenoideus, Anchura ruida,* and *A. prolabiata.*

He described from the Henry's Fork Group at the head of Waterpocket canon, Southern Utah, *Plicatula hydrotheca;* from Lower Potato Valley and Upper Pine creek, Utah, *Inoceramus howelli;* from Middle Park, south of Grand river, Colorado, *Avicula parkensis.*

He described, from the Bitter Creek Group at Black Buttes, Wyoming, *Unio petrinus, U. propheticus, U. brachyopisthus, Neritina rolvilineata, Viviparus plicapressus, Leioplax turricula;* from Almy coal mines, near Evanston, *Pisidium saginatum, Hydrobia recta;* from Point of Rocks, *Corbula subundifera;* from south base of Pine Valley Mountains, Utah, *Helix peripheria,* and from Musinia plateau, *Hydrobia utahensis.*

---

* Geo. of Uinta Mountains.

Dr. F. V. Hayden* said the Dakota Group is composed of massive beds of sandstones, intersected with layers of clay, and forms some of the most conspicuous ridges or "hogbacks" along the eastern base of the Front or Colorado range. Its importance, however, varies in different localities as much as its texture; sometimes it is' scarcely seen, and then again it forms one or more of the most important ridges. Its aggregate thickness is never great, varying from 200 to 400 feet, and may be represented by a very narrow belt on the map. West of the 100th meridian it has yielded very few organic remains, although it has a very extended geographical range. It is hardly ever wanting along the margins of the mountain ranges east of the Wasatch Mountains, in Utah. From its structure in the far West he regarded it as a sort of transitional group between the well-defined Cretaceous and the Jurassic below.

Dr. A. C. Peale measured a section of the Dakota Group beneath station 73, north side of Gunnison river, that presented a thickness of 536 feet, and another section at station 60, that presented a thickness of 651 feet.

The Fort Benton and Niobrara Groups are found in the valleys of Grand and Gunnison rivers, and on the North Fork of the Gunnison. A partial section between station 38 and station 80 gave a thickness of 753½ feet, and another section on Gunnison river, opposite Roubideau's creek, measured 687 feet. The estimated thickness, however, including the Fort Pierre Group, is from 1,500 to 2,000 feet.

On Coal creek there is a bluff, in the face of which are exposed 1,500 feet of light-gray and yellowish sandstones and shales, referred to the Fox Hills Group. And on the North Fork of the Gunnison the exposures are of greater thickness. On the ridge dividing Oh be Joyful creek from Anthracite creek, near station 32, a section of sandstones occurs 883 feet in thickness. Most of these sandstones have a metamorphosed appearance, and the ridge, in which they are exposed, is intersected with dikes. Below the strata of this section there are probably 1,000 feet of shales and sandstones to a series of coal-bearing strata on Oh be Joyful creek. The latter, according to Mr. Holmes' estimates, is about 2,000 feet above the Dakota Group.

Above these beds there is a series referred to the Lignitic Group from 7,000 to 8,000 feet in thickness, covering a large area extending from the Grand river to the Gunnison, beneath the basaltic plateaus west of Roaring Fork. The strata are conformable to the underlying

---

Fox Hills Group, and it is difficult to determine where one formation ends and the next begins. From Dr. Peale's examination and study he deduced the following conclusions:

1. The lignite-bearing beds east of the mountains in Colorado are the equivalent of the Fort Union Group of the Upper Missouri, and are Eocene-Tertiary; also that the lower part of the group, at least at the locality 200 miles east of the mountains, is the equivalent of a part of the lignitic strata of Wyoming.

2. The Judith river beds have their equivalent along the eastern edge of the mountains below the Lignitic or Fort Union Group, and also in Wyoming, and are Cretaceous, although of a higher horizon than the coal-bearing strata of Coalville and Bear river, Utah. They form either the upper part of the Fox Hills Group, or a group to be called No. 6.

3. That the upper part of the Fox Hills Group is wanting in many parts of Eastern Colorado, and when present seems to be thin and destitute of coal.

F. M. Endlich surveyed the San Juan mining district, where he found the Dakota Group resting unconformably upon carboniferous sandstone. It consists of sandstones with occasional remains of plants, and has an estimated thickness of 800 to 1,000 feet. The Fort Benton Group, consisting of dark-gray shales, subject to considerable erosion from the action of water, is found from 400 to 600 feet in thickness. It contains beds of coal.

These groups are also developed on the San Miguel and on the Rio Dolores. A creek flowing scarcely five miles has at the junction with the San Miguel a canon 1,005 feet in depth. The entire canon is cut out of the strata of the Dakota Group, and yet the whole thickness is not exposed.

Prof. Leo Lesquereux found the flora of Point of Rocks related to that of Black Butte by nine identical forms or one-third of its known species, notwithstanding that there are two to three thousand feet of interposed measures. The distance between the two localities is only eleven miles, and the superposition of the strata is exposed so that the vertical thickness of the intervening rocks may be easily ascertained. He explained the scarcity of the bones of animals in the lower beds of the Lignitic, by the fact that, no animal, not even man, if once imbedded in soft peat, can get out of it, and also by the further fact that the coriaceous, ligneous plants of the bogs are not food for mammals.

He described, from the Lignitic at Point of Rocks, *Fucus lignitum*,

*Salvinia attenuata, Selaginella falcata, S. laciniata, Sequoia biformis,
Widdringtonia compianata, Pistia corrugata, Ottelia americana,
Dryophyllum crenatum, D. subfalcatum, Populus melanarioides,
Trapa microphylla, Laurus præstans, Viburnum rotundifolium.
Greviopsis cleburni, Rhus membranacea;* from Alkali Station, *Alnites
inæquilateralis, Juglans alkalina, Carpites viburni;* from Black Butte,
*Sphæria rhytismoides, Sequoia acuminata, Diospyros ficoidea, Vibur-
num platanoides;* from South Park, near Castello ranch, *Hypnum
haydeni;* from Grand Eagle Junction, *Lygodium marvinei;* from
Golden, *Zamiostrobus mirabilis, Arundo obtusa, Palmacites goldianus.*
now *Geonomites goldanus, Sabal communis;* from Middle Park.
Myrica insignis, Castanea intermedia; from Fort Fetterman, *Betula
rogdesi;* from Pleasant Park, Plum creek, *Ficus ovalis;* and from
Evanston, *Ficus psendo-populus.*

He described, from the Dakota Group, near Fort Harker, Kansas,[*]
*Sequoia condita, Myrica cretacea, Dryophyllum latifolium, Ficus dis-
torta, F. laurophylla, Laurus proteæfolia, Daphnogene cretacea,
Aralia saportanea, Hedera schimperi, H. platanoidea, Cissites acum-
inatus, C. heeri, Ampelophyllum attenuatum, Menispermites populi-
folius, Aspidiophyllum trilobatum, Protophyllum crednerioides;* from
Clay Center, Kansas, *Aralia concreta, A. towneri, Menispermites
ovalis, M. cyclophyllus, Sterculia lineariloba;* from the Fort Benton
Group, near the San Juan river, in southwest Colorado, *Dryophyllum
salicifolium,* and *Ilex strangulata;* and from Spring Canon, *Andromeda affinis.*

Prof. F. B. Meek[†] described, from the Dakota Group, southwest of
Salina, Kansas, *Trigonarca salinaensis;* from the Big Sioux river,
Arcopagella macrodonta; from the Fort Benton Group, at the head
of Wind River Valley, Wyoming, *Mortoniceras shoshonense;* from the
Fort Pierre Group on Cherry creek, near the mouth of Sage creek,
Dakota, *Odontobasis ventricosa;* from the Fox Hills Group, Moreau
river, *Microstizia millepunctata, Ostrea subalata, Pyropsis bairdi,* var.
rotula, and *Scaphites conradi,* var. *intermedius;* from the base of the
Black Hills, *Sphæriola warrenana;* from 90 miles below Fort Benton
on the Missouri, *Sphæriola endotrachys;* from Yellowstone river, 150
miles from its mouth, *Fasciolaria gracilenta;* from the Fort Union
Group, at Clear Fork of Powder river, Montana, *Hydrobia eulimoides;*
from the Judith River Group, at the mouth of Judith river, Montana,
Hydrobia subconica, and *Valvata montanensis.*

---

* 7th Rep. Hayden's U. S. Geo. Sur. Terr.
† Invert. Cret. and Tert. Foss., vol. ix., Hayden's Sur.

R. P. Whitfield* described, from the Judith River Group, at the mouth of Judith river, *Tapes montanensis, Mactra maia, Sanguinolaria oblata,* and *Thracia grinnelli.*

W. M. Gabb† described, from the Cretaceous of New Jersey, *Pentacrinus bryani, Goniaster mammillata, Scalpellum conradi, Nautilus bryani, Surcula strigosa, Opalia thomasi, O. cyclostoma, Laxispira lumbricalis, Ostrea bryani, Paliurus triangularis;* from the Ripley Group, of North Carolina, *Exilifusus kerri, Fasciolaria kerri, F. obliquicostata, ·Gyrotropis squamosus, Alaphrus kerri, Idonearca carolinensis;* from Patula creek, Georgia, *Drillia georgiana, Tritonium edentatum, Nassa globosa, Fasciolaria crassicosta, Ptychosyca inornata, Aporrhais bicarinata, Bixonia cretacea, Pholadomya littlei, Schizodesma appressa, Tellina georgiana, Gari elliptica, Peronœoderma georgiana, Trigonia angulicosta, Idonearca littlei, Trigonarca cuneata, Ostrea littlei, O. exogyrella:* and from Alabama, *Idonearca alabamensis,* and *Neithea complexicosta.*

J. W. Spencer‡ examined the country between the Upper Assineboine river and lakes Winnipegosis and Manitoba, and found rocks of Cretaceous age on Thunder Hill, at a height of nearly 800 feet above Swan lake. Following the course of Swan river below Thunder Hill, there are numerous exposures of these rocks for about thirty miles, which, with those of Thunder Hill, furnish a thickness of from 550 to 650 feet. There are also numerous exposures along the Bell river in the Porcupine mountains. They repose on rocks of Devonian age. Mr. G. M. Dawson, from the calcareous character, the microscopic forms, and the presence of *Inoceramus* and *Ostrea congesta,* referred the rocks to the Niobrara Group.

Prof. J. F. Whiteaves§ described, from the Cretaceous of the Queen Charlotte Islands, *Ammonites perezianus, A. logananus, A. richardsoni, A. skidegatensis, A. carlottensis, A. luperonsianus, A. filicinctus, A. crenocostatus, Amauropsis tenuistriata, Pleurotomaria skidegatensis, Martesia carinifera, Pleuromya carlottensis, Pholadomya ovuloides, Callista subtrigona, Trigonia diversicostata, Meleagrina amygdaloidea,* and *Syncyclonema meekanum.*

Prof. E. D. Cope‖ described, from the Judith River Group of Montana, *Amblysodon lateralis, Lælaps incrassatus, L. explanatus, L. falculus,*

* Carroll to Yellow Stone Nat. Park.
† Proc. Acad. Nat. Sci.
‡ Geo. Sur. of Canada, 1876.
§ Mesozoic Foss., Pt. 1.
‖ Proc. Acad. Nat. Sci.

Dysganus encaustus, D. haydenanus, D. bicarinatus, D. peiganus. Diclonius pentagonus, D. perangulatus, D. calamarius, Monoclonius crassus, Paronychodon lacustris, Compsemys imbricarius, C. variolosus, Polythorax missouriensis, Hedronchus sternbergi, Ceratodus cruciferus, C. hieroglyphus, Myledaphus bipartitus, Lælaps hazenanus, L. lævifrons, Zapsalis abradens, Champsosaurus profundus, C. annectens, C. brevicollis, C. vaccinsulensis, Scapherpeton excisum, S. favosum, S. laticolle, S. tectum, and *Hemitrypus jordananus ;* and from the Fox Hills Group, of Montana, *Uronautus cetiformis.*

Prof. O. C. Marsh* described, from the upper Cretaceous of Western Kansas, *Ichthyornis victor, Hesperornis gracilis, Lestornis crassipes, Pteranodon comptus, P. ingens, P. longiceps, P. occidentalis, P. velox,* and *P. gracilis,* now *Nyctosaurus gracilis.*

In 1877, Arnold Hague† estimated the thickness of the Cretaceous on the outlying ridges and foot-hills, east of the Colorado range, as follows: Dakota Group, 300 feet; Colorado Group, 1,000 feet; Fox Hills Group, 1,500 feet; and Laramie Group, 1,500 feet.

The Dakota beds are essentially a sandstone formation, and as they are usually hard and compact, frequently almost a quartzite, they form a well-defined horizon. Lying between the easily-eroded Jurassic marls and clays below, and the overlying blue shales, clays and crumbling rocks of the Colorado Group above, the Dakota beds are usually a conspicuous feature in the ridges, which form the foot-hills of the main range. In approaching the mountains from the Great Plains, the Dakota beds are especially prominent, as they form the outlying member of the series of upturned sedimentary beds, which rise so abruptly above the plain; for although the overlying Colorado group is perfectly conformable, they never occur high up on the long ridges, which form a sort of barrier between the level country and the mountain region beyond.

The Colorado Group is used to represent the Fort Benton, Niobrara, and Fort Pierre Groups. The Fort Benton Group is only exposed along the base of the abrupt ridges, and consists of dark, plastic clays, at times distinctly bedded, and frequently occurring as thinly-laminated paper shales. The lower beds are always more or less arenaceous, with interstratified beds of purer clay, while the upper beds sometimes carry thin seams of argillaceous limestone, which, in many places, can not be distinguished from similar beds in the Niobrara. Along the Laramie

---

* Am. Jour. Sci. and Arts, 3d Ser., vol. xi.
† Geo. Sur. 40th parallel.

Hills, this group is somewhat difficult to recognize, but in Colorado it may be traced for long distances in well defined north and south lines.

The Niobrara Group, although much thinner, is more easily recognized. It frequently blends so completely with the overlying Fort Pierre Group that it is extremely difficult to separate them.

The Fox Hills Group, east of the Colorado Range, is characterized throughout by great uniformity in texture and physical habit, and consists of a coarse sandstone formation, showing only variations in color from reddish brown to reddish yellow. The strata pass by imperceptible gradations, into the Laramie series, offering no well-defined line of separation, both formations from top to bottom consisting of coarse sandstone. The Laramie Group may be traced along the Big Thompson and Cache la Poudre valleys, and then eastward up the valleys of the northern tributaries to the South Platte. The sandstones form the exposed banks along Crow and Lone Tree creeks, and may be traced northward, passing under the Tertiary of Chalk Bluffs. This group includes the valuable coal deposits at Erie, and the Marshall and Murphy mines, north of Golden, extending from within one-half mile of the base of the range far out upon the plains into Eastern Colorado.

The Laramie beds form the uppermost members of the great series of conformable strata that lie upturned against the Archaean mass of the Rocky mountains; all overlying strata resting unconformably upon the older rocks.

The Cretaceous rocks are distributed over the surface of the Laramie Plains. On Rock creek, a branch of Medicine Bow river, north of the Little Laramie, and near Rock Creek Station, the Fort Benton Group is exposed from 350 to 400 feet in thickness. In the North Park, the Dakota Group is estimated at 350 feet in thickness, and here the Fort Benton, Niobrara and Fort Pierre Groups have a combined thickness roughly estimated at from 1,500 to 2,000 feet.

The Medicine Bow river, after leaving the mountains, runs almost exclusively through beds of Cretaceous age, its course being guided by the clays and marls, and the overlying Fox Hills sandstone.

On the northern slopes of Elk Mountain, the most northern point of the Medicine Bow Range, are found all the beds from the coal measures to the Fox Hills sandstone, uplifted at high angles, lying against the Archaean formation. All the geological divisions are well represented. In the valley of the North Platte river the Fox Hills Group has an estimated thickness of between 3,000 and 4,000 feet.

The strata containing the coal beds, at the town of Carbon, 656 miles west of Omaha, Mr. Hague supposed to be Upper Cretaceous.

S. F. Emmons,* geologist of the division west of North Platte, said that Bridger's Pass, which connects the valleys of the Upper Sage creek and the south fork of the Little Muddy, has been eroded out of the soft beds of the Colorado Cretaceous. Along the northern and western borders of this valley extends a ridge of white massive sandstones of the Fox Hills Group, standing at angles of 10° to 25°, and curving in strike approximately with the shape of the ridge. To the north of the gap, they form a continuous ridge about 15 miles in length, showing a bluff face to the southwest toward Bridger's Pass, at the base of which are exposed the clayey beds of the Colorado Group. A thickness of 3,000 to 4,000 feet of heavy-bedded sandstones, mostly white and buff, with a few included beds of shale, and some thin seams of coal, dipping to the northward at an angle of 10° to 20°, is exposed.

In going northward from a point on the Little Muddy, about five miles west of the Sulphur Springs, a thickness of between 3,000 and 4,000 feet of beds of the Laramie Group, dipping northwest at an angle of 20°, is crossed. Of these, the lower 2,000 feet are composed of massive white and yellow sandstones, in which the shale beds are of subordinate importance. The upper sandstones are stained and striped in red, by iron oxide, and form ridges with considerable clayey valleys between. In the upper 800 feet are several coal seams, and near the top is a prominent bed of bright vermilion color, only a few feet in thickness, of fine-grained, hard, argillaceous material, abounding in well preserved impressions of leaves. This is overlaid by a white sandstone, about 200 feet in thickness, carrying a coal seam, which in turn is capped by a thin-bedded brown sandstone, which weathers into flags about three inches in thickness; the dip of these upper beds has shallowed to 10°, and to the north the beds of the Laramie Group are practically horizontal.

The exposures of the Fox Hills Group, as seen in Bear Ridge, near the valley of the Upper Tampa river, show a series of massive, white, fine-grained sandstones of several thousand feet in thickness.

The Cretaceous of the Uinta Mountain region consists of over 10,000 feet of beds of sandstones and clays, carrying coal seams, which are most abundant in the upper part of the series. The Dakota Group consists of about 500 feet of rather thinly-bedded sandstones, with some clay beds, having at its base the persistent conglomerate carrying small pebbles of black chert. The Colorado Group, about 2,000 feet in thickness, is made up mostly of clays and yellow marls, with

---

* Geo. Sur. 40th parallel.

some sandstones at the base, which inclose one prominent coal-seam; the outcrops of this group are generally occupied by valleys. The Fox Hills Group consists of about 3,000 feet of heavily-bedded white sandstones, with a few coal-seams and comparatively little clay. The Laramie Group, whose actual thickness is not definitely ascertained, consists also of gray and white sandstones, often iron-stained, containing a greater development of clay beds, and very rich in coal seams. It is overlaid by an unconformable series of beds. The fauna of this group is brackish, and, locally, even fresh water forms are found associated with marine types.

In the valley of Bitter creek, the Fox Hills Group is estimated at 3,000 feet in thickness, and the Laramie at 6,000 feet. The latter is characterized by the greater development of clayey beds, and by the great number of coal seams, and by the presence of great quantities of leaves and plant remains, especially in the upper portion of the series. The beds are conformable, and were evidently deposited prior to the great period of plication and uplift in which the Rocky Mountains and the Uinta and Wahsatch ranges received their main elevation.

West of Bear River City, in Utah, along the face of the hills north of Sulphur creek, are exposed outcrops of the Fox Hills and Laramie Groups, from 5,000 to 7,000 feet in thickness, standing at angles of 85° to 90° west, and striking north 30° to 45° east, and consisting of heavy white sandstones with conglomerate beds, and passing to the westward into reddish brown sandstones. The beds of the Colorado Group west of the sandstone ridge, at the bend of Sulphur creek, expose a thickness not less than 5,000 or 6,000 feet. About two miles west of Bear River City, a railroad-cut, through a low ridge running out from the high ground forming the northeastern wall of the Sulphur Creek Valley, shows a section of about 150 feet of beds, separated by an interval, bare of outcrops, from the sandstones west of Bear River City, but corresponding with them in strike, and standing with an inclination of 70° to 80° to the southeast. It is formed of sandstones, marls and clays, with a few bituminous and gypsiferous seams, and is remarkable for the fine definition of its bedding-lines, the strata varying from half an inch up to a foot or more in thickness. The strata abound in fossils of fresh and brackish water types, viz.: *Unio, Corbula, Limnaea, Campeloma, Viviparus,* etc. They evidently belong to the conformable beds of the Laramie Group, and are overlaid a short distance to the north by horizontal strata of the Vermilion creek Eocene.

G. K. Gilbert* found the Cretaceous strata well displayed upon the flanks of the Henry Mountains, in Southern Utah, where they consist of four principal sandstones, with intervening shales, and have a thickness of 3,500 feet. They also contain thin beds of coal, one of which was observed at the foot of Mount Ellen, four feet in thickness. The lower 500 feet he referred to the Henry's Fork Group.

Dr. A. C. Peale,† geologist of the Grand river division, said that the massive, yellow silicious sandstone, in some places quartzite, at the base of the Cretaceous, is so well defined lithologically, that there has never been any difficulty in separating it from the overlying shales. Along the edge of the plains in Colorado, it is underlaid by greenish shaly beds, sometimes lignitic near the top, generally in part or wholly covered, which have always been referred to the upper part of the Jurassic. In the West these shaly beds still persist, and the massive sandstone, although still recognizable without difficulty, is much thinner, being only from 50 to 100 feet, and as we descend, in the sections carried below, we find other beds of silicious sandstone separated by shaly beds that are arenaceous, calcareous and argillaceous. In these beds, in 1874, he found a sassafras-leaf, which led him to refer them to the Lower Cretaceous. He drew an arbitrary line separating the Cretaceous and Jurassic. The beds below have the same lithological characters to the top of the red beds, with this exception, that limestones occur more frequently toward the base. In Arizona, G. K. Gilbert found Jurassic and Cretaceous fossils, associated in beds, resembling those usually referred to the Jurassic. He is of the opinion that we can not draw any line between the two formations, palæontologically, or lithologically, but for convenience in description it is best to draw an arbitrary line, which may be changed as we obtain more facts in relation to the formation.

There is a narrow outcrop of the Dakota Group on the south side of the Gunnison, above the Grand Cañon, between the breccia and the granite. It appears, and is faulted, at the head of the Uncompahgre river and on Dallas Fork, the latter stream flowing on the line of the fault. Between this creek and the San Juan Mountains it rises until it reaches the summit of the foot hills, appearing from beneath the shales. On the Uncompahgre plateau, it dips gently to the eastward, and is the surface formation until we approach Escalante creek. Between the latter and Roubideau's creek, there are some isolated

* Rep. on the Geo. Henry Mountains.
† 9th Rep. Hayden's U. S. Geo. Sur. Terr.

patches of it. It is found along the western side of the Gunnison and forms the floor of the San Miguel plateau. Going north on the San Miguel plateau, we find the massive sandstones of the Dakota Group broken, and forming the tops of mesas between the streams rising in the Uncompahgre plateau and flowing into the San Miguel and Dolores rivers. Still further north it disappears altogether, until we approach Grand river, near the mouth of the Dolores.

In the Uncompahgre valley, on both sides of the river, until the canon is reached, there are exposures of the Fort Benton and Niobrara Groups. East of the Uncompahgre Agency the thickness of the beds is about 3,000 feet.

F. M. Endlich, geologist of the southeastern division, found the Dakota Group in the San Juan region forming a ridge parallel with the Piedra river, and having a thickness of more than 1,000 feet. He also discussed the age of the Lignitic Group of the Trinidad region, which spreads over an area of 750 square miles, and with Prof. Lesquereux supposed it to be of Tertiary age.

Dr. B. F. Mudge* said the Cretaceous in Kansas covers an area of over 40,000 square miles, or more than half the surface of the State. The Fort Benton, Fort Pierre and Fox Hills Groups are entirely wanting. The Dakota Group rests upon the Permian, and is succeeded by the Niobrara Group.

The average width of the Dakota is less than 50 miles, being somewhat less than that in the north part of the State, and more on the Smoky and Arkansas rivers. The dip is to the northwest, and very slight. It is conformable to the formation above it, and has a maximum thickness of about 500 feet.

The Niobrara Group occupies a belt of country about 30 miles in width, in the northern part of the State, but gradually widens to more than twice that extent in the Smoky Hill valley. The upper part is composed of chalk and chalky shales, the lower part which is called the Fort Hays Group, consists in its higher strata of heavy bedded limestone, under which is a friable, bluish black, or slate colored shale, which abounds in concretions or septaria, of all sizes, from one inch to six feet in diameter. The body of the concretions is of hard clay-marl, with cracks lined with beautiful crystals of calc spar. The lower part has a thickness of 260 feet, and the upper part of 200 feet, making the total thickness 460 feet. It is succeeded by strata of Pliocene age.

---

* 9th Rep. Hayden's U. S. Geo. Sur. Terr.

Alfred R. C. Selwyn[*] explored the country north and northeast of Fort George near the 54th parallel. The exploration was almost wholly within the Arctic watershed, and the basin of Peace river. From "The Fork"—Smoky river—up to Dunvegan, and thence to about five miles below Hudson's Hope, the rocks which are exposed along Peace river are mesozoic; they consist of dark, earthy shales, in parts character-ized by numerous bands and septarian nodules of clay iron-stone, many of which inclose large *Ammonites*, and they are also associated with sandy calcareous layers, holding other Cretaceous fossils, among which a species of *Inoceramus* is tolerably abundant, while in the dark argillaceous shales the scales of fishes are frequently observed. Descending Peace river, these dark shales are first seen at about six miles below Hudson's Hope. They are nearly or quite horizontal, and are exposed at intervals between this point and Fort St. John, in cliffs which rise almost perpendicu-larly from the water to heights of 50 or 100 feet. Near where they are first seen, the hills at a little distance back rise to 500 or 600 feet, and toward their summits present cliffs in which some thick beds of brown fine-grained sandstone crop out. About a mile below St. John, on the left bank, a section is exposed nearly 700 feet in thickness. These rocks are exposed at intervals down to The Fork, and also on Smoky and Pine rivers. On the latter stream the exposed thickness is esti-mated at 1,700 feet, and contains four thin seams of bituminous coal.

Prof. George M. Dawson, who explored the country between the 52d and 54th parallels, in British Columbia, found the equivalent of the Shasta Group in the vicinity of Tatlayoco lake. Along the eastern shore of the lake these rocks overlie those of the porphyrite series. They dip eastward, or away from the anticlinal axis, in which the lake lies, and form, at a short distance from its eastern margin, a rampart-like wall of mountains, from 2,000 to 3,000 feet high, and twelve miles in length. The rocks are compact, bluish-gray quartzites, or hard sand-stones, and conglomerates of all grades in regard to size of particles, associated with blackish or dark colored slaty and shaly beds, which recur frequently at different horizons. The thickness of the entire Cretaceous series on the east side of Tatlayoco lake is estimated at 7,000 feet. Their geographical extension is also great. He regarded the Jackass Mountain Group as the equivalent of the Shasta Group of California.

---

[*] Geo. Sur. of Canada.

Prof. E. D. Cope* called the Judith River Group No. 6 Cretaceous. He showed its conformability with the underlying marine Cretaceous, and gave a section 332 feet in thickness, though its maximum is not less than 500 feet. His section in ascending order is as follows:

Arenaceous marl (with Dinosaurian bones near the top)....125　feet.

Sandstone, 1st........................	5	"
Sandstone............................	6	"
Impure lignite........................	2	"
Sandstone, 2d........................	10	"
Impure lignite........................	4	"
Unio bed...........................	30	"
Rusty sandstone (with fresh water shells)	25	"
Arenaceous marl (with petrified wood)................	50	"
Sandstone, 3d	15	"
Marl	20	"
Reddish shale........................	10	"
Lignite	5	"
Shale...............................	7	"
Black shale and lignite................	3½	"
Bed of *Ostrea subtrigonalis*	15	"

(left margin: SAURIAN REMAINS.)

Total.............................332½ feet.

The presence of Dinosaurians, gar fishes, turtles, *Physa*, *Viviparus* and *Unio* prove the fresh water character of the strata, while the *Ostrea* indicates a return to brackish water.

Dr. C. A. White† described, from the Judith River Group at Cow Island and Dog creek, a tributary of the Upper Missouri river, in Montana Territory, *Unio cryptorhynchus*, *U. senectus*, *U. primaevus*, *Anodonta propatoris*, *Bulinus atavus*, and *Physa copei*.

Prof. F. B. Meek‡ described, from near Laporte, Colorado, *Anomia rœtiformis;* from East Canon creek, Wasatch Range, Utah, *Cucullœa obliqua*, *Mactra emmonsi:* from Cooper creek, Laramie Plains, Wyoming, *Axinœa wyomingensis;* from Red creek, Uinta Mountains, Utah, *Mactra arenaria;* from East Canon creek, Utah, *Mactra utahensis*, *Tellina isonema*, *T. modesta*, *Gyrodes depressa*, and *Anchura fusiformis.*

Prof. C. A. White§ described, from east of Impracticable Ridge, Utah, *Ostrea prudentia;* from near Pueblo, Colorado, *Inoceramus*

* Bull U. S. Geo. Sur., Vol. 3., No. 3.
† Bull U. S. Geo. Sur., Vol. 3., No. 3.
‡ U. S. Geo. Expl., 40th parallel.
§ Wheeler's Sur. W. 100th Mer., Vol. 4.

flaccidus, Mactra incompta; from the Rio Puerco, New Mexico, *Ido-nearca depressa;* from Mount Taylor, New Mexico, *Lispodesthes lin-gulifera;* from Ojo de los Cuervas, New Mexico, *Ammonites laevianus;* from Paria, Utah, *Helicoceras pariense,* and *Serpula intrica.*

Prof. E. D. Cope* described, from the Fort Pierre Group of Kansas, *Pelycorapis berycinus;* and from the Niobrara Group of the Upper Missouri, *Elasmosaurus serpentinus,* and *Anogmius aratus.*

Prof. O. C. Marsh† described, from West Kansas, *Baptornis advenus;* from Texas, *Gracularus lentus, Diplosaurus felix;* from the Rocky Mountain region, *Nanosaurus agilis, N. victor, Apatodon mirus ;* and from the Dakota Group of Colorado, *Titanosaurus montanus.*

In 1878, Prof. C. A. White‡ surveyed a portion of Northwestern Colorado, and found the Dakota Group reaching an aggregate thickness of between 500 and 600 feet; the lower half consisting of a dark-colored, coarse, silicious, pebble-conglomerate, which is somewhat irregularly bedded and easily disintegrated; and the upper portion, consisting of a yellowish or brownish, rough, heavy-bedded sandstone, between which and the conglomerate some variegated bad-land sandstones usually exist.

The equivalent of the Fort Benton and Niobrara Groups he called the Colorado Group, which is also the equivalent of the Sulphur Creek Group. He united, under the name of the Fox Hills Group, both the Fox Hills and Fort Pierre Groups, the former of which has a thickness of 1,000 feet, and the latter of 800 feet. The strata that have been called by the name of the Fort Union Group, Lignitic Group, Bitter Creek Group, Judith River Group, and by other names, including the name of Laramie Group, proposed by Mr. King, he proposed to call Post-Cretaceous. The thickness of this group in Northwestern Colorado is at least 3,500 feet.

He described, from the Laramie Group,§ on Crow creek and Danforth Hills, in Northern Colorado, *Volsella regularis. V. laticostata, Nucu-lana inclara, Anodonta parallela, Corbicula cleburni, C. cardiniæ-formis, C. obesa, C. macropistha, Physa felix, Viriparus prudentia, Odontobasis formosa;* from Black Buttes Station, Wyoming, *Unio gontambonatus, U. aldrichi, Neritina baptista;* from Bear river, near the confluence of Sulphur creek, Wyoming, *Acella haldemani, Neri-*

* Bull. U. S. Geo. Sur., Vol. 3, No. 3.

† Am. Jour. Sci. and Arts, 3d Ser., Vol. 14.

‡ 10th Rep. Hayden's U. S. Geo. Sur. Terr.

§ Bull. U. S. Geo. Sur., Vol. 4, No. 3.

tina naticiformis, *Viviparus couesi*; from near Evanston, *Helix evanstonensis*, and *Goniobasis endlichi*.

Prof. Leo Lesquereux* described, from the Fort Union Group, at Black Buttes, Wyoming, *Sequoia acuminata*, *Vitis sparsa*, *Grewiopsis saportana*, *G. tenuifolia*, *Rhus pseudomeriani*, *Podogonium americanum*, *Carpites myricarum*, *C. glumæformis*, *C. mitratus*, *C. verrucosus*, *C. viburni*, *C. bursæformis*; from Golden South Mountain, Colorado, *Sabalites fructifer*, *Palmocarpon truncatum*, *P. corrugatum*, *P. subcylindricum*, *Populus ungeri*, *Laurus ocoteoides*, *Viburnum anceps*, *V. goldianum*, *V. solitarium*, *Fraxinus eocenica*, *Cornus suborbifera*, *Carpites oviformis*, *C. triangulosus*, *C. costatus*, *C. coffæformis*, *C. rostellatus*, *C. rhomboidalis*, and *C. minutulus*; from the divide between the source of Snake river and Yellowstone lake, *Geonomites schimperi*; from Raton Mountains, near Fischer's Peak, New Mexico, *Geonomites tenuirachis*, *G. ungeri*; from Castello's Ranch, near South Park, Colorado, *Fraxinus brownelli*, *Sapindus stellariæfolius*; from Florissant, *Carpites pealei*; from Evanston, Wyoming, *Laurus socialis*, *Carpites laurineus*, *C. utahensis*; from Bridger's Pass, Wyoming, *Laurus utahensis*; from above Spring Canon, near Fort Ellis, Montana, *Dombeyopsis platanoides*, *Celastrinites lævigatus*; from Carbon, Wyoming, *Cratægus æquidentata*; from Fort Steele, *Carpites calvatus*, and from other places, *Quercus cinereoides*.

Mazyck & Vogles† described, from the Cretaceous beds reached in artesian boring, at Charleston, South Carolina, at the depth of 1,880 feet below the surface, *Anomia andersoni*.

In 1879, F. M. Endlich‡ described the Cretaceous cast of the Wind River range in Wyoming, and separated it in ascending order into:

1. The Dakota Group, consisting of yellow and brown shales, interstratified with sand stones of the same color. In the shales, above some of the thin beds of sandstone, there are slight indications of coal. The seams are but half an inch thick, and the coal is of that variety called jet coal. Higher up the sandstones predominate, separated by thin layers of homogeneous, dark shales. Near the top there is a heavy bed of shale, which is covered by massive white, yellow and brown sandstones. A small thickness of arenaceous shales closes the group. This is the general section of the Dakota, as exposed west of the anticlinal axis. In some of the upper sandstones indis-

---

*Tert. Flora., Vol. 7, Hayden's Sur.
†Proc. Acad. Nat. Sci.
‡11th Ann. Rep. U. S. Geo. Sur. Terr.

tinct remains of plants occur, and in the higher shales a *Gryphæa*. The thickness is about 400 feet.

2. The Colorado Group, consisting of an extensive series of dark gray, slightly calcareous shales. They are thinly laminated, easily eroded, and become light gray or white upon exposure. Covering the highest portions of the region lying between Sheep Mountain and the base of the third chain, they present comparatively steep bluffs parallel to their strike, and rounded surfaces along their dip. A few banks of argillaceous limestone may be found within them. Within the upper third the shales are more arenaceous than lower down. A cold sulphur spring near Camp Brown seems to take its rise in these shales which must be regarded as a very prolific source for alkaline compounds of a highly soluble nature. Within the shales there are small inclusions of pyrite. Upon decomposition of this and the shales various salts are formed. The thickness of this group is about 600 feet, increasing southerly to 900 feet.

3. The Fox Hills Group, consisting in the lower part of brown and yellow shales, interstratified with thin beds of sandstone. Some of the shales are very dark and carbonaceous. Above this alternating series there is a considerable thickness of yellow and brown shales. As a rule, they are arenaceous, but some of them quite free from sand. Small particles of mica occur throughout. Higher up, sandstones set in again, containing, together with thin seams of shales, small deposits of coal. The upper part is formed by thinly-bedded, micaceous and argillaceous sandstones, covered by a thick stratum of the same material. The thickness is estimated at 500 feet.

About two miles west of Camp Brown, a very interesting hot spring occurs, which rises in the beds of this group. It is known as the Hot Sulphur Spring. The temperature is from about 100° to 110°, and varies but little with the weather. The bright green and blue water is contained within an elliptic basin 315 feet long and 250 feet wide. A constant bubbling up of carbonic-acid gas gives it the appearance of boiling. The mineral constituents held in solution by the water are iron, lime, magnesia, soda and potash. They seem to be contained in the form of sulphates, carbonates and chlorides. The heat which supplies the warmth of the water is supposed to be due to chemical changes going on within the strata through which the moisture finds its way. A petroleum spring also occurs near Camp Brown, originating probably in the same rocks.

The Laramie Group consists of a succession of shales and yellow

sandstones, forming low, long-continued bluffs.   The thickness is estimated at 400 feet.

The yellow and white sandstones of the Dakota Group occur in the
northern portion of the Sweetwater Hills.   East of Elkhorn Gap they
are much folded and plicated.   In Whisky Gap, the strata curve around
the Western base of the Seminole Hills, with a partiversal dip, and a
short distance farther west they take part in an anticlinal upheaval.
The thickness is estimated at about 700 feet.

The Colorado Group occurs also near Elkhorn Gap, and in Whisky
Gap.   At the latter place the shales are dark gray, finely laminated, and
have a thickness of 650 to 700 feet.

The Fox Hills Group, in Whisky Gap, forms sharp, low ridges, par-
ticipates in the stratigraphical disturbances, and has an estimated
thickness of 1,000 feet, which increases toward the south.   Near Salt
Wells, this group is well developed, and occupies a prominent position.
A valley of approximately semicircular shape, lies directly north of
the railroad, bordered by steep brown bluffs of shales and sandstones
of this group.   Dipping off in every direction, they present a most
typical partiversal arrangement of the strata.   Near the base, they are
composed of thinly-bedded sandstones.   These are followed by yellow
and brown shales, more or less arenaceous and micaceous.   Above these
there is a succession of sandstones and shales, containing carbonaceous
strata.   A recess in the bluffs is caused by the higher series of shales.
The latter are covered by sandstone strata of varying thickness, separated from each other by shales.   Some good coal is found in this
horizon.   Near the top, massive yellow sandstones are overlaid by thin
beds of shale and white sandstone.   On every side the beds are conformably overlaid by strata of the Laramie Group.   The thickness is
from 1,200 to 1,300 feet.

The Laramie Group has a wide distribution in the southern area of
this territory.   On the west side of the anticlinal it can be traced nearly
to Whisky Gap, and probably juts against the granite of the Sweetwater Hills.   From the stratigraphical structure of the entire region it
is ascertained that this group forms a basin, upon which the younger
strata are conformable.   It is composed of sandstones, shales, marls,
clays and coals.   Near the base, heavy sandstones set in, soon super--
seded, however, by shales.   These contain strata of sandstones at varying intervals.   A number of coal-beds overlie the sandstones.   The
coal is generally covered by a comparatively thin stratum of sandstone,
upon which follow clays, shales and arenaceous marls.   Higher up a

succession of sandstones is interstratified with shale. Selenite is common in the shale. The higher members of the group are composed of yellow and white sandstones, containing beds of coal, and dark and often carbonaceous shales. Sandstones mediate the transition into the lower Tertiary groups. The lower coal-horizon is the most productive. The total thickness of this group west of Rawlings Springs, and from there northward, is estimated at 1,600 feet.

The decomposition of pyrite in dumps from coal banks, produces a spontaneous combustion of the coal which changes the color of the shales to a brilliant red. In the same manner probably the coal at places in the bank has taken fire and burnt as long as the supply of oxygen could sustain a flame. Through this process of metamorphosis by heat the overlying beds, containing more or less hydrated ferric oxide, were changed to a bright vermilion color. Sandstones occur, the faces and edges of which have been literally glazed by the long continued action of heat. Fragments are firmly baked together, and resemble cinders from a furnace. Purely argillaceous shales and clays have been thoroughly fritted and altered into very hard, compact porcelain jasper. Throughout the area covered by the Laramie Group, and in some of the Wasatch beds red colored strata occur which have been produced by these causes.

Dr. A. C. Peale,* estimated the thickness of the Laramie Group on Smith's Fork, and in the Bear River region, near the western shore line of the Wahsatch lake, at 5,000 feet.

Geo. M. Dawson,† explored the Cretaceous in British Columbia, on the headwaters of the Skagit, west of the main axis of the range, which forms the watershed, between that river and the Similkameen. The trail traverses the area in a general northeast direction for nearly thirteen miles. A section occurs on the trail immediately east of the crossing of the north branch of the Skagit, representing a thickness of 4,429 feet. The rocks are much disturbed, are lying at all angles up to vertical, and have suffered considerable hardening and alteration. They consist, generally, of sandstones, conglomerates and argillites.

Still further north-westward, from the vicinity of the mouth of Anderson river and Boston Bar, they were found to extend, in a long, narrow trough, nearly coinciding, in the main, with the Frazer river, with a general bearing of about N. 70° W., to the vicinity of Lillooet and Fountain, a distance of about 80 miles. The estimated thickness is

---

* 11th Ann. Rep. U. S. Geo. Sur. Terr.
† Geo. Sur. of Can.

5,000 feet. They were also found on the Thompson, below its junction with the Bonaparte. The thickness on Tatlayoco, 220 miles north-eastward from Skagit valley is estimated at 7,000 feet. These rocks are regarded as of the same age as the Shasta Group of California.

Prof. C. A. White* described, from the Fox Hills and Fort Pierre Group, at Cimarron, New Mexico, *Caryophyllia johannis*, *C. egeria*, *Crassatella cimarronensis*; from Hilliard Station, U. P. R. R., Wyoming, *Placanopsis hilliardensis, Neritina incompta*; from Coalville, Utah, *Neritina patelliformis*, var. *weberensis*; from Monument creek, near Colorado Springs, *Palinurus pentangulatus*; from the mouth of the Saint Vrains, Northern Colorado, *Baroda subelliptica*, *Pachymya herseyi, Actaeon woosteri, Actaeonina prosocheila*; from west of Greeley, Colorado, *Tancredia coelionotus, Glycimeris berthondi* and *Anchura haydeni*; from the Cretaceous, at Salado, Bell County, Texas, *Exogyra walkeri*; from Dennison, Texas, *Anchura mudgeana*; from Helotes, Bexar County, Texas, *Turritella marnochi*; and from the Cretaceous, at the head of Waterpocket Canon, Southern Utah, *Cardium trite*.

He† described, from the Cretaceous, on Fossil Creek, 16 miles west of Greeley, and 6 miles south of Fort Collins, Colorado, *Chetetes (?) dimissus*, and *Beaumontia (?) solitaria*.

Prof. J. F. Whiteaves‡ described, from the Cretaceous rocks of the Sucia Islands, *Nautilus suciensis, Ammonites selwynanus, Surcula suciensis, Cerithium lallierianum*, var. *suciense, Amauropsis suciensis, Cirsotrema tenuisculptum, Stomatia suciensis, Cinuliopsis typica, Teredo suciensis, Linearia suciensis, Veniella crassa, Laevicardium suciense, Inoceramus cripsi*, var. *suciensis*; from Vancouver Island, *Ptychoceras vancouverense, Opis vancouverensis, Discina vancouverensis, Smilotrochus vancouverensis*; and from Hornby Island and Nanaimo river, *Potamides tenuis*, var. *nanaimoensis*, and *Periploma suborbiculatum*.

In 1880, Prof. C. A. White§ said that the geographical limits of the Laramie Group are not yet fully known, but strata bearing its characteristic invertebrate fossils have been found at various localities within a great area, whose northern limit is within the British Possessions, and whose southern limit is not further north than Southern Utah and Northern New Mexico. Its western limit, so far as known, may be stated as approximately upon the meridian of the Wahsatch

---

* 11th Rep. Hayden's U. S. Geo. Sur. Terr.
† Bull. U. S. Sur., Vol. 5, No. 2.
‡ Mesozic Foss., Part 2.
§ Cont. to Pal. No. 4, 12th Rep. U. S. Geo. Sur. Terr.

range of mountains, but extending as far to the southwestward as the southwest corner of Utah, and its eastern limit is far out on the great plains, east of the Rocky Mountains, where it is covered from view by later formations and the prevailing debris of the plains. These limits indicate for the ancient Laramie sea a length of about 1,000 miles north and south, and a maximum width of not less than 500 miles. Its real dimensions were no doubt greater than those here indicated, especially its length; and we may safely assume that this great brackish-water sea had an area of not less than 500,000 square miles. The present range of the Rocky Mountains, which has been entirely raised as a mountain range since the close of the Laramie period, traverses almost the entire length of this great area, and far the greater part of the other extensive and numerous displacements which the strata of the different geological ages have suffered within that great area, have also taken place since all the Laramie strata were deposited, although some of those changes thus especially referred to began before the close of the Laramie period.

The invertebrate fauna consists almost wholly of brackish-water, fresh-water and land mollusca. Species belonging to all three of these categories are often found commingled in the same strata, but it is also often the case that certain strata, sometimes only thin layers, which contain the fresh-water and land molluscs alternate with those which contain the brackish-water species. All the species of fresh-water and land mollusca which prevailed during the Laramie period, seem to have ceased with the disappearance of their contemporary brackish-water forms, although they were succeeded by other fresh-water and land species.

He described from 'Point of Rocks' station, Bitter Creek valley, Wyoming, *Acinœa holmesana;* from the mouth of Sulphur creek, Bear river valley, Wyoming, *Rhytophorus meeki;* from the Cretaceous of Collin county, Texas, *Ostrea blacki, Exogyra winchelli, Pteria* (?) *stabilitatis;* from Bexar county, Texas, *Exogyra forniculata;* from Bell county, Texas, *Pachymya compacta, Thracia myæformis;* from the estuary strata of the age of the Fox Hills Group at Coalville, Utah, *Anomia propatoris;* from the Fox Hills Group at Cimarron, New Mexico, *Barbatia barbulata;* from Dodson's Ranch, near Pueblo, Colorado, *Lispodesthes obscurata;* from the Dakota Group, Saline county, Kansas, *Pteria salinensis, Gervillia mudgeana;* from the Fort Pierre Group at Fort Shaw near Muscleshell river, Montana, *Tessarolax hitzi;* and from the Cretaceous of Yellow Stone river, Montana, *Fasciolaria alleni.*

Prof. R. P. Whitfield described, from near San Antonio, Texas, *Paramithrax (?) walkeri*. And Prof. O. C. Marsh[*] described, from the Cretaceous chalk of Kansas, *Holosaurus abruptus*.

To conclude this cursory review of the growth of our knowledge of the Cretaceous formation of North America, I will add a few observations upon the present state of the science. The Cretaceous is found either exposed upon the surface or covered by the Tertiary, forming a border of variable width, on the eastern coast, from New York to Florida. It constitutes the surface rocks, or is overlaid with the Tertiary at all places south of the 33d parallel, with the exception of limited areas in the mountain regions. It extends up into Tennessee, spreads over all Mississippi, and reaches southern Illinois. West of the 97th meridian from the 33d parallel to the Arctic ocean, the whole country is covered with this formation, with the exception of limited areas in the mountain regions, or inconsiderable extensions of land, where it has been swept away, and an area of some magnitude north and west of Hudson's Bay. This of course includes the area covered by the Tertiary. It is found east of the 97th meridian, extending into Iowa, Minnesota, and some parts of British America. Or approximately stated, the Cretaceous now forms the surface rock, or is overspread by the Tertiary, over more than half the area of the North American Continent, and from the extensive denudation which it has evidently suffered, we may fairly presume, that at the commencement of this formation the continent was an island of less than one third its present dimensions.

In the east and south the formation is exclusively a marine deposit, but in the west, over great areas, the marine Cretaceous is succeeded by a brackish or fresh water Cretaceous deposit. In the east it never exceeds half a mile in thickness, but in the west the marine Cretaceous sometimes exceeds a mile in thickness, and is followed by the brackish and fresh water deposits, which are also more than a mile and sometimes even two miles in thickness. This formation is, therefore, preeminently the building deposit or land making deposit of the North American Continent.

The brackish and fresh water deposits were first named the Fort Union or Lignitic Group, and there is no reason known to the author, why these deposits, wherever found, should bear any other geological name. It is true that the name Bear River Group was given to a group of rocks lower than those first named the Fort Union Group, but

---

[*] Am. Jour. Sci. and Arts, 3d ser., Vol. xix.

the separation has not been maintained, and the authors, instead of extending the Fort Union Group to include these rocks, have called both the Fort Union and Bear River Groups the Laramie Group. Prof. White, and some other authors, call the rocks Post-Cretaceous. This is not objectionable, because it is treating them with reference to their geological position, and not proposing a new name for a group of rocks. If it is desirable to retain the name Bear River Group, it should be applied to the rocks below the Fort Union Group, and in no event can the Fort Union Group be swallowed up by another name for the same group of rocks. A great many synonyms have been proposed for this Group, some of which it is difficult to wipe out, and others will burthen the science for a longer or shorter period, but, finally, we may hope for their burial in oblivion. Any one can propose to call an exposure of rocks, at any place, by a new name, but it requires a palæontologist to determine the age of the rocks and to refer them to their proper position in the geological column. A little reflection, therefore, will satisfy the reader, that proposing a new name for a group of rocks, wherever exposed, without giving the palæontological reasons for so doing, is an evidence of ignorance, and most frequently we find those who do it are suffering from downright stupidity.

The plants which have been described, from the Cretaceous rocks in question, have been referred to about 150 genera, and number about 500 species. About 50 of these genera are now extinct, and about 100 are living. The larger part are from the Fort Union Group of the West, and from their intimate relation with living forms, the great palæo-botanist, Prof. Lesquereux, referred the rocks to Eocene age. The testimony, however, of the animal remains, which Prof. Cope was the first to discover, has proven that they must be referred to the upper or later Cretaceous. This determination has, if we may trust investigations of our fossil botanists, specifically united the Cretaceous era with the present time. For the living plants, *Corylus americana, C. rostrata, Davallia tenuifolia,* and *Onoclea sensibilis* have been identified among the fossils from the Fort Union Group. It is likely that too much confidence in this identification may lead to error, for as yet we may fairly suppose that we know but little of the vegetable life of this vast period of time in comparison with what will be known in a few decades. And better specimens than those upon which the identifications have been made may show specific distinctions. It is sufficient that the forms so much resemble the living as to be mistaken for them, to show how closely the living forms are connected with the ancient dead.

The relation between the invertebrate kingdom of the Cretaceous period and the living invertebrates is shown (according to present identifications) by the survival of more than one third of the Cretaceous genera, though all Cretaceous species have become extinct. The survival, however, in different classes, is by no means uniform. In the class Polypi, of sixteen Cretaceous genera, six are living. In the class Echinodermata, of twenty-two genera, eight are living. In the class Bryozoa, of thirty-two genera, nine are living. In the class Brachiopoda, of six genera, five are living. In the class Gasteropoda, of one hundred and seventy-four genera, ninety-six are living. In the class Lamellibranchiata, of one hundred and sixty-four genera, seventy are living. But in the class Cephalopoda, where there were more than thirty genera and subgenera, all have become extinct except a single genus, the *Nautilus.*

The connection between the vertebrates of the Cretaceous period, and the living vertebrates, is, seemingly, much farther removed. No Cretaceous genera of birds or mammals survive. In the class Reptilia, where more than seventy-five Cretaceous genera have been determined, only three genera are known to have survived, *Crocodilus*, *Trionyx* and *Emys.* A few species of fishes, found in the Cretaceous, have been referred to living genera, and probably some of them are correctly so referred; but from the great differentiation observed in the vertebrates, during the long period of time which has transpired, we can not expect to find many forms preserving unchanged their ancient outlines, though we may be able to trace backward the living genera into what we call distinct ancestral genera or families.

This closes our remarks upon the Mesozoic period, and we will now take up the Cænozoic. There is no great break in animal or vegetable life in passing from the Mesozoic to the Cænozoic, as early geologists, from very limited observations, supposed. Indeed, it may be said to be a most propable hypothesis that there are no breaks in genealogical trees. All organic life has descended from ancestral forms, and among the vertebrates, in the later geological periods, profitable accretions or accessions of important parts or functions have been developed in successive generations. This will become more apparent as we pass from one group of rocks to another in the Tertiary period.

THE CÆNOZOIC AGE, OR TERTIARY PERIOD.

When the words Primary, Secondary and Tertiary are used to distinguish geological subdivisions, the rocks are so comprehended as to leave none to which the word Quaternary can be properly applied. The organic remains of the Tertiary are likewise so completely blended with the living organisms, that we can not distinguish a Quaternary age or period. The subdivision of the Tertiary, with reference to the survival of conchological species, into Eocene, Miocene, Pliocene and Post-pliocene, brings us to the living species as gradually as the species are found to change within any of the subdivisions of geological time, or within any of the minor subdivisions of the strata into groups. It is, therefore, evidently a mistake to use the word Quaternary, in a geological subdivision, with reference either to the rocks or their organic contents.

The Tertiary rocks, generally, consist of marls, clays, sands, or other friable material, filling depressions in the underlying rocks, and, though widely distributed, seldom form hard continuous strata. This condition of the rocks in Europe made it very difficult to determine the order of superposition, and led Deshayes to suggest, after having examined 1,122 species of fossil shells from the Paris basin, and having identified only thirty-eight with the living, that a subdivision of the Tertiary might be based upon the relative proportion of the extinct and living species of shells. He drew up, in tabular form, lists of all the living shells known to him as occurring in Tertiary rocks, and submitted the same to Mr. Lyell. The number of species of fossil shells examined by Deshayes was about three thousand, and the living species with which they were compared about five thousand. With this assistance, and that furnished by the works of Basterot and some Italian authors, Mr. Lyell, in 1833, estimated that, in the lower Tertiary strata of London and Paris, 3½ per cent. of the species are identical with the living; that, in the middle Tertiary of the Loire and Gironde, about 17 per cent. are living; that in the upper Tertiary, or Subappenine beds, from 35 to 50 per cent.; and that, in strata still more recent, in Sicily, from 90 to 95 per cent. He proposed to call the lower Tertiary " Eocene," which signifies the dawn of the present state of things; the middle Tertiary " Miocene," which implies less recent; and the upper Tertiary " Pliocene," which means more recent. The

Pliocene he subdivided into the Older Pliocene and Newer Pliocene. In the latter, out of 226 fossil species of shells, he found 216 to be living. He afterward proposed the name Post-pliocene for rocks having all the imbedded fossil shells identical with living species, though they may contain extinct mammalian remains. We now include in this group strata which belong to more modern time, and which are frequently called "Recent."

This subdivision of the Tertiary, with reference to the survival of conchological species, and the subdivision of the strata, or rocks, into groups, have made a double system of nomenclature, which does not prevail in the older geological periods. The determination of the North American equivalents of the European strata, by the per cent. of living species, was soon ascertained to be impracticable, and, instead of that method, the age is determined by the extinct species. Certain species have come to be regarded as types of Eocene age, or Miocene, as the case may be, and, from the presence of these, the rocks are referred to the proper subdivision of the Tertiary.

I have not found time to separate the consideration of the Tertiary, into the groups into which it has been subdivided, and preserve the chronological order, or history of our knowledge of it. For this reason, I will follow the order of discovery in matters relating to the Tertiary, separating only that part relating to the fresh water drift of the central part of the continent, which will form the conclusion of this essay; nor will I dwell upon the few vertebrate fossils mentioned prior to 1820.

In 1824, Prof. Silliman* noticed the Tertiary exposed at Martha's Vineyard and the Elizabeth Islands. Prof. Olmstead,† in the first report ever made, as it is said, in any country, upon geology, with State or Government funds, described the country through which the Beauport canal was excavated, and separated the strata into: 1st. A black mould; 2d. Potters' clay, of a yellowish brown color; 3d. A thin layer of sand, full of sea shells and the remains of land animals, particularly of the mammoth, from three to eight feet deep; and, 4th. A soft blue clay.

Thomas Say‡ described, from strata now referred to the Miocene of Maryland, *Turritella plebeia, Natica interna, Buccinum porcinum,* now *Ptychosalpinx porcina, B. aratum, Fusus cinereus,* now *Urosalpinx cinereus, F. 4-costatus,* now *Ecphora quadricostata, Calyptraea*

---

* Am. Jour. Sci. and Arts, vol. vii.
† Rep. on the Geo. of North Carolina.
‡ Jour. Acad. Nat. Sci. vol. iv., pt. 1.

grandis, now *Dispotæa grandis*, *Fissurella redimicula*, *Ostrea compressirostra*, *Pecten jeffersonius*, *P. madisonius*, *P. clintonius*, *P. septenarius*, *Plicatula marginata;* and from strata now referred to the Pliocene of Maryland, *Arca arata*, *A. centenaria*, *A. incile*, *Pectunculus subovatus*, *Nucula concentrica*, *N. lævis*, *Venericardia granulata*, now *Cardita granulata*, *Crassatella undulata*, *Isocardia fraterna*, *Tellina acquistriata*, *Lucina anodonta*, *L. contracta*, *L. cribraria*, *L. subobliqua*, *Venus deformis*, *Astarte undulata*, *A. vicina*, *Amphidesma subovatum*, *Corbula cuneata*, *C. inaequalis*, *Panopæa reflexa*, *Serpula granifera*, and *Dentalium attenuatum.*

In 1825, Dr. Richard Harlan* described, from Bigbone Lick, Kentucky, *Cervus americanus*, *Bos bombifrons*, now *Ovibos bombifrons*, *B. latifrons*, now *Bison latifrons;* from a cave in Greenbriar county, Virginia, *Megalonyx jeffersoni*, and from Skidaway Island, Georgia, *Megatherium cuvieri.*

In 1828, Dr. J. E. Dekay† described, from the Post-pliocene at New Madrid, on the Mississippi river, *Bos pallasi.*

In 1829, Dr. Morton‡ arranged, from the notes of Lardner Vanuxem, some geological observations on the Tertiary and Alluvial formations of the Atlantic coast of the United States, showing their great extent and inclination from Nantucket and Martha's Vineyard, on the coast of New England, to the Mississippi river. The modern alluvial was divided into vegetable mould and river alluvium; the ancient alluvial into white siliceous sand and red earth; the Tertiary formation into beds of limestone, buhrstone, sand and clay. He described, from strata now regarded as Pliocene, *Crepidula costata.*

In 1830, Mr. Timothy A. Conrad§ showed that Tertiary deposits occupy all that part of Maryland south of an irregular line, running from the vicinity of Baltimore to Washington City, between the Potomac river and Chesapeake bay, though most of the surface is covered with a diluvial deposit of sand and gravel; and from the presence of *Turritella mortoni*, *Cucullæa gigantea*, and *Venericardia planicosta*, he regarded the deposits in the vicinity of Fort Washington as contemporaneous with the London clay of England, which now constitutes part of the Eocene of Europe. This was the first announcement of the existence of strata, of this age, in America.

---

* Fauna Americana.
† Am. Lyc. Nat. Hist., N. Y., vol. ii.
‡ Jour. Acad. Nat. Sci., vol. vi., pt. 1.
§ Jour. Acad. Nat. Sci., vol. vi., pt. 2.

He described, from Maryland, in strata now regarded as of Pliocene age, *Murex acuticosta, Voluta solitaria, Cassis cœlata, Trochus humilis. T. reclusus, Pyrula sulcosa, Turritella laqueata, T. variabilis, Cancellaria lunata*; and from strata now referred to the Miocene, *Natica fragilis, Pleurotoma communis, P. dissimilis, P. parva, P. rotifera, Marginella denticulata, Nassa quadrata, Terebra simplex, Actaeon melanoides, A. ovoides, Mactra ponderosa, Venus alveata, Amphidesma carinatum, Arca maxillata,* and *Cardium laqueatum;* and from strata now referred to the Eocene, *Monodonta glandula, Turritella mortoni, Cucullœa gigantea,* now *Latiarca gigantea, Crassatella alaeformis,* and *Venericardia blandingi.*

In 1832, Prof. Edward Hitchcock[*] described the *alluvium* as that fine, loamy deposit, which is yearly forming from the sediment of running waters, chiefly by the inundations of rivers. It is made up of the finest and richest portions of every soil over which the waters have passed. No extensive alluvial tracts occur in Massachusetts; although limited patches of this stratum exist, not infrequently, along the banks of every stream. The *diluvium,* he said, occupied more of the surface of the State than any other stratum. It is not generally distinguished from alluvium; but it is usually much coarser, being made up, commonly, of large pebbles, or rounded stones, mixed with sand and fragments of every size, which are often piled up in rounded hills to a considerable height, and under such circumstances as preclude the probability that it could have resulted from existing streams. The Tertiary formation is represented as most perfectly developed on Martha's Vineyard, though found on the Connecticut river and in the vicinity of Boston, and in limited patches in other parts of the State. He said the difference between this formation and the diluvium is, that in the diluvium, the sand, pebbles and clay are confusedly mixed together; but in the Tertiary, these materials are arranged in regular, and generally, in horizontal layers, one above another. Hence, when the sandy stratum happens to lie uppermost, the soil will be too sandy; but if this be worn away, so that the clay lies at the surface, the soil will be too argillaceous; or if the gravel stratum be exposed, the soil can not be distinguished from diluvium.

In 1833,[†] he treated of the coast alluvium, which is produced by tides and currents in the ocean, that frequently transport large quantities of soil from one place to another, and cause it to accumulate in those

* Rep. on the Geo. of Mass., 1832.
† Rep. on the Geo. of Mass., 1833.

situations where the force abates or is destroyed. The Salt Marsh alluvium, which results from the decay of salt marsh plants; the silt brought over the marsh by the tides; and from the alluvial soil brought down by streams which empty through these marshes. He mentioned the submarine forests on the coast, and on Martha's Vineyard, and numerous deposits of peat, and the processes by which it is produced. He observed how rapidly the New Red Sandstone disintegrates and unites with the soil, giving a decidedly red hue to extensive tracts of land; and likewise the gneiss, which is found disintegrated to a depth of from six to ten feet, and thus covers the earth and obscures the rocks even in the hilly districts. Some varieties of trap, sienite, mica, talcose and argillaceous slates are similarly affected, and even quartz-rock is shown to slowly decompose by the action of the weather. As evidencing the latter fact, it is mentioned that the name of John Gilpin had been painted upon a smooth boulder of granular quartz within the past 150 years, and that the paint had so protected the surface beneath it, while the decomposing process went on over other parts of the rock, that the name is now found perceptibly elevated on rubbing the fingers over the stone. Three causes—rains, frost and gravity—are said to be constantly operating to degrade the hills and the mountains. In precipitous trap-ridges, water penetrates fissures, freezes, and breaks asunder the masses which constitute the slopes of broken fragments or *debris* of rocks, which arrest the attention on the mural faces of the greenstone ridges in the Connecticut valley. The gneiss rock, in Worcester county, abounds with sulphuret of iron, which is continually undergoing decomposition by the action of heat, air and moisture, and becoming changed into an oxide and sulphate. The oxide imbibes carbonic acid from the atmosphere, and is changed into a carbonate which is soluble in water; or this oxide is washed into cavities, where it meets with water containing carbonic acid, by which it is dissolved. Once dissolved, it is transported to ponds and swamps, where it is deposited by evaporation, and forms the well known bog iron ore. Rocks containing manganese are likewise undergoing decomposition, and producing, in a similar manner, the oxide of manganese.

The ridge of bowlders on the margin of some ponds, where the bottom is free from them to a considerable extent, is accounted for by the expansion of the ice in lifting them from the bottom and crowding them out, while there is no force on the melting of the ice to draw them back.

The encroachments of the sea upon the land, and the gain of the land upon the sea, are discussed. The dunes or downs are described. The Connecticut river is shown not to have excavated its own valley entirely, though proofs are offered to show that it has cut out the last ninety feet in depth, or all below the upper terrace which forms the great valley of the Connecticut. The terraces found in the river valleys are described, and their origin accounted for on the supposition that they were produced by the rivers when they run upon a higher elevation than they do at present. The action of ice floods which continue to operate energetically in the Connecticut valley, and more powerfully in the mountain torrents, are considered in relation to their effects in modifying the surface and excavating the beds of rivers. It is shown that the Connecticut river may have excavated its own valley above Mount Toby, in Sunderland, but that this is the only valley in the State which is strictly a valley of denundation.

He separated the Tertiary into the most recent Tertiary and the plastic clay. The newest Tertiary is found in the Connecticut valley, and at Cambridge, Charleston, and other places. At Deerfield, it is found more than sixty feet in thickness, and near Boston, from seventy to one hundred and twenty feet. The plastic clay is found at Nantucket and the southeastern part of the State. He considered the extensive beds of hydrate of iron in the limestone valleys of Berkshire county, and the claystones of the Connecticut valley as of Tertiary age. The latter are concretions of carbonate of lime mixed with clay, such as that in which they are found, consisting of alumina and fine sand, with occasional fine scales of mica. These concretions are round, lenticular or oblate, and frequently joined together. The diameter is from the thickness of a pigeon shot to two or more inches, and the thickness is that of a single layer of clay, which rarely exceeds one half an inch. The Tertiary of the Connecticut valley, and other interior places, he supposed to be of fresh water origin, and the plastic clay a marine formation. The latter he separated into its mineralogical characters, and described white pipe clay, blood red clay, red and white clay, bluish gray plastic clay, white siliceous sand, white micaceous sand, green sand, lignite, osseous conglomerate, and other conglomerates and minerals. He noticed the organic remains, and called attention to the fossil vegetables and animals.

In the same year, Mr. Isaac Lea* described the Tertiary at Clai-

---

* Contributions to Geology.

borne, Alabama, and referred it to the age of the London clay, of England, and the *Calcaire Grossier*, of Paris. Having received Lyell's Principles, wherein the Tertiary was subdivided into Eocene, Miocene and Pliocene, he was enabled to class the Claiborne strata with the Eocene. This seems to have been the first application of the word "Eocene" to American rocks, though as above remarked, Conrad had compared American rocks with the London clay and *Calcaire Grossier*, which were afterward made the type of the Eocene period.

He described, from the Eocene at Claiborne, *Lunulites bouei*, now *Discoflustrellaria bouei*, *L. duclosi*, now *Heteractis duclosi*, *Orbitolites interstitia*, now *Lunulites interstitia*, *O. discoidea*, now *Cupularia discoidea*, *Turbinolia goldfussi*, now *Platytrochus goldfussi*, *T. maclurei*, now *Endopachys maclurei*. *T. nana*, *T. phareta*, *T. stokesi*, now *Platytrochus stokesi*, *Siliquaria claibornensis*, *Dentalium alternatum*, *D. turritum*, *Spirorbis tubanella*, *Serpula ornata*, *Teredo simplex*, *Solecurtus blainvillei*, *Anatina claibornensis*, now *Periploma claibornensis*, *Mactra dentata*, *M. grayi*, *M. pygmaea*, *Corbula alabamensis*, *C. compressa*, *C. gibbosa*, *C. murchisoni*, *Byssomya petricoloides*, *Egeria bucklandi*, *E. inflata*, now *Sphaerelia inflata*, *E. nana*, *E. nitens*, *E. ovalis*, *E. plana*, now *Tellina plana*, *E. rotunda*, *E. subtrigona*, *E. triangulata*, *E. veneriformis*, *Lucina compressa*, *L. cornuta*, *L. impressa*, *L. lunata*, *L. papyracea*, *Gratelupia moulinsi*, *Astarte minor*, now *Micromeris minor*, *A. nicklini*, *A. parva*, now *M. parva*, *A. recurva*, *A. minutissima*, now *M. minutissima*, *Cytherea comis*, *C. globosa*, *C. hydi*, *C. minima*, *C. subcrassa*, *C. trigoniata*, *Venericardia parva*, *V. rotunda*, *V. sillimani*, *V. transversa*, *Hippagus isocardioides*, *Myoparo costatus*, *Arca rhomboidella*, now *Anomalocardia rhomboidella*, *Pectunculus broderipi*, *P. deltoideus*, *P. minor*, *P. ellipsis*, now *Limopsis ellipsis*, *Nucula brongniarti*, *N. carinifera*, *N. magna*, now *Nuculana magna*, *N. media*, now *Nuculana media*, *N. ovula*, now *Nuculana ovula*, *N. pectuncularis*, now *Limopsis pectuncularis*, *N. plicata*, now *Nuculana plicata*, *N. pulcherrima*, now *Nuculana pulcherrima*, *N. sedgwicki*, *N. semen*, now *Nuculana semen*, *Avicula claibornensis*, *Pecten deshayesi*, *P. lyelli*, *Plicatula mantelli*, *Ostrea alabamensis*, *O. divaricata*, *O. linguacanis*, *O. pincerna*, *O. semilunata*, *Fissurella claibornensis*, *Hipponyx pygmaea*, now *Concholepas pygmaea*, *Infundibulum trochiforme*, *Crepidula cornuarietes*, *Bulla sthillairi*, *B. dekayi*, now *Cylichna dekayi*, *Pasithea aciculata*, *P. claibornensis*, *P. elegans*, *P. guttula*, *P. lugubris*, *P. minima*, *P. notata*, *P. secale*, *P. umbilicata*, *P. striata*, now *Actæonella striata*, *P. sulcata*, now *A* sulcata, *Natica gibbosa*, now *Neverita gibbosa*, *Natica magnoum-bilicata*, *N. mamma*, *N. minima*, now *Lunatia minima*, *N. minor*, *N. parva*, *N. semilunata*, *N. striata*, *Actaeon elevatus*, *A. laevis*. *A. lineatus*, *A. magnoplicatus*, *A. punctatus*, *A. wetherilli*, *A. melanellus*, now *Obeliscus melanellus*, *A. striatus*. now *O. striatus*, *A. pygmaeus*, now *O. pygmaeus*, *Scalaria carinata*, *S. planulata*, *S. quinquefasciata*, *Delphinula depressa*, now *Solariorbis depressus*, *D. plana*, now *Architectonica plana*, *Solarium bilineatum*, now *Architectonica bilineata*, *S. cancellatum*, now *A. cancellata*, *S. elegans*, now *A. elegans*, *S. granulatum*, now *A. granulata*, *S. henrici*, now *A. henrica*, *S. ornatum*, now *A. ornata*, *Orbis rotella*, now *Discohelix rotella*, *Planaria nitens*, now *Solariorbis nitens*, *Turbo naticoides*, *T. nitens*, *T. lineatus*, now *Solariorbis lineatus*, *Tuba alternata*, *T. striata*, *T. sulcata*, *Turritella carinata*, *T. lineata*, *Cerithium striatum*, *Pleurotoma beaumonti*, now *Surcula beaumonti*, *P. calata*, now *S. calata*, *P. childreni*, now *S. childreni*, *P. desnoyersi*, now *S. desnoyersi*, *P. hoeninghausi*, *P. lesueuri*, *P. lonsdalei*, now *Drillia lonsdalei*, *P. monilifera*, now *Surcula monilifera*, *P. obliqua*, now *S. obliqua*, *P. rugosa*, now *S. rugosa*, *P. sayi*, now *S. sayi*, *Cancellaria babylonica*, *C. costata*, *C. elevata*, *C. multiplicata*, *C. parva*, *C. plicata*, *C. sculptura*, *C. tessellata*, *Fasciolaria elevata*, *F. plicata*, now *Latirus plicatus*, *Fusus acutus*, *F. bicarinatus*, *F. conybearei*, now *Strepsidura conybearei*, *F. crebissimus*, *F. decussatus*, *F. delabechei*, *F. fittoni*, *F. magnocostatus*, *F. minor*, *F. mortoni*, *F. nanus*, *F. ornatus*, *F. parvus*, *F. pulcher*, *F. pumilis*, *F. taiti*, *Pyrula cancellata*, *P. elegantissima*, *P. smithi*, *Murex alternata*, *Rostellaria curieri*, *R. lamarcki*, *Monoceras fusiforme*, *M. pyraloides*, *M. sulcatum*, now *Pseudolica sulcata*, *Buccinum sowerbyi*, *Nassa cancellata*, *Terebra costata*, *T. gracilis*, *T. venusta*, *Mitra humboldti*, *M. lineata*, *M. minima*, *M. flemingi*, now *Caricella flemingi*, *M. fusoides*, now *Conomitra fusoides*, *Voluta cooperi*, *V. defrancei*, *V. gracilis*, *V. parkinsoni*, *V. parva*, *V. striata*, *V. vanuxemi*, *Marginella ovalina*, *M. columba*, *M. incurva*, *M. ovata*, *M. plicata*, *M. semen*, *M. biplicata*, now *Ringicula biplicata*, *Anolax gigantea*, *A. plicata*, *Oliva constricta*, *O. dubia*, *O. greenoughi*, *O. minima*, *O. gracilis*, now *Lamprodoma gracilis*, *O. phillipsi*, now *L. phillipsi*, *Monoptygma alabamensis*, *M. elegans*, *Conus claibornensis*, *Miliola marylandica*, now *Triloculina marylandica*, *Palmula sagittaria*, now *Phonemus sagittarius*, and *Rotella nana;* and from the Pliocene, *Balanus finchi*, and *Mactra clathrodon*.

T. A. Conrad* described, from the Miocene, at Yorktown, and other places in Virginia, *Mactra confraga, M. congesta, M. modicella, M. clathrodonta,* now *Rangia clathrodonta, Chama congregata, C. corticosa, Petricola centenaria, Pecten eborens, Cytherea marylandica, Fulgur incilis;* and from Choptank river, near Easton, Maryland, *Corbula idonea;* from the Eocene, at Claiborne, Alabama, *Corbula oniscus, Venerupis subregra, Cardita alticosta, Astarte tellinoides, A. ungulina, Pectunculus stamineus, P. cuneus,* now *Limopsis cuneus, P. trigonellus, Lucina dolabra, L. pandata, Nucula bella, N. caelata, Melongena alveata,* now *Cassidulus alveatus, Crepidula lirata, Solarium elaboratum,* now *Architectonica elaborata, Sigaretus bilix,* and *Typhis gracilis.*

In 1834, Mr. T. A. Conrad† identified the Eocene at Claiborne, Ala.; at Eutaw Springs and Nelson's Ferry, on the Santee river; at Shell Bluff, near Milledgeville, in Georgia; at Shell Bluff, on the Savannah river, fifteen miles below Augusta; at Fort Gaines, on the Chattahoochee, and other places; from all which he projected the continuity of the strata, commencing in Maryland, at Fort Washington, and extending in a southerly direction across Virginia, North and South Carolina, and westerly across Georgia, Alabama and Mississippi. His diagram, representing the strata composing the bluff at Claiborne, showed, in descending order: 1. Diluvium, 20 feet; 2. Whitish, friable limestone, 45 feet, containing *Scutella lyelli;* 3. Six feet indurated limestone, where the fossils occur in casts; 4. Ferruginous, siliceous sand, 14 feet, containing *Cardita planicosta, Corbis lamellosa,* and *Pyramidella terebellata;* 5. Sand, with a calcareous cement, 3 feet, containing *Ostrea sellaeformis;* 6. Soft, lead-colored limestone, 70 feet, containing *O. sellaeformis* in abundance, and rarely *Plagiostoma dumosum;* 7. Friable, lead-colored limestone, of unknown thickness, containing *Cardita planicosta,* a shell very characteristic of the Eocene. He remarked that the *Plagiostoma dumosum* passed from the cretaceous rocks to the Eocene; that the Eocene at Claiborne appeared to be older than the Eocene of Europe, and older than the deposit at Fort Washington, Md.

He described, from the Eocene of the Southern States, *Tellina scandula, Pectunculus perplanus,* now *Limopsis perplana, Fusus irrasus, F. raphanoides, F. salebrosus, F. sexangulatus, F. symmetricus, Cassis brevicostatus, C. taiti, Cerithium nassula, C. solitarium, Ancillaria*

---

tenera, Fusus cooperi, now *Clarifusus cooperi, Crepidula dumosa, Murex mantelli, M. septemnarius, Terebra polygyra, Serpula squamulosa, Cytherea nuttalli, C. mortoni, Ostrea georgiana,* and *Scutella lyelli,* now *Mortonia lyelli;* and from more recent Tertiary of the Southern States, *Anatina antiqua,* now *Periploma antiqua, Saxicava pectorosa, Pandora arenosa, Tellina declivis, T. egena, Cytherea obovata, C. pandata, C. reposta, Amphidesma subreflexum, Astarte concentrica, A. lunulata, A. obruta, A. symmetrica, Balanus proteus, Fasciolaria mutabilis, Turbinella demissa, Cancellaria perspectiva, C. plagiostoma, Trochus bellus, T. labrosus, T. lapidosus, T. mitchelli, T. philantropus, Pleurotoma biscatenaria, P. incilifera, P. pyrenoides, P. tricatenaria, P. virginiana, Turbo caperatus, Marginella eburneola, M. limatula, Solarium nuperum,* now *Architectonica nupera, Delphinula lyra,* now *Carinorbis lyra, Actæon novellus, Dentalium thallus, Fissurella alticosta, F. griscomi, Infundibulum gyrinum, Capulus lugubris, Turritella alticostata, T. octonaria, Cancellaria alternata, Pecten decemnarius, P. rogersi, Lepton mactroides,* and *Tellina biplicata.*

He also mentioned the following Pliocene fossils, which are to be found living on the coast of the United States, to wit: *Arca transversa, Cytherea sayana, C. gigantea, Pholas costata, Ostrea virginiana, Solen ensis, Amphidesma inequale, Saxicava rugosa, Venus mercenaria, Panopea reflexa, Mactra leilinoides, Pandora trilineata, Cardita tridentata, Lucina contracta, L. crenulata, L. divaricata, Corbula contracta, Crepidulata convexa, C. glauca, C. plana, Lutraria canaliculata, Fusus cinereus, Nassa trivittata, N. lunata, Natica duplicata, N. heros, Fulgur carica, F. canaliculatus, Mactra lateralis, Scalaria clathrus,* and *Vermetus lumbricalis.* This list does not include fossils of the newer Pliocene.

In 1836, Prof. Edward Hitchcock[*] described, from the Miocene at Portland, Maine, *Nucula portlandica.*

Dr. Samuel G. Morton[†] described, from a Miocene or Pliocene deposit, near Marietta, Ohio, *Unio petrosus, U. saxulum, U. terrenus, U. tumulatus,* and *Anodonta abyssina.*

In 1837, Wm. B. and Henry D. Rogers[‡] described the Tertiary in the counties of Elizabeth City, Warwick, York, James City, Va., and the lower extremities of New Kent and Charles City, having a total length

---

[*] Bost. Jour. Nat. Hist., vol. i., pt. 3.
[†] Am. Jour. Sci. and Arts, vol. xxix.
[‡] Trans. Am. Phil. Soc., vol. v.

of about fifty miles, and a mean breadth of fourteen miles. The superficial stratum is an argillaceous and ferruginous sand, of a yellow or reddish color, with an occasional pebble or small bowlder of sandstone, or a white, silicious sand. Beneath this superficial layer, occasionally argillaceous beds of clay are found, of a yellow, blue, green, red or variegated color. In some places this clay is from twelve to fifteen feet in thickness. Below this stratum there is usually found a red ferruginous layer, from an inch to a foot in thickness. Beneath this layer there is a yellowish brown sand, frequently containing a large proportion of clay, all of which is barren of shells. Below these superficial layers occur the various shell beds of Miocene sand and clay, from which these authors described *Turritella quadristriata*, *T. terstriata*, *Natica perspectiva*, *Fissurella catilliformis*, *Arca protracta*, *Lucina speciosa*, and *Venus cortinaria*. They also described, from the Eocene greensand, *Nucula cultelliformis*, now *Nuculana cultelliformis*, *N. parva*, and *Cytherea ovata*, now *Dione ovata*.

In 1838, Mr. Conrad* said that the most northern locality known to be decidedly of Medial Tertiary age, is in Cumberland county, N. J., from whence the deposits extend southward in a very connected series, and are spread over a large portion of the Atlantic seaboard. The eastern shore of Maryland is chiefly composed of this and the superior formations, but the greensand occasionally appears. The Medial Tertiary occupies all that portion of the western peninsula south of a line running from Annapolis to Fort Washington, on the Potomac, and nearly all that part of Virginia which lies east of a line running through Fredericksburg, Richmond and Petersburg, to Halifax, in North Carolina, in which State the formation expands to its greatest breadth. The lowest stratum of the Medial Tertiary is clay; the upper stratum sand; and the intermediate strata are composed of sand and clay, either pure or intermixed. The general surface of the country is level, and it was originally covered with a forest of pine trees. The western limit is bounded by a narrow strip of the lower, or Eocene Tertiary, which reposes upon Cretaceous strata. He described, from the Miocene, *Mya producta*, *Pandora crassidens*, *Pholadomya abrupta*, *Panopæa americana*, *Corbula elevata*, *Venus tetrica*, *V. ducateli*, now *Mercenaria ducateli*, *V. rileyi*, *Cytherea metastriata*, *Sphærella subrexa*, *Saxicava bilineata*, *Mactra incrassata*, *M. subcuneata*, *Cardium acutilaqueatum*, *Lucina crenulata*, *Venus latisulcata*, now *Euloxa latisulcata*, *Astarte arata*, *A. cuneiformis*, *A. perplana*, *A. coheni*,

---

Pecten virginianus, Ostrea percrassa, O. subfalcata, O. sculpturata, O. disparilis, Myoconcha incurva, Modiola ducateli, now *Volsella ducateli, Byssoarca marylandica,* and *Arca callipleura.*

Prof. Emmons[*] described the Tertiary of Lake Champlain as consisting of clays and sands, embracing, to some extent, marine shells of recent age—the whole formation in Essex county, New York, not exceeding fifty feet in thickness, and averaging only from twenty to twenty-five feet. From above, downward, the strata are, first, a fine white, or yellowish white, marine sand; second, a yellowish clay; and third, a blue clay. The yellowish clay abounds with argillo-calcareous concretions, of all shapes and forms, which appear to have been formed by molecular attraction, since the deposition of the beds. On the New York side of the lake, it does not form a continuous deposit from the head of the lake to its outlet, but interruptions occur where the older strata reach the lake shore. On the Vermont side, it covers a much greater extent of surface, and reaches from the lake to the base of the Green Mountains, or from six to twelve miles. The height above the level of the lake to which it extends, is about two hundred feet. This ancient sea occupied the Champlain basin, and the Hudson forming a continuous arm from the Gulf of St. Lawrence to the mouth of the Hudson, at New York.

In 1839, Prof. Charles T. Jackson[†] mentioned a recent marine Tertiary deposit, at Augusta, Maine, eighty-two feet above the level of the Kennebec river, where it is said to form the substratum of a large portion of the valley.

Wm. B. and Henry D. Rogers[‡] described the Tertiary in the counties of Lancaster, Northumberland, Richmond, Westmoreland, King George, and the eastern part of Stafford, in Virginia; thus including the peninsula between the Potomac and Rappahannock rivers. This area forms the northern portion of the Tertiary of Virginia. The Miocene extends from near the bay shore, westward over the larger portion of the peninsula, while the Eocene occupies the remaining area on the west. They described from the Miocene, *Turritella fluxionalis* and *Fasciolaria rhomboidea;* and from the Eocene, *Cytherea lenticularis,* now *Dosiniopsis lenticularis, Crassatella capricranium, Cucullaea ononcheila,* now *Latiarca ononcheila, C. transversa,* now *L. transversa* and *Venericardia ascia.*

---

[*] Geo. Rep. N. Y., 1838.
[†] Third Annual Rep. Geo. of Maine.
[‡] Trans. Am. Phil. Soc., vol. vi.

Wm. Wagner* described, from the Miocene and older Pliocene, of Maryland and North Carolina, *Venus inoceriformis*, *Pecten marylandicus*, *Panopœa goldfussi*, *Mysia nucleiformis*, and *Trochus eboreus*.

In 1840, Mr. Conrad† described, from the Miocene at Chapel Hill, North Carolina, *Fulgur excavatus*, *F. contrarius*, *Conus adversarius*, and *Voluta carolinensis*, now *Mitra carolivensis*.

The Tertiary extends from the lower limit of the Cretaceous, in Connecticut,‡ to the lower part of Kent county, and has a thickness of 125 feet.

It is found§ at Gay Head, Martha's Vineyard, and occupying Long Island and the eastern part of the Atlantic States from New Jersey to Florida, and the southern part of the Mississippi valley.

Henry C. Lea‖ described, from the Eocene at Claiborne, Alabama, *Pasithea cancellata*, *P. elegans*, *P. minima*, *Actœon lœvis*, *A. magnoplicatus*, *Scalaria elegans*, *S. venusta*, *Turbo parvus*, *Trochus planulatus*, *Turritella monilifera*, *T. gracilis*, *Turbinella fusoides*, *Pleurotoma cancellatum*, *Triton pyramidatum*, *Terebra constricta*, *T. multiplicata*, *Cancellaria pulcherrima*, *Buccinum parvum*, *Mitra eburnea*, *M. elegans*, *M. gracilis*, *Conus parvus*, and *Voluta dubia*.

T. A. Conrad¶ described, from the Middle Tertiary at the Natural Well, Duplin county, North Carolina, *Amphidesma constrictum*, now *Fabella constricta*, *Buccinum interruptum*, *B. multirugatum*, now *Ptychosalpinx multirugata*, *Cardita perplana*, *Cassis hodgei*, now *Galeodia hodgei*, *Cerithium carolinense*, now *Terebra carolinensis*, *C. unilineatum*, now *T. unilineata*, *Cyprœa carolinensis*, *Dispotœa dumosa*, *D. multilineata*, *Gnathodon minor*, now *Rangia minor*, *Infundibulum centrale*, *Lucina radians*, *L. trisulcata*, *Lunulites denticulatus*, now *Discoporella denticulata*, *Mactra crassidens*, *M. subparilis*, *Natica caroliniana*, *N. percallosa*, *Pectunculus carolinensis*, and *P. quinquerugatus*; from Wilmington, North Carolina, *Amphidesma nuculoides*, *A. protextum*, *Cardium sublineatum*, *Cardita abbreviata*, *Pectunculus carolinensis*, and *P. aratus*.

Prof. Emmons** found the direction of the drift scratches and scorings of rock, in the eastern part of New York, conforming to that of

Jour. Acad. Nat. Sci., vol. viii., pt. 1.
† Am. Jour. Sci. and Arts, vol. xxxix.
‡ Geo. Sur. of Delaware, 1841.
§ Geo. of Massachusetts, 1841.
Am. Jour. Sci. and Arts, vol. xl.
¶ Am. Jour. Sci. and Arts, vol. xli.
** Geo. 2d Dist. N. Y., 1842.

the great valleys. In the Champlain valley, it is nearly north and south; and in the St. Lawrence valley, northeast and southwest. The marine Tertiary of Champlain, though deposited in quiet waters, always overlies the scored and grooved surfaces. The bowlders succeed this Tertiary or are mixed with it. It is mineralogically composed in ascending order, of first, a stiff blue clay; second, a yellowish brown clay; and third, a yellowish brown sand. The second owes its color to weathering rather than to any important difference in its composition from the lower clay. Sand begins to appear in the yellowish clay, and increases gradually until it predominates, and finally becomes a pure siliceous sand. No fossils had then been discovered in the clay, but in the clay and sand and upper part of the group fossils are found as if in their native habitat, exceedingly frail, preserving their markings and edges entire, forbidding the idea that they could have been drifted into their present position. In protected places, as at Port Kent and Beauport, the thickness of the group is about 100 feet. In unprotected places, the larger part of the group has been swept away.

Commencing at Whitehall, at the head of Lake Champlain, it may be traced continuously not only the entire length of the lake, but also to Quebec and far toward the Gulf of St. Lawrence. It lines the St. Lawrence river as far as Ogdensburg. And from Whitehall south it lines the Hudson river for a long distance. The Albany clay belongs to this group, and is therefore one of the most recent of our marine formations.

Mr. Conrad* described, from the Miocene at Calvert Cliffs, Maryland, *Venus latilirata, Cytherea subnasuta, Lucina foremani, L. subplanata, Cardium leptopleura, Astarte cuneus, A. exaltata, Lima papyria, Arca subrostrata, Pleurotoma marylandicum, P. bellicrenatum, Trochus peralveatus, Scalaria pachypleura, Solarium trilineatum,* now *Architectonica trilineata, Infundibulum perarmatum, Fissurella marylandica, Dispotæa ramosa, Cancellaria biplicifera, C. engonata, Bonellia lineata, Turritella indenta, T. exaltata, T. perlaqueata, Marginella perexigua.* And also *Astrea marylandica,* incrusting *Pecten madisonius* on James river, Virginia; *A. bella,* from Newbern, North Carolina; *Cardium nicolletti,* now *Protocardia nicolletti,* from the Lower Tertiary or Jackson Group, on the Washita river, Monroe county, Louisiana; and *Fusus pachyleurus,* from the Lower Tertiary of Alabama.

Edmund Ravenel† described, from a Pliocene calcareous deposit on

Cooper river, about seventeen miles from Charleston, South Carolina. *Scutella caroliniana*, now *Mellita caroliniana*, and *S. macrophora*, now *Encope macrophora*.

In 1843, Mr. Conrad* described, from the Miocene at Newbern, North Carolina, Cliffs of Calvert, Maryland, Petersburg, Virginia, and other places, *Carditamera carinata*, *C. protracta*, *Arca triquetra*, *Nucula liciata*, *Pectunculus parilis*, *Pecten biformis*, *P. tricenarius*, *P. vicenarius*, *Tellina laevis*, *Lucina multistriata*, *Amphidesma aequalum*, *Crassatella turgidula*, *Crepidula spinosa*, *Fulgur rugosus*, *Buccinum biles*, *B. filicatum*, *B. fossulatum*, *B. lienosum*, *B. praeruptum*, *B. protractum*, *B. sexdentatum*, *Cancellaria corbula*, *Oliva duplicata*, *Pyramidella arenosa*, *Fusus migrans*, *F. devexus*, *Voluta mutabilis*, *Ovula iota*, *Monodonta exoleta*, *Echinus improcerus*, *E. philanthropus*, *Venus cribraria*, *Plicatula densata*, *Crepidula densata*, *Arca propatula*, now *Granoarca propatula*, a subgenus of *Barbatia*, *A. scalaris*, *Cyrena densata*, *Mactra triquetra*, *Venus capax*, *Artemis elegans*, *Loripes elevata*, *Solen directus*, *S. ensiformis*, *Turritella bipartita*, *Scalaria procera*, *Pleurotoma multisectum*, *Buccinum harpuloides*, *Fusus cannabinus*, *Terebra curvilirata*, *Turbinolia pileolus*, *Spatangus orthonotus*, now *Amphidetus orthonotus*; from the Eocene at Chapel Hill, North Carolina, *Tellina arctata*; from Pamunkey river, Virginia, *Anomia ruffini*; and from the Jackson Group, *Anomia jugosa*. He said, that in a few hours' examination of the Miocene marl, in the vicinity of Petersburg, Va., he was enabled to collect about 100 distinct species. This locality is the western limit of the Miocene, which is here based on granite, and is the spot, in which, to search for the estuary and fresh water shells of the Miocene period. The elevation is considerably more than 100 feet above tide, and as the rise decreases toward the sea, it is probable that the primary rocks continued to be uplifted even after the era of the Miocene; indeed, how can we otherwise account for the elevation of fossiliferous beds, even of those of the Post-pliocene period.

It is an interesting fact that the Miocene estuaries were inhabited by two species of bivalves, now extinct, of the same two genera which still occur in similar situations in Florida and Alabama, that is at the confluence of rivers and bays, where the water is nearly fresh. These genera are *Gnathodon* and *Cyrena*, both of the family *Cyrenidae*. The extinct Gnathodon has a considerable resemblance to the recent species, but the *Cyrena* is widely different from the living shell. These

---

* Proc. Acad. Nat. Sci., vol. i.

fossils are frequently water-worn, always with disunited valves, and appear to have been transported. Occasionally a specimen occurs not in .the least abraded, a circumstance which indicates the vicinity of the Petersburg deposits to the mouth of the river. The strata occur in a meadow, and consist of blue marl, of a sandy texture, often intermixed with small gravel and ferruginous sand, full of shells ; there is here also a proportion of gravel, of rounded quartz, occasionally of large size. Water-worn fragments of bivalves are abundantly intermingled with entire shells, and many species occur with connected valves. This is particularly the case with the burrowing shells, as *Panopœa*, but also, though less frequently, with the large *Venus tridacnoides. Crassatella undulata, Astarte concentrica, Cytherea albaria,* two species of *Chama,* and even two species of *Ostrea* are not uncommon; but there is nothing like an oyster bed in these strata which might indicate shoal water. The proportion of oysters to the other bivalves is about the same which the dredge furnished at the mouth of Cape Fear river, North Carolina, at the depth of eight fathoms.

In 1844, Prof. J. W. Bailey* identified numerous living Infusorial forms with the fossil Infusoria, from the Miocene at Petersburg, Va., and Piscataway, Md., and described several new species.

Mr. Conrad† described, from the Miocene, at Petersburg, Va., *Crepidula cymbœformis;* from the Eocene at Marlbourne, Hanover county, Va., *Cytherea eversa, C. liciata, C. subimpressa;* from Stafford county, Va., *C. pyga;* from Claiborne, Ala., *Cardita densata;* and from near Santee, South Carolina, *Pecten elivatus.*

Dr. Edmund Ravenel described, from the Miocene of South Carolina, *Pecten mortoni;* from the Eocene, *Terebratula canipes,* and *Scutella pileussinensis,* now *Mortonia pileussinensis.* And Dr. Robert W. Gibbes described, from a bed of green sand near the Santee canal, and about three miles from the head waters of Cooper river, South Carolina, *Dorudon serratus.*‡

In 1845, Prof. James Hall§ described, from Tertiary, slaty, bituminous limestone, on the dividing ridge between the waters of Muddy river flowing eastward, and those of Muddy creek flowing into Bear river on the west, in long. 111 deg., lat. 40 deg., *Mya tellinoides,*

---

* Am. Jour. Sci. and Arts, vol. xlvi.
† Proc. Acad. Nat. Sci., vol. ii.
‡ This species was erroneously mentioned as Cretaceous on page 15, vol. iii., of this Journal, or page 51 of this article.
§ Fremont's Expl. Exped.

now *Unio tellinoides, Pleurotomaria uniangulata, Cerithium fremonti, C. tenerum,* now *Goniobasis tenera, Natica* (?) *occidentalis,* and *Turbo paludinæformis,* now *Viviparus paludinæformis.*

William Lonsdale* described, from the Miocene of Virginia, *Columnaria sexradiata, Heteropora tortilis,* now *Multicrescis tortilis, Escharina tumidula,* now *Cellepora tumidula, C. quadrangularis,* now *Reptocelleporaria quadrangularis, C. informata,* now *R. informata, C. similis,* now *R. similis, C. umbilicata,* now *Multiporina umbilicata.* From the Eocene, *Ocellaria ramosa, Flabellum cuneiforme, Dendrophyllia lævis, Cladocera recrescens, Caryophyllia subdichotoma, Idmonea commiscens, I. maxillaris, Hippothoa tuberculum,* now *Pyriflustrella tubercula, Eschara incumbens, E. petiolus, E. tubulata, E. viminea, E. linea,* now *Escharinella linea, Lunulites distans, L. sexangulatus,* and *L. contiguus.* Lyell and Sowerby described *Terebratula wilmingtonensis,* now *Rhynchonella wilmingtonensis,* and *Cerithium georgianum.* And Edward Forbes described, *Scutella jonesi,* now *Clypeaster jonesi.*

In 1846, Mr. Conrad† demonstrated that the white limestone of Southern Alabama and Mississippi, which had been previously classed with the upper Cretaceous rocks, belongs more properly with the lower Eocene, and described *Dentalium arciforme, Fistulana larva, Lutraria lapidosa,* now *Pteropsis lapidosa, Crassatella rhomboidea, C. palmula, Amphidesma tellinula, Tellina sillimani, T. raveneli,* and *Lucina modesta.*

He found evidences of the Eocene‡ and Miocene in East Florida, and described the Tertiary of Warren county, Mississippi, and stated, that it marks a distinct era in the American Tertiary system intermediate to the Eocene and Miocene, but more nearly allied to the former. He described the Eocene at Vicksburg, and in the bluffs on the Mississippi river, and defined, from the Upper Eocene limestone of Tampa Bay, *Bulimus floridanus, Bulla petrosa, Nummulites floridanus, Cristellaria rotella, Venus penita,* now *Cryptogramma penita, V. floridana,* now *C. floridana, Nucula tellinula, Cytherea floridana,* and *Balanus humilis.*

Dr. Dickeson§ described, from the blue clay that underlies the diluvial drift east of Natchez, Mississippi, a fossil, *os innominatum,* that once belonged, as he supposed, to a young man about 16 years of age.

---

* Quar. Jour. Geo. Soc. Lond., vol. i.
† Am. Jour. Sci. and Arts, 2d ser., vol. i.
‡ Am. Jour. Sci. and Arts, 2d ser., vol. ii.
§ Proc. Acad. Nat. Sci., vol. iii.

It was found beneath the fossil bones of the *Megalonyx jeffersoni*, and *Mastodon giganteum.*

In 1847, W. E. Logan[*] found marine testacea along the valley of the Ottawa, in the clays and sands that form the superficial deposits. These deposits cover the whole valley of the south Petite Nation and its tributaries ; and occur in Templeton, Hull, Nepean, Packenham, and Fitzroy, to the mouth of the Mississippi and Madawaska. They were found in Fitzroy, 330 feet above the level of the sea, and in Nepean, 410 feet above the sea, where *Saxicava rugosa* occurs in the gravel. At the mouth of Gattineau, near Bytown, not only marine shells were discovered, but in nodules of indurated clay the *Mallotus villosus*, or common capeling, a small fish, which still frequents the shores of the Gulf of St. Lawrence, was obtained in vast numbers.

Grooves and scratches on the surfaces of the rocks were met with on the Gattineau, between Farmer's and Blasdell's mills, having a direction S. 36° E ; on Glen's creek in Packenham, N. and S. ; on the Allumettes Lake, at Montgomery's clearing, S. 25° E. The shores of Lake Temiscamang, which is long and narrow, and has banks bold and rocky, rise into hills 200 to 400, and sometimes 500 feet above its surface. The general valley of the lake thus bounded presents several gentle turns, the directions connected with two of which, reaching down to the mouth of the Keepawa river (35 miles), are 158°, 191°, 156°, numbering the degrees from north as zero around by east. The parallel grooves in these reaches of the valley turn precisely with them, as if the bounds of the valley had been the guiding cause of their bearings, and they are registered on various rounded and polished surfaces projecting into the lake, and sometimes rising to 30 and 40 feet over its level. These projecting points did not deflect the grooved lines in the slighest degree. In one case, where the projecting point is 35 feet high, the furrows were observed to move over it without any deflection whatever ; so that, whatever body, moving downward in the valley, may have caused the grooves, it was not deflected by meeting an obstacle 35 feet higher than the surface of the lake. On the top of this projecting point, the grooves are crossed by another parallel set at an angle of 15°.

The Company's Post stands on a point on the east side, which cuts the lake nearly in two, at about 18 miles from the head, and it is opposite a less prominent point on the other side. These points approach to within a quarter of a mile of one another. Both are composed of

---

* Geo. Sur. Can.

sand and gravel, which, on the east, form a hill 130 feet high. The southern face of this hill runs in the bearing 65°, and the gravel toward the eastward rests on flat sandstone strata, which have a smooth and partially rounded surface. The gravel and the rock constitute the north side of a deep bay. The polished rock surface exhibits well marked grooves, which come from beneath the gravel hill, nearly at right angles to the margin of the water. There is here, as in some other instances, more than one set of parallel scratches. Two of these sets cross one another in the directions 140° and 196°. In the eastern bay, at the head of the lake, near the mouth of the Otter river, parallel grooves were remarked running in the bearing 105°, which is the upward direction of the valley of that stream; and about a mile westward of the Blanche, in the same bay, in the bearing 130°, partaking of the direction of the valley, bounded by the escarpment of the limestone described as running back into the interior. On the east side of the lake, three bowlders were remarked, which had been moved by the ice the previous winter. One of them measuring 32 cubic feet, had been moved nine feet in the direction 90°; another 100 cubic feet, had been moved twelve feet in the direction 350°; another 80 cubic feet, had been moved 14 feet in the direction 350°; each had left behind it a deep, broad furrow through the gravel of the beach down to the clay beneath. In front of the first was accumulated a heap of gravel, one foot high, with an area of 9 square feet; in front of the second was an accumulation of small bowlders weighing from 80 to 10 lbs. each. To move the second and third, the progress of the ice must have been up the lake, and the first across it. Had the gravel rested on the surface of a rock instead of clay, parallel scratches would have been the result in each case.

There are deep, water-worn holes on the banks of the Ottawa, at heights considerably above the highest level it has ever been known to attain. One of these, 18 inches in diameter, near Chenaux, is 60 feet above the existing surface of the water; another, on the island at Portage Dufort, 25 feet above the water, and 12 or 13 feet over the great flood of the preceding spring, is more than 5 feet deep, measuring 2 by 2½ feet in diameter.

Alexander Murray found Tertiary deposits on the eastern peninsula of the Province, between the Bay Chaleur and the Gulf of St. Lawrence, consisting of clay, generally of a blue color, with sand or gravel over it, and forming the banks at the mouths of the rivers. Over the clay in some cases, as at the mouth of the Chat, marine shells were found deposited in layers, 30 feet above high-water mark. At the mouth of the Matan the clay and gravel banks are upward of 80 feet high.

Robert W. Gibbes* described, from the Eocene of South Carolina, *Pristis agassizi.*

In 1848, M. Tuomey† said that the Tertiary rocks of South Carolina are composed of beds of loose sand, clay, gravel and sandstone, together with strata of limestone, of great thickness, and beds of soft or pulverulent marl.

A line drawn from the mouth of Stevens' creek, on the Savannah, north of Hamburg, crossing the Saluda and Broad rivers, near their junction; the Wateree, at the canal; Lynch's creek at Evan's Ferry; and Thompson's creek, at the point where it enters the State, in Chesterfield district, will approximately mark the northern boundary. Wherever the rivers, in their downward course, enter this boundary, they wash away the more yielding Tertiary rocks, and expose the metamorphic, and very frequently the granitic rocks; and hence it is, that, at these points, in ascending the rivers, we meet with the first falls.

The Eocene, in South Carolina, has a thickness of 1,000 or 1,100 feet, and consists of three well-defined groups. 1. The Buhr-stone group, composed of thick beds of sand, gravel, grit, clay and buhr-stone, amounting to at least 400 feet in thickness, and underlying the calcareous beds. Its upper portions are characterized by beds abounding in silicified shells, for the most part identical with the Claiborne fossils. As these are littoral shells, they probably occupied the coast, while the Santee beds were forming in deep water. The materials of which this group is composed are the ruins of the granitic and metamorphic rocks of the upper districts. Good exposures occur at the ferry below Augusta, in the high red cliffs overlooking the town of Hamburg, between Aikin and Graniteville, on Horse creek and Cedar creek, and at the head of Congaree creek. It may be traced from Barnwell to Sumter, a distance of 100 miles, and it occurs on Huspa creek, in Beaufort district, and at many other places.

2. The Santee beds, consisting of thick beds of white limestone, marl and green sand. These are best seen on the Santee, where, interstratified with the green sand, they dip gently toward the south. The coralline marl of Eutaw is found near the upper edge of these beds. The irregular area occupied by these beds, is about 75 miles long, and 60 miles wide.

---

* Jour. Acad. Nat. Sci., 2d ser., vol. i.
† Tuomey Rep. Geo. of South Carolina.

3. Next in order above the Santee beds, are the Ashley and Cooper beds, which are the newest Eocene beds of this State. The marl of these is characterized by its dark gray color and granular texture, while the remains of fishes and mammalia give its fossil remains a peculiar character. These, together with the Santee beds, have a thickness of from 600 to 700 feet.

Artesian boring has shown that the Ashley marl occurs at a depth of about 300 feet, at the City of Charleston.

The Eocene is succeeded, in South Carolina, by isolated patches of highly fossiliferous beds of sand and marl, in which Tuomey estimated the proportion of living species to amount to 40 per cent, and for this reason referred the beds to the age of the Older Pliocene. On the Waccamaw and Peedee, this older Pliocene is found super-imposed upon Cretaceous rocks, and, in general, the strata appear to have been deposited on a plane that rises gently from the Atlantic till it reaches its greatest elevation in Darlington district. The protected patches may be traced, at short intervals, from Horry to Darlington, and from thence by Lynch's creek to Sumter. It occurs on Cooper river, and at various other places.

The Post-pliocene, of South Carolina, is confined to a belt along the coast of about 8 or 9 miles in breadth. The fossils are nearly all referable to living species now inhabiting the coast ; a few, however, belong to the fauna of Florida and the West Indies. There appears to have been a slight elevation of the coast during this period.

T. A. Conrad* separated the Eocene into the Upper or Newer Eocene, found at Vicksburg, Miss., and including the white limestone of St. Stephens. and of Claiborne, Ala., and part of that in Charleston county, South Carolina, characterized by *Scutella lyelli, S. rogersi, Pecten poulsoni*, and *Nummulites mantelli*; and the limestone in the vicinity of Tampa bay, Florida, characterized by *Nummulites floridana, Cristellaria rotella,* and *Ostrea georgiana ;* and into the Lower or Older Eocene, consisting of the fossiliferous sands of Claiborne, and St. Stephens, Ala., of the Washita river, near Monroe, La. ; of Pamunky river, at Marlborne, and the greensand on James river, below City Point, Va., and at Fort Washington, Piscataway, and Upper Marlborough, Maryland, characterized by *Cardita planicosta, C. blandingi, Crassatella alta, Ostrea sellæformis,* and *Turritella mortoni.* He described,* from the Eocene. in the vicinity of Vicksburg, Mississippi, *Dentalium mississippiense, Fissurella mis-*

---

sissippiensis, Solarium triliratum. now *Architectonica trilirata, Bulla crassiplica, Cypræa sphæroides, C. lintea, Natica mississippiensis, Sigaretus mississippiensis, Natica mississippiensis, N. vicksburgensis. Scalaria trigintinaria, Turritella mississippiensis, Terebra divisura, T. tantula, Pleurotoma porcellanum, P. abundans, P. cochleare, P. congestum, P. cristatum, P. declive, P. eboroides, P. mississippiense, P. rotadens, P. serratum, P. tantulum, P. tenellum, Phorus humilis, Buccinum mississippiense, Typhis curvirostris, Murex mississippiensis, Melongena crassicornuta, Fusus mississippiensis,* now *Ficopsis mississippiensis, F. spiniger, F. vicksburgensis, Chenopus liratus,* now *Aporrhais liratus, Ringicula mississippiensis, Actæon andersoni, Cancellaria funerata, C. mississippiensis, Triton crassidens, T. abbreviatus. T. mississippiensis, Cassidaria lintea,* now *Sconsia lintea, Cassis cælatura, C. mississippiensis, Oniscia harpula, Fulgoraria mississippiensis, Oliva mississippiensis, Mitra conquisita,* now *Fusimitra conquisita, M. mississippiensis,* now *F. mississippiensis, M. cellulifera,* now *F. cellulifera, M. staminea,* now *F. staminea, M. vicksburgensis, Caricella demissa, Scobinella cælata, Turbinella perexilis, T. protracta, T. wilsoni, Panopea oblongata, Mactra funerata, M. mississippiensis. Amphidesma mississippiense, Psammobia lintea,* now *Gari lintea, P. papyria,* now *G. papyria, Crassatella mississippiensis, Cardium eversum, C. diversum,* now *Protocardia diversa, C. vicksburgense, Tellina pectorosa, T. serica, T. vicksburgensis, Donax funerata, Cytherea astartiformis, C. imitabilis C. mississippiensis, C. sobrina, C. perbrevis, Corbis staminea, Lucina mississippiensis, L. perlævis. Loripes eburnea, L. turgida, Corbula alta, C. engonata, C. interstriata, Chama mississippiensis, Pectunculus arctatus, Nucula sericea, N. vicksburgensis, Arca mississippiensis, Byssoarca lima, B. mississippiensis, B. protracta. Avicula argentea, Modiola mississipiensis,* now *Volsella mississippiensis, Pinna argentea, Lima staminea, Ostrea vicksburgensis, Pholas triquetra, Madrepora mississipiensis, M. vicksburgensis, Turbinolia caulifera,* now *Osteodes cauliferus, Lunulites vicksburgensis,* now *Oligotresium vicksburgense.* From the Eocene, at Claiborne, Alabama, and other places, *Ampullaria* (?) *perovata, Turbinolia elaborata,* now *Osteodes elaboratus, Madrepora vermiculosa,* now *Dendrophyllia vermiculosa :* and from St. Matthews Parish, Orangeburg District, South Carolina, *Nucula calcar ensis, N. carolinensis, Cardita bilineata, C. carolinensis, C. vigintinaria, C. subquadrata, C. subrotunda, Turbo biliratus. Cerithium siliceum, C. bicostellatum, Infundibulum carinatum, Tellina subæqualis, Madrepora punctulata, Nautilopsis canuxemi.* From the

Miocene of Suffolk and Yorktown, Virginia, and other places, *Eulima eborea, E. migrans, Odostomia limnia, O. protexta, Delphinula arenosa,* now *Angaria arenosa, Bulla subspinosa.* From the Eocene of the Southern States,* *Kellia oblonga, Tellina perovata, Cytherea lenis, Nucula impressa,* now *Yoldia impressa, N. claibornensis, N. parilis, Lithodomus claibornensis,* now *Lithophagus claibornensis, Cerithium claibornense, Amphidesma peroratum,* and *Psammobia mississippiensis,* now *Gari mississippiensis.* From the Columbia river,† near Astoria, *Nucula abrupta, N. cuneiformis, N. divaricata, N. penita, Mactra albaria, Tellina oregonensis, T. obruta, Loripes parilis, Solen curtus, Cytherea oregonensis, C. respertina, Bulinus petrosus,* now *Cylichna petrosa, Pyrula modesta,* and *Fusus oregonensis.*

Dr. Joseph Leidy‡ described, from the Miocene of Nebraska and the west, *Pœbrotherium wilsoni,* and *Merycoidodon culbertsoni,* now *Oreodon culbertsoni.* Dr. S. G. Morton described, from the Eocene of Washington county, Alabama, *Cidaris alabamensis,* and *Galerites agassizi.* And Dr. Robert W. Gibbes described, from the Eocene of South Carolina, *Carcharodon mortoni, C. acutidens, C. lanciformis, Oxyrhina sillimani, Otodus lævis,* and *Glyphis subulata.*

In 1849, T. A. Conrad§ described, from the Upper Eocene of Vicksburg, Mississippi, *Clavella vicksburgensis,* now *Fasciolaria vicksburgensis, Fulgur nodulatum,* and *Triton subalveatus.* And Robert W. Gibbes described, from the Eocene of South Carolina, *Galeocerdo contortus,* and *Oxyrhina wilsoni.*

In 1850, W. E. Logan‖ said that in the valleys of the Gouffre and the Murray Bay rivers, as well as along the margin of the St. Lawrence between them, there are, at various parts, great accumulations of clay and sand, with some gravel; and it is very perceptible that while they often present a confused aggregation of hummocks in the lower grounds, at higher levels, lying in horizontal beds, they are arranged into a succession of opposite terraces of equal height along the sides of the valleys, and corresponding terraces at intervals along the St. Lawrence, all probably marking ancient beaches or periods of retrocession of a Tertiary sea by the elevation of the land. One of these terraces, in the valley of the Gouffre, has a height, as indicated by a spirit level, of 130 feet above the Bay St. Paul, and another has a height of 360 feet.

---

* Jour. Acad. Nat. Sci., 2d ser., vol. i.
† Am. Jour. Sci. and Arts, 2d ser., vol. v.
‡ Proc. Acad. Nat. Sci., vols. iii. and iv.
§ Jour. Acad. Nat. Sci., 2d ser., vol. i.
‖ Geo. Sur. of Canada.

The deposits in which these terraces have been worn consist of clay, containing marine shells, among which are *Tellina groenlandica*, *T. calcarea*, *Saxicava rugosa*, *Nucula*, *Venus*, *Mytilus*, and *Balanus*. These shells were found as high as 390 feet above the bay. At Little Malbaie there are six terraces, plainly distinguishable, one above another.

T. A. Conrad[*] described, from the Eocene of Georgia, *Mitra georgiana*, *Catopygus conradi*, now *Cassidulus conradi*, *Holaster mortoni*, *Nucleolites lyelli*, *Discoidea haldemani*, and *Cidarites mortoni*. Robert W. Gibbes described, from the Eocene of Ashley river, *Myliobates holmesi*. And Zadock Thompson[+] described, from the drift in Vermont, exposed in excavating for the Rutland and Burlington railroad, *Delphinus vermontanus*, now *Beluga vermontana*.

In 1851, Philip T. Tyson[‡] described the Sacramento Valley as a long prairie, occupying the space between the flanks of the Sierra Nevada and those of the Coast Range, closed in on the north by the terminal spurs of the Cascade mountains, and on the south by the junction of the Coast Range with the Sierra Nevada. Its greatest width is less than 60 miles, but it maintains a mean width of nearly 50 miles throughout almost its entire length. The surface strata are not older than the Eocene or Miocene, and rest immediately upon the metamorphic and hypogene rocks.

Prof. James Robb[§] showed the direction of the Drift striæ in New Brunswick to be, generally, about 10 deg. W. of true north to 10 deg. E. of south, but that some striæ have a direction N. 30 deg. E. Others N. 45 deg. W., and still others east and west.

T. T. Bouve[||] described, from the Eocene of Georgia, *Catopygus patelliformis*, now *Cassidulus patelliformis*, and *Hemiaster conradi*.

In 1852, Mr. J. Evans[¶] explored that region of the Upper Missouri country, lying high up on White river, called the "*Mauvaises Terres*," or "Bad Lands." He said that from the high prairies, which rise in the back, by a series of terraces or benches toward the spurs of the Rocky Mountains, the traveler looks down into an extensive valley, that may be said to constitute a world of its own, and which appears to have been formed partly by an extensive vertical fault, and partly by the long continued influence of the scooping action of denudation.

---

* Jour. Acad. Nat. Sci., 2d ser., vol. ii.
+ Am. Jour. Sci. and Arts, 2d ser., vol. ix.
‡ Geo. and Ind. Resources of Cal.
§ Proc. Am. Ass. Ad. Sci., 4th Meeting.
|| Proc. Bost. Soc. Nat. Hist., vol. iv.
¶ Geo. Sur., Wis., Iowa and Minn.

The width of this valley may be about 30 miles, and its whole length about 90, as it stretches away westwardly toward the base of the gloomy and dark range of mountains known as the Black Hills. Its most depressed portion, 300 feet below the general level of the surrounding country, is clothed with scanty grasses, and covered by a soil similar to that of the higher ground.

To the surrounding country, however, the Mauvaises Terres present the most striking contrast. From the uniform, monotonous, open prairie, the traveler suddenly descends, one or two hundred feet, into a valley that looks as if it had sunk away from the surrounding world, leaving, standing all over it, thousands of abrupt, irregular, prismatic, and columnar masses, frequently capped with irregular pyramids, and stretching up to a height of from one to two hundred feet or more.

So thickly are these natural towers studded over the surface of this extraordinary region, that the traveler threads his way through deep, confined, labyrinthine passages, not unlike the narrow, irregular streets and lanes of some quaint, old town of the European continent. Viewed in the distance, indeed, these rocky piles, in their endless succession, assume the appearance of massive, artificial structures, decked out with all the accessories of buttress and turret, arched doorway and clustered shaft, pinnacle, and finial and tapering spire.

One might almost imagine oneself approaching some magnificent city of the dead, where the labor and the genius of forgotten nations had left behind them a multitude of monuments of art and skill.

On descending from the heights, however, and proceeding to thread this vast labyrinth, and inspect, in detail, its deep, intricate recesses, the realities of the scene soon dissipate the delusions of the distance. The castellated forms, which fancy had conjured up have vanished; and around one, on every side, is bleak and barren desolation.

Then, too, if the exploration be made in midsummer, the scorching rays of the sun, pouring down in the hundred defiles that conduct the wayfarer through this pathless waste, are reflected back from the white or ash-colored walls that rise around, unmitigated by a breath of air, or the shelter of a solitary shrub.

The drooping spirits of the scorched geologist are not permitted, however, to flag. The fossil treasures of the way, well repay its sultriness and fatigue. At every step, objects of the highest interest present themselves. Embedded in the debris, lie strewn, in the greatest profusion, organic relics of extinct animals. All speak of a vast fresh water deposit of the early Tertiary period, and disclose the former existence of most remarkable races that roamed about in bygone ages high up in the valley of the Missouri, toward the source of its western tributaries, where now pasture the big-jhorned *Ovis montana*, the shaggy buffalo, or American bison, and the elegant and slenderly constructed antelope.

A section of the Tertiary of the "Bad lands," or, "Mauvaises Terres," in descending order, is as follows : 1. Ash colored clay, cracking in the sun, containing siliceous concretions, 30 feet. 2. Compact, white limestone, 3 feet. 3. Light gray, marly limestone, 8 feet. 4. Light gray, indurated, siliceous clay (not effervescent), 30 feet. 5. Aggregate of small angular grains of quartz, or conglomerate, cemented by calcareous earth, slightly effervescent, 8 feet. 6. Layer of quartz and chalcedony (probably only partial), 1 inch. 7. Light gray, indurated, siliceous clay, similar to No. 4, but more calcareous, passing downward into pale flesh colored, indurated, siliceous, marly limestone, turtle and bone bed, 25 feet. 8. White and light gray, calcareous grit, slightly effervescent, 15 feet. 9. Similar aggregate to No. 5, but coarser, 8 feet. 10. Light green, indurated, argillaceous stratum (slightly effervescent); Palæotherium bed, 20 feet.

Dr. Joseph Leidy described, from the Eocene of Nebraska, *Eucrotaphus auritus*, and from the Miocene of Virginia,[*] *Crocodilus antiquus*, now *Thecachampsa antiquus*. Prof. F. Unger[†] described, from the Tertiary of Texas, *Sillimania texana*, *Rœmeria americana*, and *Thuioxylon americanum*.

In 1853, Alexander Murray[‡] informed us that the clays on the Ottawa, in the vicinity of Bytown, at the mouth of the Gatineau on the north, and of Green's creek on the south side, in addition to marine shells, yield, in the latter locality, two species of fish, the *Mallotus villosus*, and *Cyclopterus lumpus*, or lump-sucker, the former now living and frequenting the Gulf of St. Lawrence in great numbers, and the latter abounding on the northern shores of Scotland and America. The fossils are enclosed in nodules of indurated clay of reniform shapes, and occupy a bed nearly on a level with the water of the Ottawa, and about 118 feet above the tide level of Lake St. Peter. The same sort of nodules frequently enclose fragments of wood, leaves of trees, and portions of marine plants ; among the last is one of the species of littoral algæ still found near the coasts of Arctic seas. Beside the stratified deposits of clay and sand, there is a deposit of clay drift, holding pebbles and bowlders, sometimes angular, but generally

---

[*] Jour. Acad. Nat. Sci., 2d ser., vol. ii.

[†] Kreid. von Texas.

[‡] Geo. Sur. of Canada.

rounded, showing no decided lines of stratification, but irregularly associated with isolated beds of gravel and sand, among which great quantities of marine shells of comparatively recent origin occur. One of these localities is on the Prescott Road, about a mile and a half from Kemptville, where a vast accumulation of *Tellina grœnlandica* overlays a two feet bed of limestone gravel, the latter resting on gravel of a still coarser quality, and of more angular fragments, and irregularly mixed up with sand and clay, some of the bowlders being from 6 to 10 inches in diameter. The height of this locality is about 350 feet over Lake St. Peter. At another locality, near Armstrong's Mills, the shells consist chiefly of *Saxicava rugosa*, mixed with sand and loam, at a height of about 300 feet above Lake St. Peter. In Kenyon, on the Garry river, these shells occur at the height of 270 feet above Lake St. Peter. On the road between the 5th and 6th concessions of the township, on the 19th and the 21st lots, these shells occur at the height of 330 or 340 feet above Lake St. Peter. Two localities occur in Lochiel, one of them on the 15th lot of the 1st concession, at the height of 264 feet, and the other on the 5th lot of the same concession, 280 or 290 feet above Lake St. Peter, where the marine shells are mixed with the sand, and where bowlders and fragments of limestone and sandstone abound.

Prof. Edward Hitchcock* described the brown coal deposit in Brandon, Vermont, and referred it to the Pliocene or Newer Tertiary. He found it abounding in fruits and lignites, which appear to have been transported by water, and probably accumulated in an ancient estuary. It abounds in white and variegated clays, water-worn beds of sand and gravel, beds of carbonaceous matter not bituminous, and deposits of iron and manganese.

T. A. Conrad† described, from the Miocene of Upper California, *Gnathodon lecontei*, now *Rangia lecontei*, and *Ostrea vespertina.*

In 1854, Dr. Leidy‡ described, from the Post-pliocene, of Ashley river, South Carolina, *Arctodus pristinus*, from Kansas, *Camelops kansanus ;* and from the mouth of Pigeon creek, below Evansville, Ind., *Canis primævus*, now *C. indianensis.* From the Pliocene, on Bijou Hill, east of the Missouri river, *Hippodon speciosus*, now *Hippotherium speciosum*, and *Merycodus necatus*, now *Cosoryx necatus;* from the Miocene of Nebraska, *Deinictis felina.*

Evans and Shumard described, from the Tertiary (White Riv. Gr.),

* Am. Jour. Sci. & Arts, 2d ser., vol. xv.
† Jour Acad. Nat. Sci., 2d ser., vol. ii.
‡ Proc. Acad. Nat. Sci., vol. vii.

in Nebraska, in the vicinity of Peno creek, a small tributary of Teton, or Little Missouri river, in a thin bedded, light gray, siliceous limestone, near the summit of the elevated plateaux which border the Mauvaises Terres, *Planorbis nebrascensis, Limnœa diaphana, L. nebrascensis, Physa secalina,* and *Cypris leidyi.*

T. A. Conrad* described, from the (Jackson Group) Greensand Marl-bed of Jackson, Mississippi. *Astarte parilis, Umbrella planulata, Corbula bicarinata, C. densata, Leda multilineata,* now *Nuculana multilineata, Navicula aspera, Crassatella flexura, Glossus filosus,* now *Axinœa filosa, Ostrea trigonalis, Pecten nuperus, Capulus americanus, Clavelithes humerosus, C. varicosus, C. mississippiensis,* now *Papillina mississippiensis, Trochita alta, Mitra dumosa,* now *Lapparia dumosa, Conus tortilis, Volutilithes symmetricus, V. dumosus, Rostellaria velata,* now *Calyptrophorus velatus, R. Staminea,* now *C. stamineus, Caricella subangulata, C. polita, Natica permunda, Rostellaria extenta,* now *Platyoptera extenta, Mitra millingtoni,* now *Fusimitra millingtoni, Teredo mississippiensis, Architectonica acuta, A. bellistriata, Cyprœa pinguis, C. fenestralis, Gastridium vetustum, Phorus reclusus, Turritella alveata, Galeodia petersoni,* and *Strepsidura dumosa.*

Dr. J. W. Dawson,† describing the drift of Nova Scotia, in 1855, said, that in the low country of Cumberland there are few bowlders, but of the few that appear, some belong to the hard rocks of the Cobequid hills to the southward; others may have been derived from the somewhat similar hills of New Brunswick. On the summits of the Cobequid hills, and their northern slopes, we find angular fragments of the sandstones of the plain below, not only drifted from their original sites, but elevated several hundreds of feet above them. To the southward and eastward of the Cobequids, throughout Colchester, Northern Hants, and Pictou, fragments from these hills, usually much rounded, are the most abundant traveled bowlders, showing that there has been great driftage from this elevated tract. In like manner, the long ridge of trap rocks, extending from Cape Blomidon to Briar Island, has sent off great quantities of bowlders across the sandstone valley which bounds it on the south, and up the slopes of the slate and granite hills to the southward of this valley. Well characterized fragments of trap from Blomidon may be seen near the town of Windsor, and unmistakable fragments of similar rock from Digby neck, on the Tusket river, may be seen, thirty miles from their original position. On the

---

* Wailes' Geo. of Miss.
† Acadian Geology.

other hand, numerous bowlders of granite have been carried to the northward from the hills of Annapolis, and deposited on the slopes of the opposite trappean ridge; and some of them have been carried round its eastern end, and now lie on the shores of Londonderry and Onslow. So, also, while immense numbers of bowlders have been scattered over the south coast from the granite and quartz rock ridges, immediately inland, many have drifted in the opposite direction, and may be found scattered over the counties of Sydney, Pictou and Colchester. These facts show that the transport of traveled blocks, though it may have been principally from the northward, has, by no means been exclusively so; bowlders having been carried in various directions, and more especially from the more elevated and rocky districts to the lower grounds in their vicinity.

The surface of the country was greatly modified by the drift; the ridges of Cumberland, the deep valleys of Cornwallis and Annapolis, the great gorges crossing the Cobequid mountains, and the western end of the North mountains in Annapolis and Digby counties, such eminences as the Greenhill in Pictou county, and Onslow mountain in Colchester, are due in great part to the removal of soft rocks by denuding agencies of this period, while the harder rocks remained in projecting ridges. The surface of the rocks are frequently found polished, scratched or striated. The striæ at different places have different courses, and sometimes they are found to cross each other as at Gore mountain, where one set is S. 65 deg. E., and the other S. 20 deg. E. At Gay's river, Musquodoboit Harbor, and near Guysboro the direction is from S. to N. At Polson's Lake, from N. to S., and near Pictou, E. & W. Bowlders or traveled stones are often found in places where there is no other drift. For example, on bare granite hills, about 500 feet in height, near the St. Mary's river, there are large, angular blocks of quartzite, derived from the ridges of that material which abound in the district, but are separated from the hills on which the fragments lie by deep valleys.

The only evidence of organic life during the bowlder period, or immediately before it, noticed by Dr. Dawson, consists of a hardened, peaty bed, which appears under the bowlder clay on the northwest arm of the River of Inhabitants. It rests upon gray clay, similar to that which underlies peat bogs, and is overlaid by nearly twenty feet of bowlder clay. Pressure has rendered it nearly as hard as coal, though it is somewhat tougher and more earthy than good coal. It has a glossy appearance when rubbed or scratched with a knife, burns with considerable flame, and approaches in its characters to the brown coals, or more imperfect varieties of bituminous coal. It contains many small roots and branches, apparently of coniferous trees allied to the spruces. The vegetable matter composing this bed, must have flourished before the drift was spread over the province, so that it belongs to some part of the great Tertiary group of rocks, of which the drift is the latest member.

Dr. Dawson accounted for the drift phenomena of Nova Scotia, in this manner. Let us suppose the surface of the province, while its projecting rocks were uncovered by surface deposits, exposed for many successive centuries to the action of alternate frosts and thaws—the whole of the untraveled drift might have been accumulated on its surface. Let it then be slowly submerged, until its hill-tops should become islands or reefs of rocks in a sea loaded in winter and spring with drift ice, floated along by currents, which, like the present Arctic current, would set from N. E. to S. W., with various modifications produced by local causes. We have, in these causes, ample means for accounting for the whole of the appearances, including the traveled blocks and the scratched and polished rock surfaces.

The *stratified sand and gravel* rests upon and is newer than the unstratified drift. This may often be seen in coast sections or river banks, and occasionally in road-cuttings. In Pictou county there occurs a very thick bed of conglomerate, of the age of the Coal Measures, the outcrop of which, owing to its comparative hardness and great mass, forms a high ridge extending from the hill behind New Glasgow, across the East and Middle rivers, and along the south side of the West river, and then crossing the West river reappears in Rogers Hill. The valleys of these three rivers have been cut through this bed, and the material thus removed has been heaped up in hillocks and beds of gravel, along the sides of the streams, on the side toward which the water now flows, which happens to be the north and northeast. Accordingly, along the course of the Albion Mines railway, and the lower parts of the Middle and West rivers, these gravel beds are everywhere exposed in the road-cuttings, and may in some places be seen to rest on the bowlder-clay, showing that the cutting of these valleys was completed after the drift was produced. The stratified gravels do not, like the older drift, form a continuous sheet spreading over the surface. They occur in mounds, and long ridges, sometimes extending for miles over the country. They are supposed to have been distributed when the country was being elevated, while the bowlder drift was deposited when the land was subsiding beneath the sea.

T. A. Conrad separated the Eocene of Mississippi and Alabama, in ascending order, into : 1. Claiborne group, characterized by *Cardita densata*, *Ostrea selliformis*, *Crassatella alta*, *Pectunculus stamineus*, *Meretrix æquorea*, *Gratelupia hydi*, *Leda cælata*, and *Crepidula lirata*. 2. Jackson group, characterized by *Umbrella planulata*, *Cardium nicollettii*, *Conus tortillis*, *Cyprœa fenestralis*, *Galeodia petersoni*, and *Rostellaria extenta*. 3. St. Stephen's group, characterized by *Pecten poulsoni*, and *Orbituliies mantelli*. 4. Vicksburg group, characterized by *Corbula alta*, *Crassatella mississippiensis*, *Arca mississippiensis*, *Meretrix sobrina*, *M. imitabilis*, and *Turbinella wilsoni*. He described,* from Jackson, Miss., and Claiborne, Ala., *Endopachys expansum*, *E. triangulare*, *E. alticostatum*, *Flabellum wailesi*, *Osteodes irroratus*, *Turbinolia lunulitiformis*, *Chiton antiquus*, *C. eocenensis ;* and from White river, Arkansas, *Petrophyllia arkansasensis;* from the Miocene, of Colorado and the West, *Anomia subcostata*, *Mercenaria perlaminosa*, *Pecten heermani*, *Pandora bilirata*, *Astarte thomasi*, *Turritella secta*, *Ostrea contracta*, and *Idmonea californica*.

Prof. Wm. P. Blake noticed Miocene strata, containing the remains of Infusoria and Polythalamia, near Monterey, California. The strata are white, porous, light, resemble chalk, and are situated about two miles southeast from the center of Monterey, and form part of a hill which fronts the bay, and rises on the east side of the stage-road to San Francisco to the height of 500 or 600 feet.

Tuomey and Holmes† described, from the Pliocene of South Carolina, *Cellepora formosa*, *C. depressa*, *C. radiata*, *C. tessellata*, *Membranipora lacinia*, *Placunanomia plicata*, *Ostrea ravenelana*, *Janira affinis*, *Pecten comparilis*, *P. peedeensis*, *Mytilus inflatus*, *Arca hians*, *A. rustica*, *Pectunculus lævis*, *Lucina costata*, *Crassatella gibbesi*, *Psammocola pliocena*, *Dentalium pliocenum*, *Hipponix bulli*, *Monodonta kiawahensis*, *Trochus armillatus*, *T. gemma*, *Terebellum striatum*, *T. burdeni*, *T. etiwanensis*, *Voluta trenholmi*, *Porcellana oliviformis*, *Purpura tridentata*, *Cancellaria depressa*, *C. venusta*, *Busycon conradi*, *Cassidulus carolinensis*, and *Fasciolaria tuomeyi*.

J. McCrady described, from the same strata, *Psammechinus excoletus*, *Agassizia porifera*, *Amphidetus ampliflorus*, *A. gothicus*, *Brissus spatiosus*, *Plagionotus holmesi*, and *P. ravenelanus*.

Dr. Trask‡ described, from the Pliocene of Santa Barbara, California, *Chemnitzia papillosa*, *Tornatella elliptica*, *Murex fragilis*, *Fusus barbarensis*, *F. robustus*, and *F. rugosus*.

---

* Proc. Acad. Nat. Sci., vol. vii.
† Tuomey and Holmes' Fossils of South Carolina.
‡ Proc. Cal. Acad. Sci., vol. i.

In 1856, W. P. Blake* described the Tertiary rocks of the vicinity of San Francisco, California. They consist of fine-grained, compact sandstone, associated with shales, and underlie the city of San Francisco, and are exposed along the shores of the bay, both north and south of the city, forming the principal promontories and points, and several islands. On entering the bay from the Pacific, they are first seen at Point Lobos, the outer point, and again at North and Tonquin points. They border part of the Golden Gate on the north, and form the shores of Richardson and Saucelito bays. Angel, Yerba Buena and Alcatrazes Islands, are of the same age. In some places, hills and ridges of 200 or 300 feet in elevation are formed entirely of this sandstone. Rocks of the same age are found at Benicia, New Almaden, and between San Juan and Monterey.

On the south end of the Island of Yerba Buena, a section, 200 feet thick, shows the sandstone layers, varying from a few inches to six and eight feet, and alternating with beds of argillaceous slates and shales. All the weathered surfaces of this series of beds are of a rusty-brown or drab color, which extends throughout the rock to a depth of from ten to twenty feet, down to the limit of atmospheric influences. There are parts, however, of the upper beds that have not been reached and changed by decomposition; these parts are found in the condition of spherical or ellipsoidal masses, from which the weathered parts scale off in successive crusts. These nuclei have the appearance of great, rounded bowlders, and have accumulated, in great numbers, at the base of the cliff. They are of various sizes, but are smallest in the upper parts of the strata, near to the surface.

This spherical or globular condition does not appear to be the result of any peculiar arrangement of the material of the strata, a concretionary action, such as takes place in the igneous rocks, but is probably due to decomposition, the result of the absorption of infiltrating waters charged with impurities. A solid and homogeneous cube of sandstone thus exposed, under conditions favorable for absorption of the water on all its sides, would decompose most rapidly on the angles, producing a succession of curved surfaces gradually approaching a sphere.

The color of the sandstone is dark, bluish green, inclining to gray. It is exceedingly compact and tough, and does not break so readily as the fine-grained, red sandstone of the Connecticut river and New Jersey quarries.

---

* Explorations and Surveys for a railroad from the Mississippi river to the Pacific ocean, vol. v.

A section at Navy Point, Benicia, exposes a thickness of conformable beds of sandstone conglomerate and shales a little more than 1,000 feet in thickness. The strata are uplifted, being inclined at an angle of from 20 to 60 deg., and dipping toward the southwest. The trend of the outcrops is 75 deg. west of north, and the strata underlie, or rather form the hill upon which the government buildings are erected. The ridge of conglomerate is the hardest and most unyielding of all the strata, and its resistance to abrasion and atmospheric influences has determined the form of the hill and the shape of Navy Point. It is prominent at several points, along the surface of the ground, and is almost the only rock that appears above the soil in that vicinity. The bed is about twenty-five feet thick, and is composed of pebbles and gravel, very round, much water-worn, and chiefly derived from the wear of volcanic or eruptive rocks. Their colors are generally dark; and porphyries, agates and carnelians are abundant. Their average diameter does not exceed an inch, and many are about the size of beans and peas. They are closely united by a small portion of finer materials. The strata on both sides of the conglomerate consist of alternate beds of soft and friable argillaceous shales, with an occasional layer of gravel and pebbles.

The wide development of the formation, and the great thickness which it attains—probably 2,000 or 3,000 feet—and the even grain of the thick beds of sandstone over large areas, together with the remarkable uniformity of the strata, indicate that they were formed in a wide spread ocean or sea, and the thick beds of shale attest the depth and comparative quiescence of the water.

He found the Miocene rocks extending in a continuous belt along the base of the Sierra Nevada, from White creek to Ocoya creek and beyond it for many miles to the southward, forming high banks on both sides of Posuncula or Kern river, and even extending in a narrow strip to the Tejon.

Although by far the greater portion of the materials composing the formation are extremely light, fine and unconsolidated, there are, in some places, layers of sandstone and conglomerate, which offer more resistance to the action of the weather than the other strata, and that slightly modify the rounded contour of the hill sides. The principal constituent of the formation is a fine gray sand, mingled, in some of the beds, with a considerable portion of clay, and alternating with layers in which clay predominates. Volcanic materials, or sands derived from their abrasion, constitute a large part of the strata. Thick beds are formed almost wholly of white pumice stone, in rounded masses, or in a fine powder, like fine sand, regularly bedded. The color of these beds is white, but the lines of stratification are rendered very distinct by the stains produced by the percolation of impure waters; also, by layers of the same ingredients, differing in their fineness, and by occasional seams of charcoal, in fragments. Thin layers of pebbles are also numerous, even among the strata of the finest materials. The inclined stratification, called diagonal stratification, is very common, and in many cases is beautifully shown by multitudes of the finest layers of sand, inclined in different directions.

He also identified this Group on Chico creek, in the valley of the Sacramento, at the foot of the Hills of the Sierra Nevada, on Carrizo creek, near San Diego, at Williamson's Pass, Los Angeles and San Pedro. Near Monterey it contains a bed of microscopic organisms, 50 feet in thickness ; and he supposed it to underlie the alluvium of the Colorado desert.

He described the Post-pliocene deposits of Monterey, San Pedro, and San Diego, and showed a comparatively recent elevation of the strata. The low hills around the bases of the mountains in the Colorado desert, and the elevation of the Coast Mountains, he supposed to be of the same age, because they are composed in great part of Tertiary strata, thrown into great wave-like flexures, with here and there a granitic axis of limited extent, but with serpentine abundant. In the auriferous regions, a similar serpentine abounds, and has in all cases the aspect of an intrusive rock. The movements which attended the uplift and plication of the Coast Mountains, must have affected the whole western slope of the Sierra Nevada. He expressed the opinion that the impregnation of the rocks with gold, and the formation of the Coast Mountains, were nearly synchronous.

He described, from the Miocene, at Point Lobos, near San Francisco, *Scutella interlineata*, and from a brown calcareous sandstone at Volcano Ridge, *Leda subacuta*, now *Nuculana subacuta*. Prof. Agassiz described, from Ocoya creek, at the western base of the Sierra Nevada, *Echinorhinus blakei, Scymnus occidentalis, Galeocerdo productus, Prionodon antiquus, Hemipristis heteropleurus, Carcharodon rectus, Oxyrhina plana, O. tumula, Lamna clavata,* and *L. ornata.* T. A. Conrad described, from the same locality, *Natica geniculata, N. ocoyana, Bulla jugularis, Pleurotoma transmontanum, Sycotypus ocoyanus, Turritella ocoyana, Colus arctatus, Tellina ocoyana, Meretrix decisa, Pecten nevadensis, P. catilliformis ;* from the San Diego Mission, *Cardium modestum, Corbula diegoana, Nucula decisa, Tellina diegoana, T. congesta, Mactra diegoana, Natica diegoana,* now *Vanikoro diegoana, Crucibulum spinosum, Trochita diegoana;* from Monterey county, eighteen miles south of Tres Pinos, *Meretrix uniomeris;* from the Tulare valley, *Meretrix tularana, Arca microdonta, Purpura petrosa;* from Carmello, *Lutraria traskei;* from sixteen miles south of Tres Pinos, *Modiola contracta,* now *Volsella contracta;* from Carrizo Creek, *Pecten deserti,* and *Ostrea heermanni.*

He described, from the Post-pliocene at San Pedro, *Tellina pedroana, Saxicava abrupta, Petricola pedroana, Schizothærus nuttalli, Mytilus pedroanus, Penitella spelæa, Nassa interstriata, N. pedroana, Oliva pedroana, Littorina pedroana,* and from Santa Barbara, *Crepidula princeps.*

W. P. Blake estimated the thickness of the Eocene, at the southern end of the Tulare valley, at 2,000 feet. The strata are chiefly argillaceous sandstone. T. A. Conrad described, from the Canada de las Uvas, *Cardium linteum,* now *Cymbophora lintea, Dosinia alta, Meretrix californiana, M. uvasana, Crassatella alta, C. uvasana, Mytilus humerus, Natica alveata,* now *Ampullina alveata, Turritella uvasana, Volutilithes californianus, Busycon blakei,* now *Perissolax blakei,* and *Clavalula californica.* It is quite likely that part or all of the strata referred to the Eocene from which these fossils were collected, belong to the Cretaceous.

He described,[*] from the Miocene, at Santa Barbara, California, *Janira bella, Mulinia densata, Arca canalis, A. trilineata, Axinæa barbarensis;* from Santa Clara, *Schizopyga californiana;* from Estrella valley, *Pallium estrellanum,* now *Lyropecten estrellanum, Spondylus estrellensis;* from Monterey county, *Pallium crassicordo,* now *Lyropecten crassicordo, Thracia mactropsis, Mya montereyana, Arcopagia medialis, Cryptomya ovalis, Cyclas tetrica, Dosinia alta, D. longula, Tamiosoma gregaria, Astrodapsis antiselli;* from San Raphael Hills, *Pecten meeki, P. altiplicatus;* from Santa Inez Mountains, *Pachydesma inezanum;* from Ranche Triumpho, near Los Angelos, *Lutraria transmontana;* from other places in California, *Arca congesta, Tapes linteatum;* and from Texas, *Mellita texana.*

Dr. Leidy described, from the Bad Lands of Nebraska, *Hipparion occidentale,* now *Hippotherium occidentale, Hyopotamus americanus, Leptauchenia decora, L. major, Leptochærus spectabilis, Steneofiber nebrascensis, Ischyromys typus, Palæolagus haydeni, Eumys elegans, Amphicyon gracilis, Agriochærus major, Enteledon ingens,* now *Elotherium ingens, Palæochærus probus,* now *Perchærus probus;*

---

* Proc. Acad. Nat. Sci., vol. viii.

from Bear creek, *Protomeryx halli;* from the Pliocene of Ashley river, South Carolina, and from the Miocene of New Jersey and Virginia, *Manatus antiquus, Phoca debilis;* from the Miocene of Cumberland county, Md., *Macrophoca atlantica,* now *Squalodon atlanticus, Sphyræna speciosa;* from North Carolina, *Orycterocetus cornutidens, Pliogonodon priscus;* from Salem county, New Jersey, *Chelonia grandæva;* from the Eocene of the Neuse river, North Carolina, *Ischyrhiza antiqua;* from Green river, Missouri, *Clupea humilis,* now *Diplomystus humilis;* from the Upper Tertiary of the Bijou Hills, on the Upper Missouri, *Merychippus insignis,* now *Protohippus insignis* and *Leptarctus primus.*

In 1857, Dr F. V. Hayden* made an estimated vertical section, showing the order of superposition of the different beds of the Bad Lands of White river, in Nebraska, referred to the Miocene, in ascending order as follows:

Bed A.—Light gray, calcareous grit, passing down into a stratum composed of an aggregate of rather coarse, granular quartz; underlaid by an ash-colored, argillaceous, indurated bed, with a greenish tinge. *Titanotherium* bed. Best developed at the entrance of the Basin from Bear creek. Seen also in the channel of White river. Thickness, 50 feet.

Bed B.—A reddish, flesh-colored, argillo-calcareous, indurated material, passing down into a gray color, containing concretionary sandstone, sometimes an aggregate of angular grains of quartz, underlaid by a flesh-colored, argillo calcareous, indurated stratum, containing a profusion of mammalian and chelonian remains. Turtle and Oreodon bed. Revealed on both sides of White river and throughout the main body of the Bad Lands. Thickness, 80 feet.

Bed C.—Light gray, siliceous grit, sometimes forming a compact, fine-grained sandstone. Seen on both sides of White river. Also at Ash Grove Spring. Thickness, 20 feet.

Bed D.—Yellow and light yellow, calcareous marl, with argillo-calcareous concretions, and slabs of siliceous limestone, containing well-preserved fresh-water shells. On the south side of White river. Seen in its greatest thickness at Pina's Spring. Thickness, 40 feet.

Bed E.—Yellowish and flesh colored, indurated argillo-calcareous bed, with tough argillo-calcareous concretions, containing *Testudo, Hipparion, Steneofiber, Oreodon* and *Rhinoceros.* Seen along the White river valley, on the south side. Thickness, 30 feet.

---

* Proc. Acad. Nat. Sci., vol. ix.

Bed F.—Grayish and light gray, rather coarse-grained sandstone, with much sulphate of alumina (?) disseminated through it. Along White river valley, on the south side. Thickness, 20 feet.

Bed G.—Yellowish-gray grit, passing down into a yellow and light yellow argillo-calcareous marl, with numerous calcareous concretions, and much crystalline material, like sulphate of baryta. Fossils : *Hipparion*, *Merychippus* and *Steneofiber*. Bijou Hills, Medicine Hills, Eagle Nest Hills, and numerous localities on south side of White river, also at the head of Teton river. Thickness, 50 feet.

Bed H.—Gray and greenish-gray sandstone, varying from a very fine compact structure to a conglomerate. Bijou Hills, Medicine Hills, and Eagle Nest Hills. Thickness, 20 feet.

T. A. Conrad described,[*] from the Miocene, in Monterey county, California, *Mya subsinuata ;* from San Pablo bay, *Pecten pabloensis ;* from Santa Margarita, Salinas valley, California,[†] *Hinnites crassus ;* from between La Purissima and Santa Inez, *Pecten discus ;* from Santa Inez mountains, *Pecten magnolia*, *Tapes inezensis*, *Crassatella collina*, *Mytilus inezensis*, *Turritella inezana*, *T. variata*, *Natica inezana ;* from Estrella valley, *Cyclas estrellana*, *Ostrea panzana*, *Glycimeris estrellanus*, *Astrodapsis antiselli ;* from San Buenaventura, *Tapes montana*, *Perna montana ;* from Pajaro river, Santa Cruz, *Venus pajaroana ;* from the shore of Santa Barbara county, *Arcopagia unda;* from Sierra Monica. *Cyclas permacra*. *Ostrea subjecta ;* from San Luis Obispo valley, *Arca obispoana ;* from Salinas river, Monterey county. *Dosinia montana*, *D. subobliqua ;* and from Gaviote Pass, *Mactra gaviotensis*, and *Trochita costellata*.

He described[‡] from supposed Miocene strata at Rancho Helena, below Salado, *Ostrea veleniana*, and from Western Texas, in strata supposed to be Eocene, *Venus vespertina*. I am inclined to believe, however, that this is a Cretaceous species.

He described,[§] from the Eocene of Alabama, *Calyptrophorus trino-diferus*.

Dr. Thomas Antisell‖ made the following section of the Miocene strata of California, viz :

1. Upper Miocene, consisting of bituminous and foraminiferous beds, trappean conglomerate, soft, yellow sandstone, foraminiferous layers and argillite beds. Thickness, 400 feet.

* Expl. and Sur. R. R. Miss. Riv. to Pacific ocean, vol. vi.
† Expl. and Sur. R. R. Miss. Riv. to Pacific ocean, vol. vii.
‡ U. S. and Mex. Bound, Sur. vol. i.
§ Pro. Acad. Nat. Sci., vol. ix.
‖ Expl. and Sur. R. R. Miss. Riv. to Pacific ocean, vol. vii.

2. Middle Miocene, consisting of grits and calcareous sandstones, as at Panza and Santa Margarita. Thickness, 360 feet. And the San Antonia sandstones with Dosinia. Thickness, 250 feet.

3. Lower Miocene, consisting of the gypseous and ferruginous sandstones of Santa Inez, Panza, and Gavilan, containing *Ostrea*, and *Turritella*. Thickness, 1,200 feet. Total thickness of the Miocene, 2,211 feet, but part of this has since been referred to Eocene age.

He supposed that the elevation of the Coast Range, in California, above the water level, was an event much later in time than that of the Sierra Nevada. During the Eocene period, the latter range must have had its crest considerably above water, and was uplifted, finally, after the Miocene period ; but it is probable that during the whole of the Miocene period, the Coast Range was altogether beneath the sea level. Anterior to the Post-pliocene period, the erupted rock tilted up their strata, which, perhaps, did not reach the level of the ocean surface, and upon these smoothed edges, were deposited the unconsolidated clays and local drift. They had not, however, fully appeared above the surface of the ocean until the close of the Post-pliocene period. The elevated sea beaches found distributed over so large an extent of country, from north to south, at a level of from 100 to 150 feet above the sea, and containing species, all of which are now existing, show how comparatively recent is the final elevation of the lower lands of the State, and places the time of elevation of this range in the early portion of the Post-pliocene period. The plutonic rocks of the coast hills, also attest the comparative newness of the land : pumice, obsidian, felspathic lava, trachyte, amygdaloidal greenstone, and serpentine. Volcanic rocks, of the latest kind, are those which are commonly distributed both in the form of axes and veins, or seams. Granite is also found, though not so extensive as a disturbing agent, or an elevator of a mountain ridge. When found in place, it is an older rock than those above mentioned, being cut through and injected by them, in many places ; but the granite, in the Coast Mountains, is a modern granite, being either highly felspathic, passing into leucite, and even trachyte in many places, or it is hornblendic, and passes into a hornblende porphyry ; micaceous granite is very sparingly distributed in Southern California. The elevation of the Coast Range must have taken place from two points, one in the north, and one in the south ; the latter force commencing in the southern part of San Luis Obispo, and the eastern part of Santa Barbara counties, and thence extending north ; as the upheaving force passed northward, its power became spent, and unable to lift the imposed strata ; a similar action from the north, acting in a southerly direction with less vigor, produced an uplift, whose action ceased between latitude 37° and 38°. So that while the consolidated crust of the State was uplifted at each end, it was quiescent, or nearly so, in the middle ; and the two forces acting against each other may have produced a rupture of the superficial strata, and even a depression of the surface below the sea level, in which the waters of San Pablo, Suisun, and San Francisco, have taken their resting place.

Depressions of the strata and fissures from east to west across the line of the mountain ranges are common along the Pacific, north of this point, latitude 38 deg., and extend inland even east of the Sierra Nevada. In the course of these depressions rivers run. The Klamath and the Columbia are examples; which rivers might possibly never have emptied their waters into the Pacific, but for this fracturing effect produced by opposing volcanic forces.

The upheaval of the Coast Ranges have brought to view only Tertiary strata of the Miocene, and beds of clay of the Post-pliocene periods. These beds are thicker and more extensively distributed in a connected series than anywhere else on this continent. In this respect they rival or even excel the strata on the shores of the Mediterranean. It is interesting to trace the resemblance of form and outline of hills produced by similarity of geological circumstances, whether of formation or upheaval. Many of the scenes of California resemble those on the shores of northern Greece, Roumelia, northern Syria and the Calabrian peninsula.

There are no phenomena in California referable to the period of the polar drift or ancient alluvium, when the transport of large blocks or bowlders occurred. Over the extensive plains east of the Sierra Nevada, in Tulare valley, in the pleasant little oak valleys of the Coast Ranges, or on the terrace plains of the shore, not a single bowlder is to be met with—not a stone from which the plough might turn aside. This period, was, apparently, one of quiet in this State. Yet the mountain chains were elevated at this time. The topography was almost the same as at present, save the whole plain country was below the water level ; there were, therefore, elevated ranges from which the counties along the coast might have had scattered over their surface these blocks; but the Sierra Nevada has contributed no bowlders upon these plains, nor is there any stone included in the terraces which may not be classed as belonging to those ranges immediately bounding the deposit.

Not that the whole Post-pliocene epoch was passed without produc-ing its effects: denudation on an extensive scale, lacustrine deposits, immense deposits of clay, sands, and gravels attest the long period alike of action and of repose which characterize the later Post-pliocene period, when the effects were more local, and every valley and plain had its beds of gravel and clay formed from its mountain margins.

Considerations founded on the zoological characters of the mollusks of the Miocene period of Europe, have led to the belief that the tem-perature of that epoch approached very much to that of Spain and Italy at the present time, or a mean temperature about 66 deg. Fahren-heit. As that temperature is almost the exact figure for a great por-tion of the area observed, it follows that there is little, if any, difference between the climate of the Miocene of Europe, and the present period in those places; and since the drift of California is local, and not gen-eral, and there are no traces on the surface of rocks exposed, of scratch-ing or grooving, no moraines, no polished rocks (roches moutonnees), no traces of glacier action, perhaps it may be asserted with safety that the climate and temperature of this region, from the Miocene period to the present time, has preserved a constancy and equality, which latitudes more polar than 40 deg. never possessed.

Artesian boring through the Post-pliocene beds, in the Los Angelos valley, showed :

1. Alluvium, 6 feet.
2. Blue clay, 30 feet.
3. Drift gravel, 22 feet.
4. Arenaceous clay, 16 feet.
5. Tenaceous blue clay, 300 feet.

Such a thickness of deposit might be attributable to the local circum-stances, namely, a deep trough in the sandstone strata under an eleva-tion, almost vertical, close by ; yet that these incoherent beds are usually of great depth is evident from the smooth surface of the whole plain, which preserves its gradual slope from the Cordilleras to the ocean, independent of the dip or upheaval of the strata beneath. Again, when looking from the south entrance of the Cajon Pass toward San Bernardino, at an altitude of 2,000 feet, there may be perceived a broad terrace at the base of the mountain, consisting of loose conglomerates, gravel and clay beds, lying at an elevation nearly 200 feet above the present level of the plain in its neighborhood, and which are the only remains of a series of beds which have been removed from the lower and more exposed parts of the plain. Its average thickness, perhaps, might be about 200 feet; the other beds would preserve throughout a pretty uniform thickness ; of these, bed 4, an arenaceous, yellow clay, is described as containing small marine shells. The brownish loamy clay (bed 1), is exposed by every creek, and in the sections produced by the Los Angelos river, several feet of the bluish clay (2) are exposed ; the beds are deposited almost perfectly horizontal, and are, therefore, unconformable to the soft sandstones of the San Pedro hills and the Sierra Monica, which in the former case have a dip of 20 deg., and in the latter are in places almost vertical ; they have, therefore, been deposited posterior to the upheaval of these soft Tertiary sandstones, and the surfaces have undergone no material alteration of contour since, the only change being that of elevation of the whole region out of the bed of the sea.

An investigation into the mineral nature of these various deposits shows that the alluvial covering, to the depth of six or seven feet, is aluminous in its finer parts, and granitic pebble in its coarser, and has been the result of the degradation of granitic and felspathic rocks. The soil of the plain is rarely quartzose, except when close to some of the low Tertiary hills, which alteration may therefore be due to the wash of these latter.

The blue clay is generally assigned by geologists to a slow deposit of mud, produced by the sifting action of the tide in estuaries or gulfs where matter is not transported by current actions ; it is the evidence of a calm condition of the waters during the period of deposit, and a cessation of upheavals of the land contiguous ; the two beds of bluish clay are separated by nearly forty feet of gravel and sand.

The drift gravel (bed 3) consists not only of rounded granitic pebbles, but also those of syenite, hornblende schists, metamorphic, brown sandstones, trap and amygdaloid ; and the underlying sandy bed is chiefly quartzose, and probably is the detritus of the sandstones at the base of the Cordilleras.

There have been no very large stones seen in the drift beds, there are no loose bowlders or erratic blocks, nor is there, either on the surface or in the deposits, any stone which can not be traced to masses of similar mineral constitution in the ranges bordering the plain. The period of general or polar drift, therefore, which was one of the earliest of the Post-pliocene epoch, passed by without affecting California ; and it was during the later periods of drift that the processes of wearing down continents and depositing them in the seas around took place, and were carried out on an immense scale, and over an immensely extended period of time.

Los Angelos plain is not the only one in California, where these deposits of clays and gravels are of great depth. The borings which have been made in Sacramento and San Joaquin plains have revealed a similar structure of basin, while that in Santa Clara valley, Santa Clara county, shows that the deposit has not been to so great a depth in that plain. Thus, at the Stockton well boring, after passing through red clays, sands, and gravel, the blue clay was met with at the depth of 400 feet. On the Sacramento valley, between the city and Pit river, the lava clays and sands cover the blue clay to the depth of 358 feet. In the Santa Clara valley, the covering of clay and light sand above blue clay is from 80 to 115 feet.

Blue clays are found 465 feet below the surface at Los Angelos, and, therefore, below the present sea level; while the surface of the terrace on San Bernardino is somewhat above 2.000 feet in altitude, and as the beds are horizontal, or nearly so, it follows that near Los Angelos the deposit took place when the water was over 2.000 feet deep at that point. All the low Tertiary hills were ledges of rock, several hundred feet below low water. The ocean then rolled up east of the Cordilleras, occupying the Colorado desert and the Mohave valley; and the Cordilleras stood up like a peninsula in the great mass of waters, with its crests from 3,000 to 5,000 feet above the surface, and with a breadth not more than 60 miles from S. W. to N. E. From the wearing down of the felspathic rocks, the granitic porphyries, and the dark colored shales, arose the blue clays, while the trappean and hornblende rocks formed the material of the coarser drift, transported by currents produced by the elevations. The carriage of such coarse matter would inevitably remove large portions of the Tertiary hills of the plain, and form the breaks which now occur in what was once a continuous chain, the denuded matter itself going to form the bed of arenaceous clay.

It has been calculated that the deposit going on at present in the Gulf of Mexico, produced both by the alluvium of the Mississippi and the transported mud of the Amazon, does not exceed more than half an inch yearly. There is nothing in the topographical condition of Southern California to warrant the belief that the slow deposit could have occurred to a greater depth in the same space of time; for there is no evidence of the double influence of a large river and a strong current of sea water coinciding. Admitting, however, that the same rate of deposit occurred then as now in the two localities, the period of deposit of the lower blue clay bed would be 7,20 years, and of the upper blue clay and gravels above 1,600 years, making a total of 8,800 years of perfect repose. If to this we add the periods of elevation, both rapid and slow, the total period occupied by the deposit of Post-pliocene beds, would equal the period occupied by some deposits of the Secondary age. Yet such a calculation would scarcely give the total period accurately, since neither has the base of the lower blue clay yet been reached, nor should the present alluvial surface be looked upon as the last deposit of that epoch, or the prelude of the modern period; since, the slopes of San Bernardino display a series of conglomerates and gravels 200 feet above the level of the nearest stream (Cajon creek). These are coarse accumulations of primary pebbles and granitic clays, which have been removed from every portion of the plain where it is exposed. In the gorges and canons it still remains ; and wherever a pass has been traveled, there it is found, as the superficial covering, between 200 and 300 feet deep ; this, the last evidence of deposit of the Post-pliocene period, has not been considered in calculation of duration. Yet such a deposit must have existed over the plain, and must have been removed afterward; so that two additional periods would still require to be added to make the calculation complete, namely, the period occupied by the last deposit, and the period occupied by its removal.

Prof. J. W. Dawson* said that the mountain of Montreal, in Canada, which rises 700 feet, forms a tide-gauge of the Post-pliocene sea, marking, on its sides by a series of sea cliffs and elevated beaches, the stages of gradual or intermittent elevation of the land as it rose to its present level. The most strongly marked of these sea margins, are at heights of 470, 440, 386, and 220 feet above Lake St. Peter, on the St. Lawrence, or 450, 420, 366, and 200 feet above the river at Montreal.

The highest of these beaches contains sea shells of existing species. Below the lowest, and at an elevation of about 100 feet above the river, spreads the great Tertiary plain of Lower Canada, everywhere containing marine shells, and presenting a series of deposits partly unstratified and partly assorted by water. In this vicinity, the regular sequence is as follows : 1. Fine, uniformly grained sand, in some places underlaid or replaced by stratified gravel. Marine shells in the lower part. 2. Unctuous, calcareous clay, of gray, and occasionally of brown and reddish tints. A few marine shells. 3. Compact, bowlder clay, filled with fragments of various rocks, usually, partially rounded, and often scratched and polished.

The thickness of these beds is at least 100 feet, of which the lower or bowlder clay constitutes the greater part, but the sand often attains

* Can. Nat. and Geo. vol. ii.

the thickness of 10 feet, and the fine clay 20 feet. The City of Montreal is built upon this deposit. The bowlders are not confined to the bowlder clay, but are also found in the stratified clays and sand. The bowlders derived from the mountain, have been drifted to the southwest, in which direction they have been traced to the south shore of Lake Ontario, 270 miles distant. The terraces are best seen on the northeast side of the mountain. The rocks beneath the bowlder clay, are striated here S. 70° W. and S. 50° W. In some places the surface of the bowlder clay has been deeply cut into furrows by the currents which deposited sand and gravel upon it. In like manner the surface of the stratified clay, is sometimes cut into trenches filled by the overlying sand.

All the beds above referred to belong to the close of the Tertiary period, and they are all marine ; but they may have been deposited at distant intervals of time, and in waters of very various depths and area. From the abundance of the *Saxicava rugosa* in the upper bed, it was named the Saxicava sand, and from the abundance of the *Leda portlandica*, in the middle bed, it was called the Leda clay.

Dr. Albert C. Koch[*] stated that he collected, in 1839, in Gasconade county, Missouri, about 400 yards from the bank of Bourbense river, where there was a spring, the remains of a Mastodon under such circumstances as to show that it had been burnt to death, and while undergoing this punishment had also been struck with bowlders and shot with arrows. The animal had evidently been mired as its legs were in an upright position with the toes preserved. The ashes was from 2 to 6 inches in thickness, showing that the fire had been kept up for a considerable length of time. In the ashes he found stone arrow heads, a stone spear head, and some stone axes. The whole was covered by strata of alluvial deposits consisting of clay, sand and soil from 8 to 9 feet thick. He also stated that about one year later he found another Mastodon which he called the " Missourium," in Benton county, under about 20 feet of alluvial deposits, and with the bones were several stone arrow heads, one of which lay underneath the thigh bone, and in contact with it.

---

[*] Trans. St. Louis. Acad. Sci., vol. i.

In 1858, Dr. F. V. Hayden* prepared a vertical section, showing the order of superposition of the different beds of the Tertiary Basin of White and Niobrara rivers. The Miocene, he divided, in ascending order, as follows:

1. *Bed A.*—Light gray, fine sand, with more or less calcareous matter, passing down into an ash-colored plastic clay, with large quantities of quartz grains disseminated through it, sometimes forming aggregated masses like quartzose sandstone cemented with plaster; then an ash-colored clay with a greenish tinge, underlaid at base by a light gray and ferruginous silicious sand and gravel, with pinkish bands. Immense quantities of silex, in the form of seams, all through the beds. Titanotherium Bed. Found on Old Woman's creek, and in many localities along the valley of the South Fork of Shyenne. Best development on Sage and Bear creeks. Seen at several localities in the valley of White river. Thickness, 80 to 100 feet.

2. *Bed B.*—A deep flesh-colored, argillo-calcareous, indurated grit; the outside, when weathered, has the appearance of a plastic clay. Passes down into a gray clay, with layers of sandstone; underlaid by a flesh-colored, argillo-calcareous stratum, containing a profusion of Mammalian and Chelonian remains. Turtle and Oreodon Bed. Found on Old Woman's creek, a fork of Shyenne river, on the head of the South Fork of the Shyenne; most conspicuous on Sage and Bear creeks, and at Ash Grove Spring, and well developed in numerous localities in the valley of White river. Thickness, 80 to 100 feet.

3. *Bed C.*—Very fine, yellow, calcareous sand, not differing very materially from Bed D, with numerous layers of concretions, and rarely organic remains, passing down into a variegated bed, consisting of alternate layers of dark brown clay, and light gray, calcareous grit, forming bands, of which twenty-seven were counted at one locality, from one inch to two feet in thickness. Found on White river, Bear creek, Ash Grove Spring and head of Shyenne river, but most conspicuous near White river. Thickness, 50 to 80 feet.

4. *Bed D.*—A dull, reddish-brown, indurated grit, with many layers of silico-calcareous concretions, sometimes forming a heavy-bedded,

* Proc. Acad. Nat. Sci., vol. x.

fine-grained sandstone, and containing comparatively few organic remains. Found on the Niobrara and Platte rivers; well developed in the region of Fort Laramie, and in the valley of White river; and conspicuous, and composing the main part of the dividing ridge between White and Niobrara rivers. Thickness, 350 to 400 feet.

5. *Bed E.*—Usually a coarse-grained sandstone, sometimes heavy bedded and compact; sometimes loose and incoherent, and varying much in different localities. It forms immense masses of conglomerate, and contains layers of tabular limestone, with indistinct organic remains, and a few mammalian remains, in a fragmentary condition. It passes gradually into the bed below. It is most fully developed along the upper portion of Niobrara river, and in the region around Fort Laramie. It is seen also on White river, and on Grindstone hills. Thickness from 180 to 200 feet.

The Pliocene consists of 1st, dark gray or brown sand, loose, incoherent, with remains of mastodon and elephant ; 2d, sand and gravel, incoherent; 3d, yellowish-white grit, with many calcareous, arenaceous concretions ; 4th, gray sand with a greenish tinge, which contains the greater part of the organic remains ; 5th, deep yellowish-red arenaceous marl ; 6th, yellowish-gray grit, sometimes quite calcareous, with numerous layers of concretionary limestone, from two to six inches in thickness, containing fresh water and land shells, closely allied, and perhaps identical with living species, which belong to the genera, *Succinea, Limnea, Paludina* and *Helix.* It contains also, much wood of coniferous character. It covers a very large area on Loup Fork, from the mouth of North Branch to the source of Loup Fork, and occurs in the Platte valley. It is most fully developed on the Niobrara river, and extends from the mouth of Turtle river three hundred miles up the Niobrara. It occurs on Bijou hills, and Medicine hills, and is thinly represented in the valley of White river. Thickness from 300 to 400 feet.

The Post-pliocene consists of yellow, silicious marl, similar in its character to the loess of the Rhine, passing down into variegated indurated clays, and brown and yellow fine grits. It contains the remains of extinct quadrupeds, mingled with those identical with recent ones, and a few mollusca, mostly identical with recent species. It is most fully developed along the Missouri river, from the mouth of the Niobrara to St. Joseph, and occurs in the Platte valley and on the Loup Fork. Thickness from 300 to 500 feet.

Prof. G. C. Swallow* referred a formation made up of clays and

* Proc. Am. Ass. Ad. Sci.

sands and sandstone, extending along the bluffs, and skirting the bottoms, from Commerce, in Scott county, Missouri, westward to Stoddard, and thence south to the chalk bluffs in Arkansas to the Tertiary age. His section shows a thickness of 214 feet, but no fossils were obtained.

Prof. E. Emmons* described, from the Eocene of Craven county, North Carolina, *Carcharodon ferox, Cidaris carolinensis, Echinolampus appendiculatus, Echinocyamus parvus;* from near Newbern, *Carcharodon triangularis, Trygon carolinensis;* from Wilmington, *Carcharodon crassidens, C. contortidens, Cidaris mitchelli, Gonioclypeus subangulatus, Lunulites oblongus;* and from other places, *Hemipristis crenulatus.*

He described, from the Miocene at Elizabethtown, and near Cape Fear river, Bladen county, North Carolina, *Polyptychodon rugosus, Elliptonodon compressus, Fusus æqualis, F. lamellosus, F. moniliformis, Fasciolaria elegans, F. alternata, F. acuta, F. nodulosa, F. sparrowi, Cancellaria carolinensis, Buccinum moniliforme, B. multilineatum, Voluta obtusa, Paludina subglobosa;* and from the marl of other places, *Galeocerdo sub-crenatus, Pycnodus carolinensis, Terebra neglecta, Dolium octocostatum, Marginella constricta, M. elevata, Pleurotoma elegans, P. flexuosum, P. tuberculatum, Pyramidella reticulata, Chemnitzia reticulata, Eulima subulata, Cerithium annulatum, C. bicostatum, Terebellum constrictum, Scalaria curta, Littorina lineata, Delphinula quadricostata,* now *Carinorbis quadricostatus, Tornatina cylindrica, Cæcum annulatum, Pecten princepoides, Chama striata,* and *Artemis transversus.*

Prof. F. S. Holmes† described, from the Post Pliocene of South Carolina, *Nodosaria obtusa, Astræa crassa, Pectunculus charlestonensis, Lucina kiowahensis, Tapes grus, Mulinia milesi, Mesodesma concentricum, Abra angulata, Mya simplex, Carolina tuomeyi, Fusus conus, F. filiformis, F. bullata, F. rudis, Volutomitra wandoensis, Turbonilla cancellata, T. quinquestriata, T. lineata, T. subulata, T. caroliniana, T. acicula, T. subcoronata, Obeliscus crenulatus, Architectonica gemma, Angaria crassa,* and *Adeorbis nautiliformis.*

Dr. B. F. Shumard‡ described, from rocks supposed to be of Eocene age, at Port Orford and at Davis' Coal Mine in Oregon Territory, *Lucina fibrosa, Corbula evansana, Leda oregona,* now *Nuculana ore-*

* Geo. Sur. N. Carolina.
† Post Pliocene Fossils of South Carolina.
‡ Trans. St. Louis Acad. Sci., vol. i.

gona, L. willamettensis, now *N. willamettensis;* and from gray, fine-grained sandstone, at the mouth of Coose Bay, *Pecten coosensis,* and *Venus securis.*

Dr. Leidy* described, from the Pliocene of the Niobrara river, Nebraska, *Mastodon mirificus, Procamelus gracilis, P. robustus, P. occidentalis, Canis haydeni, C. sævus, C. temerarius, C. vafer, Felis intrepidus,* now *Pseudælurus intrepidus, Aelurodon ferox, Hystrix venustus, Castor tortus, Cervus warreni, Megalomerys niobrarensis, Merychyus elegans, M. major, M. medius, Hypohippus affinis, Parahippus cognatus, Equus excelsus, E. fraternus, Protohippus perditus, Merychippus mirabilis, Rhinoceros crassus, Euelephas imperator,* and from the red grit bed of Niobrara, near Fort Laramie (Miocene), *Merycochoerus proprius.*

In 1859, James Richardson† made a geological examination of the Gaspe peninsula, and observed two terraces in the drift to the west of Trois Pistoles river, at 130 and 300 feet, respectively, above the sea, and another at the mouth of the Matanne, at the height of 50 feet. Stratified clay occurs at the head of lake Matapedia, 480 feet above the sea and near the outlet at the height of about 530 feet. Marine testacea occur in the terrace on the east side of the Matanne river at the height of 50 feet above the sea; about two miles west of the Metis river, at the height of 130 feet, and eight miles up the Metis river, at 245 feet above the sea. At the St. Anne river there are five or six terraces in a height of 25 feet, abounding in fragments of marine shells. Grooves and scratches were observed a half mile below Trois Pistoles church, 60 feet above the sea, bearing S. 32 deg. E., and on the Kempt road, two miles from Lake Matapedia, 630 feet above the sea, and bearing S. 80 deg. E.

W. E. Logan‡ explored the river Rouge, a branch of the Ottawa, to the Iroquois Chute, about fifty miles from the mouth. He found an undisturbed deposit of clay on the left bank of the river, on the fourth range of Grenville, 280 feet above Lake St. Peter. In the rear of Grenville and front of Harrington, not far east of the Rouge, there spreads out a flat surface of several hundred acres in extent, which is underlaid by clay, and has a height of about 500 feet above Lake St. Peter. The plain of the three mountains has an elevation above the ordinary summer level of the river, of about 30 feet, and above Lake St. Peter of

---

* Proc. Acad. Nat. Sci., vol. x.
† Rep. of Progr. Geo. Sur. of Canada.
‡ Geo. Sur. of Canada, Rep. of Progress.

about 585 feet. It consists, in general, of sand or fine gravel at the top, with clay interstratified toward the lower part, but the sand greatly predominates. The surface of the rocks in the valley wherever examined were found to be grooved and striated. The courses of the grooves vary from S. 30 deg. E. to S. 25 deg. W., and accord in a general way, with the direction of the valley. The limits of the valley evidently guided the direction of the moving masses which produced the striæ.

Prof. Leo Lesquereux* described, from the Pliocene near Sommerville, Fayette county, Tennessee, *Salix densinervis, Quercus saffordi, Andromeda dubia,* and *Elœagnus inœqualis.*

In 1860, Prof. E. W. Hilgard† divided the Tertiary of Mississippi in ascending order into, 1st, The Northern Lignitic Group; 2d, The Claiborne Group; 3d, The Jackson Group; 4th, the Vicksburg Group; 5th, The Grand Gulf Group.

The Northern Lignitic Group occupies the central part of Northern Mississippi, and though generally covered by later deposits it outcrops at numerous places and is found at all deep borings. It consists of estuary deposits of sandstone, with marine shells; gray clays and sands, and dark brown and yellow clays and sands with lignite. Estimated thickness, including the Claiborne Group, 425 feet.

The Claiborne Group is found in the central part of the northern half of the State, in Holmes, Atala, Carroll and Choctaw counties, and in the western part of the State in Clarke, Lauderdale, Newton and Scott counties. It consists of blue and white marls, the latter always sandy and often indurate, and sandstones and claystones with sometimes lignitic clays and sands.

The Jackson Group forms a band across the central part of the State through Wayne, Clarke, Jasper, Newton, Scott, Madison and Yazoo counties. It consists of white (often indurate) and blue marls, highly fossiliferous. Estimated thickness, 80 feet.

The Vicksburg Group is the highest of the marine Eocene, and the only one which reaches the Mississippi river. It occupies a narrow belt of nearly uniform width, south of the Jackson Group, and extending across the State from Vicksburg to the Alabama line, and thence to the Tombigbee river, where it forms the bluff at St. Stephens. It consists of crystalline limestones and blue marls with ferruginous strata. It is the only one of the marine stages of the Eocene which

---

* Am. Jour. Sci. & Arts, 2d ser., vol. xxvii.
† Geo. of Miss.

exhibits crystalline limestones. It is highly fossiliferous. Estimated thickness, including the lignite at its base, 112 feet.

The Grand Gulf Group covers an immense extent of country south of the Vicksburg Group, and is composed essentially of clays and sandstones, the latter generally rather aluminous and soft, and of white-gray and yellowish-gray tints; the sand being very sharp. It takes its name from the bluff at Grand Gulf on the Mississippi river, where it is well exposed. It is overlaid near the coast by strata of Pliocene and Post-pliocene age. Estimated thickness, 150 feet.

Prof. F. S. Holmes* made three vertical sections of the Post-pliocene strata of South Carolina in descending order as follows:

1. The marine bed of the Wadmalur, consisting of yellow sand, 15 feet; ferruginous sand with casts of shells, 2 feet; red clay, 2 feet; and gray sand and mud with comminuted shells and fossils in fine preservation, 3½ feet.

2. The Ashley river beds, consisting of yellow sands with bands of ferruginous clay, 4 feet, and blue mud resting on the white Eocene marl, 1 foot.

3. The Goose creek beds, consisting of yellow sand, 12 feet; blue mud, 2 feet; ferruginous sand containing bones, 3 inches; yellow sand, 3 feet; and Pliocene marl resting on the Eocene white marl, 12 feet.

The fossil bones obtained from these strata are often in a fine state of preservation, especially those taken from the blue mud, which are generally petrified; those from the sands are likewise well preserved, but in the peaty or upper beds they are not so petrified, retain all their gelatin and appear to decompose rapidly. They consist of the bones of horses, hogs, dogs, rabbits, beavers, the tapir, and other mammalian remains.

T. A. Conrad† described, from the Eocene of Alabama and Mississippi, *Exilia pergracilis, Volutilithes limopsis, V. rugatus, Athleta leioderma, Simpulum showwalteri, S. autopsis, S. exilis, Galeodia tricarinata, Cithara nereidis, Murex morulus, Pseudoliva tuberculifera, Scala lintea, S. octolineata, S. staminea, Actæonina subvaricata, Tornatellæa bella, Cerithioderma prima, Mazzalina pyrula, Leda bella, now Nuculana bella, L. eborea, now N. eborea, Axinæa bellisculpta, Diplodonta astartiformis, D. deltoidea, Crenella latifrons;* from Texas, *Pseudoliva carinata, P. fusiformis, P. linosa, P. perspectiva,* and *Monoptygma crassiplica.*

* Proc. Acad. Nat. Sci., vol. ii., and in Post-pliocene Foss. S. Carolina.
† Jour. Acad. Nat. Sci., 2d ser., vol. iv.

Wm. M. Gabb described, from the Eocene at Wheelock, and in Caldwell county, Texas, *Belosepia ungula*, *Odontopolys compsorhytis*, *Fusus mortoniopsis*, *Neptunea enterogramma*, *Turris moorei*, *T. kelloggi*, now *Surcula kelloggi*, *T. nodocarinata*, now *Surcula nodocarinata*, *T. retifera*, *T. texana*, *Eucheilodon reticulatum*, *Scobinella crassiplicata*, *S. læviplicata*, *Distortio septemdentata*, *Phos texanus*, *Agaronia punctulifera*, now *Olivula punctulifera*, *Fasciolaria polita*, *F. moorei*, now *Cordiera moorei*, *Cymbiola texana*, *Mitra exilis*, *M. mooreana*, now *Lapparia mooreana*, *Erato semenoides*, now *Marginella semenoides*, *Neverita arata*, *Lunatia moorei*, *Architectonica meekana*, *A. texana*, *A. vespertina*, *Spirorbis leptostoma*, *Turritella nasuta*, *Eulima exilis*, *E. tenua*, *Dentalium minutistriatum*, *Ditrupa subcoarctata*, now *Gadus subcoarctatus*, *Bulla kelloggi*, *Volvula conradana*, *V. minutissima*, *Helcion leanus*, *Corbula texana*, *Tellina mooreana*, *Leda compsa*, now *Nuculana compsa*, *Noetia pulchra*, *Crassatella antestriata*, *Anomia aphippioides*, *Serpula texana;* from Alabama, *Cirsotrema megaptera*, *Leiorhinus crassilabris*, *Axinæa intercostata*, and *Pecten spillmani*.

He described, from the Miocene, near Shiloh, New Jersey, *Cantharus cumberlandana*, *Fasciolaria woodi*, *Natica hemicrypta*, *Mercenaria cancellata*, and from Maurice river, New Jersey, *Ostrea mauricensis*.

Gabb and Horn described, from the Eocene, in Caldwell county. Texas, *Flabellum pachyphyllum* and *Trochosmilia mortoni*.

Prof. Leo Lesquereux* described, from the lower Eocene or lignitic Tertiary of Tennessee and Mississippi, *Magnolia hilgardana* and *Rhamnus marginatus*.

Meek and Hayden† described, from the Miocene of the Bad Lands of White river, *Planorbis leidyi* and *P. vetulus*.

Prof. J. W. Dawson‡ described, from the Pliocene of Labrador, the foraminifer, *Nonionina labradorica*.

In 1861, Prof. C. H. Hitchcock§ said that there is not a mountain in Maine, fragments of which will not be found scattered over the country to the south or southeast. The granite of the Katahdin region is scattered over the southern part of Penobscot county, and the rocks of Mt. Abraham and Mt. Blue may be recognized among the bowlders in Kennebec county. One of the effects of the drift action is the smooth-

---

* Geo. of Ark., vol. ii.
† Proc. Acad. Nat. Sci.
C n. Nat. and Geo., vol. v.
‡ Rep. Geo. Maine.

ing, rounding, scratching and furrowing of the ledges over which the drift materials have passed, and unless these ledges have been decomposed upon their surfaces, they are covered with scratches or striæ, usually parallel to one another, and indicating the course of the drift agency. Ledges of talcose and argillaceous rocks preserve these markings the most distinctly. Were the rocks of Maine laid bare, fully half the surface would show these marks of smoothing.

The course of the striæ in Maine vary from north 70 deg. west to north 80 deg. east.

At the Lubec lead mines, a series of striæ were observed upon the side of a perpendicular wall, following the course of the wall around a corner. The course of the striæ ultimately varied at right angles from their original directions. At several places at the sea shore the striæ have been noticed below high water mark, and others were seen to run under the ocean at low-water mark. The course of the striæ upon the lakes north of the Katahdin mountains have more of an easterly course than those to the east and south of the same mountains. It looks as if the mountains formed an obstruction around which the striating agency operated, in preference to climbing the elevation. It is a curious fact, in the same connection, that the striæ are wanting on the summit of Katahdin. It appears also that there was another deflection of the course of the striæ in the valley of Sandy river. Mt. Abraham may have arrested the drift current on the north and turned it into Sandy river valley on the west, from which deflection it struck against the Saddleback mountain range, continued to Mount Blue, and was then directed toward French's Mountain in Farmington.

Drift striæ are never found upon the south side of mountains, unless for a short distance, where the slope is very small. It is common to see different courses of striæ intersecting one another, as on the south side of Chamberlin lake, where striæ north 70 deg. west and north 50 deg. west intersect, and north 17 deg. west and north 67 deg. west, intersect.

The only examples of glacial markings discovered, in Maine, are on the St. John river, in its upper portion. Above the Lake of the Seven Islands, on this river, there are no glacial markings, unless the scratches upon the pavement of bowlders are to be referred to them. The bed of the river is full of stones, and upon the banks below high-water mark they are as firmly set as paving stones in the streets of a city. The scratches are not as constant and distinct as those of the glacier below, and may possibly have been formed by ice freshets in the spring of the year. Descending the river to No. 14 we find a ledge which has been struck by a force descending the river, as the stoss and lee sides plainly show. The course of the striæ is north 65 deg. west, the stoss side being on the southeast. A similar example occurs near the mouth of Black river, where the course of the striæ is toward north 60 deg. west. The country above Black river being quite level, is not so well adapted for the existence of a glacier as the region below, where high mountains crowd the river on both sides. At the mouth of Little Black river the upper side of the ledges is uniformly the struck side. Some of the ledges are covered with both drift and glacial striæ, the former coming from north 60 deg. west, and the latter running down the river northeasterly. A mile above the mouth of the St. Francis river, the glacial striæ run down the river with the direction north 47 deg. east. Near the village of St. Francis the two sets of striæ appear again, the drift with the directions of north 60 deg. west, and north 20 deg. west, and the glacial with the direction of north 16 deg. east. This is the course of the river around a curve. The former are here the most prominent. In the township below Fort Kent, striæ appear running north 30 deg. west. One of the finest exposures of the glacial striæ is in Dionne, where the river makes a great bend and pursues a northerly course. The striæ change with the river and run north 20 deg. west, or directly opposite to the normal course of the drift in the vicinity, the force having gone northerly instead of southerly. No glacial markings were observed below this, in fact the glacial and drift markings could not be distinguished from each other below the Madawaska settlements. The evidence for an ancient glacier is not so strong on the St. John river as in the western part of New England. Some might contend that the immense ice freshets in the spring would be sufficient to explain all the phenomena. On the other hand, the objection to glaciers in northern Maine would be less than in Massachusetts, on account of the colder climate.

An unstratified mass of a stiff, dark, bluish clay, containing rounded and striated bowlders, and called bowlder clay, is found on the precipitous banks of rapid streams in narrow valleys. It underlies the finer sands and gravels of later periods, and always rests directly upon the solid rocks.

Modified drift occurs, in Maine, in the form of moraine terraces, horsebacks, sea beaches, sea bottoms, marine clays and terraces. Moraine terraces are generally accumulations of gravel, bowlders and sand,

often arranged in heaps and hollows, or conical and irregular elevations with corresponding depressions. A class of alluvial ridges found
in great abundance in Maine are called horsebacks. Sea beaches and
sea bottoms are found 150 feet higher than the ocean level, and containing littoral shells. Fossiliferous marine clays form almost a continuous belt, extending up the rivers to about this height above the
ocean. Alluvial terraces are those banks of loose materials, generally
unconsolidated, which skirt the sides of the valleys about rivers, ponds,
and lakes, and rise above one another like the seats of an amphithea-
ter.

Prof. Edward Hitchcock* collected about 300 measurements of the
drift striæ found in the State of Vermont. The course varied from
north 70 deg. east to north 80 deg. west. It seems to have been rare
to find the striæ, at any two points, exactly agreeing in direction,
though he divided the predominant courses into three divisions, viz:
1. From the northwest. 2. From the northeast. 3. From the north.
The striæ differ in size from the finest scratch visible, up to a furrow a
foot deep.

Prof. J. W. Dawson† described the Post-pliocene deposits at Murray
bay, on the St. Lawrence river, 90 miles below Quebec, where they consist of the Leda clay and Saxicava sand. There are several terraces
at this place, varying from 30 to 132 feet above the sea level, but the
highest true shore-mark observed, is a narrow beach of rounded pebbles
at the height of 326 feet. This beach appears to become a wide terrace
further to the north, and also on the opposite side of the bay. It probably corresponds with the highest terrace observed by Sir W. E. Logan, at Bay St. Paul, and estimated by him at the height of 360 feet.
The two principal terraces at Murray bay correspond nearly with two
of the principal shore-levels at Montreal and in various parts of Canada, where two lines of old sea beaches occur at about 100 to 150 feet,
and 300 to 350 feet above the sea, though there are others at different
levels.

Dr. F. V. Hayden‡ sketched the geology of the country about the
headwaters of the Missouri and Yellow Stone, and said that throughout the Wind river valley there is a series of beds of great thickness intermediate in their character between the true lignite beds
and the White river Tertiary deposits. They extend from Willow

---

* Rep. on the Geo. of Vermont, vol. i.
† Can. Nat. and Geol., vol. vi.
‡ Am. Jour. Sci, and Arts. 2d ser., vol. xxxi.

Springs on the North Platte westward toward the Sweet Water mountains, and near the divide between the North Platte and Wind river they reach a thickness of 400 feet. From this divide throughout the Wind river valley they occupy the greater portion of the country, and though inclining in the same direction with the older strata the beds do not dip more than from 1 to 5 deg. They differ from the other deposits in the great predominance of arenaceous sediments, and in the absence of vegetable remains, but they contain fragments of turtles and numerous fresh-water and land shells. The entire thickness of these deposits is estimated at from 1,500 to 2,000 feet.

The White River Tertiary beds extend southward along the Laramie mountains to Willow Springs, and up the North Platte to Box Elder creek, and beyond in small outliers, showing that much has been removed by erosion. From the source of Box Elder creek, they extend to the head of Bates Fork, and westward to the Medicine Bow mountains. These beds for the most part, hold a horizontal position, while those of the lignite age are much disturbed; moreover, their position shows that they are of much more recent origin. The White river Tertiary deposits are followed by the White river bone beds, which pass up into the Pliocene of Niobrara by a slight physical break, and the latter are lost in the yellow marl or Lacs deposits.

Meek and Hayden[*] made a vertical section of the Tertiary rocks of Nebraska, in ascending order as follows :

1. Wind river deposits, consisting of light gray and ash-colored sandstones, with more or less argillaceous layers. Thickness from 1,500 to 2,000 feet. Found in the Wind river valley and west of the Wind River mountains.

2. The White River Group, consisting of white and light drab clays, with some beds of sandstone and local layers of limestone. Thickness 1,000 feet or more. Found on the Bad Lands of White river ; under the Loup river beds, on Niobrara, and across the country to the Platte. Age of the Miocene.

3. Loup river beds, consisting of fine loose sand, with some layers of limestone. Thickness, 300 to 400 feet. Found on Loup fork of Platte river, and extending north to the Niobrara river, and south an unknown distance. Age of the Pliocene.

They described from the Wind River Group, in the Wind river valley, *Helix veterna*, and *H. spatiosa*, now *Macrocyclis spatiosa.*

---

[*] Proc. Acad. Nat. Sci., vol. xii.

W. M. Gabb[*] described, from the Eocene at Claiborne, Alabama, *Phos belliliratus;* from Vicksburg, *Tellina euryterma;* from a brown, highly ferruginous sandstone at Caddo Peak, Texas, *Meretrix yoakumi. Perna texana;* from Houston county, Texas, *Protocardia gambrina;* and from South Carolina, *Ostrea mortoni.*

He described, from the Miocene of Virginia, *Voluta sinuosa;* from Santa Barbara, California, *Turbonilla aspera,* now *Bittium asperum, Modelia striata, Rocellaria antiqua, Sphenia bilirata, Venus rhysomia, Cardita monilicosta,* and *Morrisia horni.*

Prof. Leo. Lesquereux[†] described, from the Pliocene beds at Brandon, Vermont, *Carpolithes brandonanus, C. brandonanus,* var. *elongatus, C. brandonanus,* var. *obtusus, C. fissilis, C. grayanus, C. irregularis, Carya vermontana, C. verrucosa, Fagus hitchcocki, Apeibopsis gaudini, A. heeri, Aristolochia curvata, A. obscura, A. œningensis, Sapindus americanus, Carpolithes bursaeformis, Cinnamomum novœanglia, Illicium lignitum, Drupa rhabdosperma, Nyssa complanata, N. lœvigata,* and *N. microcarpa.*

In 1862, Gabb and Horn[‡] described, from the Eocene, near Charleston, South Carolina, *Eschara texta, Reptescharella carolinensis;* from Claiborne, Alabama, *Eschara oralis, Semieschara tubulata, Cellepora cycloris, C. inornata, Escharella micropora;* from Vicksburg, Mississippi, *Reptocelleporaria glomerata.*

They described, from the Miocene of St. Mary's river, Maryland, *Eschara fragilissima;* from Petersburg, Va., *Ennalipora quadrangularis;* from the Miocene, of New Jersey, *Cellepora urceolata, Membranipora sexpunctata, Reptoflustrella tubulata;* from Santa Barbara, California, *Semitubigera tuba, Entalophora punctulata, Cellepora californiensis, C. bellerophon, Reptescharella heermanni, R. plana, Phidolopora labiata, Reptoporina enstomata, Reptescharellina disparilis, R. heermanni, R. cornuta, Siphonella multipora, Membranipora californica, Crisina serrata,* and *Lichenopora californica.*

T. A. Conrad[§] described, from the Miocene of Virginia, *Surcula engonata, S. nodulifera, Drillia impressa, D. distans, D. arata, D. bella, D. eburnea, Mangelia virginiana, Pleiorytis ovato, Busycon carinatum, B. filosum, Tritia scalaris,* now *Buccinum scalare, Astyris reticulata, Dactylus eborens,* now *Oliva eborea, Leiotrochus distans,*

---

[*] Proc. Acad. Nat. Sci., vol. xii.
[†] Geol. Vermont, vol. ii.
[‡] Jour. Acad. Nat. Sci., 2d ser., vol. v.
[§] Proc. Acad. Nat. Sci., vol. xiii.

*Pecten fraternus, Busycon tritonis, Melampus longidens, Mactra me-
dialis, Astarte bella, A virginica, Lirophora athleta, Dione densata,*
and *D. virginiana;* from Calvert cliffs, and St. Mary's county, Maryland, *Surcula rugata, Bulliopsis marylandica, B. ovata, Astyris communis, A. avara,* var. *granulifera,* and *Busycon alveatum;* from South
Carolina, *Anomalocardia trigintinaria;* from North Carolina, *Dentalium carolinense, Pecten edgecomensis, Noetia carolinensis, Dactylus carolinensis,* now *Oliva carolinensis,* and *Siliquaria carolinensis;*
from Cumberland county, New Jersey, *Turritella æquistriata, T. cum-
berlandia, Saxicava, myæformis, Carditamera aculeata,* and *Astarte
distans;* from California, *Lyropecten crassicardo;* and from the
Eocene, at Enterprise, Clark county, Mississippi, *Crassatella producta.*

Wm. Stimpson described, from the Post-pliocene at Cape Hope, on the
southeast side of Hudson's bay, *Cardium dawsoni.*

Along Lake Temiscamang, [*] the Ottawa river and Riviere Rouge,
north of the Ottawa, the furrows conform in a general way to the directions of the river-valleys, the limits of which appear to have guided
the moving masses which produced the grooves. The direction of the
grooves at a single locality is not only not uniform, but, on the contrary,
they frequently cross each other. Measurements taken at 145 different places in Canada show that there is no uniformity in the direction
of the striæ, but as in these cases they vary from S. 80° E. to S. 70° W.

Bowlders are found in great abundance in many places, especially in
the valleys, where the bowlder formation has been extensively denuded
by the action of the water, and its lighter materials swept away. On
elevations, they are often seen resting upon the unstratified drift, which,
in the adjacent depressions of the surface, is covered over by stratified sand and clay. They appear, in most instances, to have traveled
southward, but there are exceptions to this general rule. Thus in the
county of Rimonski, in the valley of the Neigette river, there are large
bowlders of limestone, one of them 40 feet in diameter, belonging to the
Gaspe series, which have been moved several miles northward or northeastward. Farther down the valley of the St. Lawrence, blocks of
trachytic granite have been carried northeastward from the Table-
topped mountain down the valley of the Magdalen. There are also
instances of the northward transportation of bowlders in Nova Scotia.

The valleys of the St. Lawrence and the Richelieu, in Canada East,
and a considerable portion of the region between the St. Lawrence and

the Ottawa, to the east of the meridian of Kingston, are occupied by stratified clays, which, unlike those of western Canada, contain abundance of marine shells, for the most part identical with species now living in the lower St. Lawrence and the gulf. The clays are in many cases overlaid by sands, occasionally interstratified with clay, which also contain marine remains. The two are regarded as forming parts of one formation, and as corresponding to the upper and lower divisions of the Champlain clay of Vermont. The lower division is called the Leda clay, and the upper the Saxicava sand. If a line be drawn from the outlet of Lake Champlain to Ottawa, and from the extremities of this, as a base, two others be carried to Quebec, there will be included a very level triangular area of about 9,000 square miles, for the greater part covered by the Champlain clays and sands. The plains on either side of the St. Lawrence below Quebec are occupied by the same formation, which is found at intervals as far down as Matanne; while on the north side it covers an extensive area in the valley of the Saguenay and around lake St. John and its tributaries. Clays belonging to the lower division are found at various levels from the surface of the sea to 600 feet above it, and in some cases they have been observed some feet below the sea-level. The river Rouge enters the Ottawa between hills of bare rock; but on its western side, in the fourth range of Grenville, a bank of clay 125 feet in thickness occurs, the summit of which is 405 feet above the sea. Again, not far east of this river, in the rear of Grenville, and in the front of Harrington, is an area of several hundred acres, underlaid by stratified blue clay, the surface of which is about 500 feet above the sea. Several similiar portions of clay occur in that vicinity. In Gaspe, at the head of Lake Matapedia, stratified clay occurs at the height of 480 feet, and near the outlet of the same lake, at the height of 530 feet above the sea. At Bay St. Paul, on the north side of the St. Lawrence, terraces occur at 130 and 360 feet above the sea. Marine fossils occur throughout the strata in which these terraces are worn, and still higher at 390 feet above the sea level. In the valley of the Saguenay, marine clays, generally overlaid by sand and gravel, are found almost everywhere between Ha-ha bay and the west side of Lake St. Johns; as well as between that bay and Chicontimi. Between Chicontimi and Ha-ha bay the clay is sometimes 600 feet in thickness. About a half mile below the falls of Bell Riviere, marine shells occur in the clay at 400 feet above the sea.

The Saxicava sand forms a belt on the north side of the St.

Lawrence, at the base of the Laurentide hills, from Ottawa to Cape Tourmente. It expands on the St. Maurice to a breadth of thirty miles. To the westward it covers much of the surface in the triangular area between the St. Lawrence and the Ottawa east of the meridian of Kingston. Marine shells occur in this sand in Nepean, at 410 feet above the sea ; in Kenyon, at 335 feet ; in Fitzroy, at 330 feet ; in Winchester, at 300 feet ; and at Pakenham mills, at 226 feet. South of the St. Lawrence these sands are found along the boundary of New York. From the east side of Missisquoi bay, a belt extends between the clay plains of the south shore of the St. Lawrence, which it partly overlies, and the more elevated region to the southeast, as far as Metis. At the Wallbridge Mills, in Stanbridge, marine shells occur at a height of 160 feet, and near Upton, on the Grand Trunk railway, at 300 feet above sea level.

In 1863, J. S. Newberry[*] described, from the Miocene of Bellingham bay, *Equisetum robustum*, *Sabal campbelli*, *Quercus coriacea*, *Q. flexuosa*, *Q. banksiæfolia;* from Birch bay, Washington Territory, *Taxodium occidentale*, *Smilax cyclophylla;* and from Bellingham bay *Quercus elliptica*, and *Populus flabellum.*

Remond[†] described, from the Pliocene near Kirkers Pass, *Cardium gabbi*, and *Ostrea bourgeoisi.*

In 1864, T. A. Conrad[‡] described, from the Eocene of Dallas county, Alabama, *Turritella præcincta ;* from Pamunkey river, Virginia, *Protocardia virginiana ;* and from 6 miles east of Washington, D.C. *Dosiniopsis meeki.*

He described, from the Miocene at Natural Well, Dauphin county, North Carolina, *Fasciolaria subtenta*, and *Lirosoma curvirostrum.*

The Miocene Strata, on the northern slope of the Monte Diablo Range, consists of heavy-bedded sandstones.

In crossing over the Santa Cruz Range from Santa Cruz, in a northerly direction to the Santa Clara Valley, before reaching the metamorphic, a mass of rocks is traversed, which is much broken and elevated, some of the ridges being fully 2,000 feet high. In rising on to this elevated ridge, however, we first pass over a belt of unaltered strata, which near the town lie nearly horizontal, and which appear to have escaped the action of the elevating forces, by which the main

---

[*] Bost. Jour. Nat. Hist., vol. vii.
[†] Proc. Cal. Acad. Sci.
[‡] Pro. Acad. Nat. Sci., vol. xiv.
[§] Geo. Sur. of California, 1865.

chain has been raised. There appears to be no doubt that these horizontal strata are the same ones which are tilted up in the mountains, and that they belong to the Miocene Tertiary. At about six miles from Santa Cruz are some singular examples of weathered sandstone, which are known as the " Ruins" or the "Ruined City." Here perpendicular tubes or chimneys of rock are found, from one to three feet in diameter, the sandstone appearing to have been hardened in concentric layers by the infiltration of ferruginous solutions, and this hardened portion has withstood the action of the elements, while the softer bands, and the interior columnar or cylindrical masses, have weathered away, leaving a pile of rocks behind, which, by some exertion of the imagination, can be construed into a resemblance to a ruined city, on a very small scale.

The whole region traversed by the trail from Pescadero to Scarsville, as far as the metamorphic on the eastern edge of the range, is bituminous shale, of Miocene age, with occasional beds of interstratified sandstone, of which the dip is irregular, but not high.

Between Petaluma and the entrance of Tomales bay, patches of Miocene sandstone occur from 250 to 300 feet thick, resting unconformably upon altered strata. The rocks are soft, yellow sandstone, with large nodules of hard, blue calcareous sandstone, imbedded in them. Between the highest points near the head of Tomales bay and Punta Reyes, there are minor ridges of Miocene sandstone, having a low southwest dip.

The sandstones of the Santa Monica and Santa Susanna Ranges, are, in large part, of Miocene age. The ridges bounding the San Fernando valley on the southwest, are made up of light bituminous slates, dipping generally to the east or north east ; they form rounded hills, bearing the marks of extensive erosion. A higher range to the west of these hills connects the two chains, and rises to a height of 3,000 feet above the sea, being made up of Miocene sandstones, highly inclined and in some places metamorphosed.

The chain of the Santa Inez Range rises to the north of Santa Barbara, a conspicuous object to those approaching this place by water. As far as known, it takes its origin at a point due north of Buenaventura, and running a little north of west (N. 84 deg. W.) for a distance of over 60 miles, it meets the sea at Point Concepcion. The chain has its greatest elevation apparently near Santa Barbara, where it is about 3,800 feet high. To the west of the Gaviota Pass it has an elevation of about 2,500 feet. The main ridge is entirely composed of Miocene sandstones, without any appearance of eruptive rock, and also with very little metamorphism.

The unaltered sandstones extending along the Gavilian Range, near the San Juan valley, and forming the San Juan hills, which extend to the Pajaro river, are referred to the Miocene. In these hills the strata are very heavy bedded, and have a dip everywhere to the south. The materials of which they are made up are often coarse, and sometimes large enough to form a conglomerate, among the pebbles of which jasper and other metamorphic rocks predominate.

In the vicinity of the Bay of Monterey the granite is flanked by Miocene sandstone. Both rocks are considerably altered, for a distance of about 20 feet from the junction; the sandstone is softened and disintegrated, and the granite discolored. The metamorphism has so affected both rocks that it is not easy to determine the exact line of junction.

The Miocene sandstones are displayed in some places in the region between the Canada de las Uvas and Soledad Pass, nearly 2,500 feet in thickness. From the summit of the higher upturned strata, a wide belt of Tertiary rocks may be seen skirting the Coast Ranges, and worn into rounded hills, which are generally barren, especially on the west side of the Tulare valley.

The Pliocene beds between Merced Lake and Mussel Point, on the peninsula of San Francisco, are made up of a bluish sandstone, of which the grains are cemented by carbonate of lime, interstratified with hard, fine conglomerates, of which the pebbles are evidently derived from the adjacent jaspery rocks of Cretaceous age. These strata contain *Scutella interlineata, Crepidula princeps*, both of which are extinct, together with several species still living on the coast.

At the head of Pleasant valley, the strata are overlaid by beds of volcanic ashes, interstratified with gravels, the whole series being conformable and dipping at a low angle to the east. They appear to be of Pliocene age, and identical in most respects with the sedimentary volcanic beds to the north of Kirker's Pass.

To the north of San Pablo are low hills of very recent strata, which are nearly horizontal, and which rest unconformably on the edges of the Tertiary. They are referred to Post-pliocene age.

From Tres Pinos, 13 miles from San Juan, to Booker's, a distance of about 13 miles in a direct line, the road follows the Arroyo Joaquim Soto, a branch of the San Benito. Along this road there are vast deposits of gravel, or entirely unconsolidated detritus, and which form a

large portion of the series of ridges between the Gavilan, on the one
side, and the Monte Diablo Range on the other. At the first exposure,
about two miles beyond Tres Pinos, the stratified detritus forms a
steep bluff about 400 feet above the creek. The gravel is made up of
pebbles of granite, red and green jaspers, and silicious slate and other
metamorphic materials. At a point a few miles below Bookers the
strata are worn into precipitous canons, with bare bluff banks or almost perpendicular walls, regularly stratified, and varying in fineness
from a coarse gravel to fine sand, with here and there a thin band of
consolidated materials, the remainder entirely in the original condition
in which it was deposited, as far as being held together by any cement
is concerned. The thickness of these deposits is enormous; one hill
was found to be 1,274 feet above the valley, and another 1,800 feet.
Both these hills are entirely made up of these unconsolidated materials.
This region gives one a most vivid idea of how recently geological
changes of magnitude have taken place in this part of the State, and
furnishes most impressive testimony to add to that obtained in other
places, in relation to the lateness of the geological epoch, during which
this portion of the chain was elevated. It would appear that the basin,
in which these strata were deposited, was drained of the water at successive intervals, by the elevation of the basin itself, judging from the
disturbed position of the strata it contains, and not by the gradual
wearing away of a barrier at its lower end.

Prof. J. W. Dawson[*] described the Post-pliocene deposits in the
country around Cacouna and Riviere-du-Loup. The depressions between the ridges are occupied by these deposits resting upon the
Quebec Group of rocks. The oldest member of the deposit, is a tough
marine bowlder clay, its cement formed of gray or reddish mud, derived from the waste of the shales of the Quebec Group, and the stones
and bowlders with which it is filled, partly derived from the harder
members of that Group, and partly from the Laurentian hills, on the
opposite or northern side of the river, more than twenty miles distant.
The thickness of the bowlder clay is variable, but at Ile Verte, it forms
a terrace 50 feet in height. The bowlder clay at Cacouna, is a deep-
water deposit. Its most abundant shells are *Leda truncata*, *Nucula
tenuis*, and *Tellina proxima*, and these are imbedded in the clay with
the valves closed, and in as perfect condition as if the animals still inhabited them. The bowlder clay is also fossiliferous at Murray bay,
St. Nicholas, and Cape Elizabeth.

* Can. Nat. and Geol. new ser., vol. ii.

Above the bowlder clay, there occurs a dark gray, soft, sandy clay, containing numerous bowlders, and above this several feet of stratified sandy clay without bowlders ; while on the sides of the ridges, and at some places near the present shore, there are beds and terraces of sand and gravel constituting old shingle beaches, apparently much more recent than the other deposits. All of the deposits are more or less fossiliferous. The surface of the rocks beneath the bowlder clay, is polished and striated in the direction of northeast and southwest, or that of the St. Lawrence valley.

W. M. Gabb[*] described, from the Post-pliocene of San Pedro and Santa Barbara, *Turcica coffea*, and *Calliostoma tricolor*.

Dr. Joseph Leidy[†] described, from the Miocene of White river, Nebraska, *Rhinoceros occidentalis ;* from Texas, *R. meridianus ;* from Calaveras county, California, *R. hesperius*. And from the Pliocene of California, *Equus occidentalis*.

R. P. Whitfield[‡] described, from the Eocene of the Southern States, *Pisania claibornensis, Pyrula juvenis, Fulgur triserialis, Fusus tortilis, Pseudoliva elliptica, Monoptygma leai, Columbella turricula, Pleurotoma capax, P. nasutum, P. persa, P. adeona, Voluta newcombana, Mitra halcana, M. biconica, Natica erecta,* now *Lacunaria erecta, N. perspecta, N. reversa, N. onusta, N. alabamensis,* now *Lacunaria alabamensis, N. aperta, Velutina expansa, Cerithium vinctum, Potamides alabamensis, Turritella eurynome, T. multilira, T. alabamensis, Cucullæa macrodonta, Crassatella tumidula.*

T. A. Conrad[§] described, from the Jackson Group, at Enterprise, Mississippi, *Corbula filosa, Dione securiformis, D. annexa, Tellina eburneopsis, T. albaria, T. linifera, Alveinus minutus, Sphærella bulla, Cyclas curta, Protocardia lima, Gouldia pygmæa, Axinæa inequistriata, A. duplistriata, Nuculana linifera, Nucula spheniopsis, Arcoperna filosa, Pecten scintellatus,* now *Camptonectes scintellatus, Doliopsis quinquecosta,* now *Galeodia quinquecosta, Turritella perdita, Mesalia arenicola.*

From divers places in Alabama, Mississippi and Texas, *Strepsidura lintea, Surcula gabbi, S. lintea, Cochlespira engonata, Moniliopsis elaborata, Drillia texana, Tortoliva texana, Monoptygma curta, Volutilithes indenta, V. impressa, Obeliscus perexilis, Architectonica*

---

* Pro. Cal. Acad. Sci.
† Pro. Acad. Nat. Sci.
‡ Am. Jour. Conch , vol. i.
§ Am. Jour. Conch., vol. i.

cælatura, Cancellaria impressa, C. tortiplica, Tornatellæa lata, Corbula filosa, Egeria donacea, Cytheriopsis hydana, Cyclas claibornensis, Mysia delloidea, Conus alveatus, C. subsauridens, Cochlespira bella, Buccitriton altum, Limatia marylandica, Cirsostrema claibornensis, Cancellaria ellapsa, Dentalium densatum; from Shark river, Monmouth county, New Jersey, *Pleurotomaria perlata, Surcula annosa, Actæonema prisca,* and *Avicula annosa.*

In 1866, Prof. J. W. Dawson[*] said the snow-clad hills of Greenland send down to the sea great glaciers, which in the bays and fiords of that inhospitable region, form, at their extremities, huge cliffs of everlasting ice, and annually "calve," as the seamen say, or give off a great progeny of ice islands, which slowly drifted to the southward by the Arctic current, pass along the American coast, diffusing a cold and bleak atmosphere, until they melt in the warm waters of the Gulf stream. Many of these bergs enter the straits of Belle-Isle, for the Arctic current clings closely to the coast, and a part of it seems to be deflected into the Gulf of St. Lawrence through this passage, carrying with it many large bergs. Mr. Vaughan, late superintendent of the light house at Belle-Isle, has kept a register of icebergs for several years. He states that for ten which enter the straits, fifty drift to the southward, and that most of those which enter pass inward on the north side of the island, drift toward the western end of the straits and then pass out on the south of the island, so that the straits seem to be merely a sort of eddy in the course of the bergs. The number in the straits varies much in different seasons of the year. The greatest number are seen in spring, especially in May and June; and toward autumn and in the winter very few remain. Those which remain until autumn are reduced to mere skeletons; but if they survive until winter, they again grow in dimensions, owing to the accumulations upon them of snow and new ice. Those that we saw early in July were large and massive in their proportions. The few that remained when we returned in September, were smaller in size, and cut into fantastic and toppling pinnacles. Vaughan records that on the 30th of May, 1858, he counted in the straits of Belle-Isle 496 bergs, the least of them 60 feet in height, some of them half a mile long and 200 feet high. Only $\frac{1}{8}$ of the volume of floating ice appears above water, and many of these great bergs may thus touch the ground in a depth of 30 fathoms or more, so that if we imagine 400 of them moving up and down under the influence of

* Can. Nat. & Geol., 2d series, vol. iii.

the current, oscillating slowly with the motion of the sea, and grinding on the rocks and stone-covered bottom, at all depths, from the center of the channel, we may form some conception of the effects of these huge polishers of the sea floor.

Of the bergs which pass outside of the straits, many ground on the banks off Belle-Isle. Vaughan has seen a hundred large bergs aground at one time on the banks, and they ground on various parts of the banks of Newfoundland, and all along the coast of that island. As they are borne by the deep seated cold current, and are scarcely at all affected by the wind, they move somewhat uniformly, in a direction from N. E. to S. W., and when they touch the bottom the striation or grooving which they produce must be in that direction.

In passing through the straits in July, we saw a great number of bergs, some were low and flat topped with perpendicular sides, others were concave or roof-shaped like great tents pitched on the sea ; others were rounded in outline or rose into towers and pinnacles. Most of them were of a pure dead white, like loaf sugar, shaded with pale bluish green in the great rents and recent fractures. One of them seemed as if it had grounded and then overturned, presenting a flat and scored surface covered with sand and earthy matter.

After describing the glaciers of Mont Blanc, he lays down the following rules :

1. Glaciers heap up their debris in abrupt ridges. Floating ice sometimes does this, but more usually spreads its load in a more or less uniform sheet.

2. The material of moraines is all local, icebergs carry their deposits often to great distances from their sources.

3. The stones carried by glaciers are mostly angular, except where they have been acted on by torrents. Those moved by floating ice are more often rounded, being acted on by the waves and by the abrading action of sand drifted by currents.

4. In the marine glacial deposits, mud is mixed with stones and bowlders. In the case of land glaciers, most of this mud is carried off by streams, and deposited elsewhere.

5. The deposits from floating ice may contain marine shells. Those of glaciers can not, except where, as in Greenland and Spitzbergen, glaciers push their moraines out into the sea.

6. It is the nature of glaciers to flow in the deepest ravines they can find, and such ravines drain the ice of extensive areas of mountain land. Icebergs, on the contrary, act with greatest ease on flat surfaces, or slight elevations in the seat bottom.

7. Glaciers must descend slopes, and must be backed by large supplies of perennial snow. Icebergs act independently, and being water-borne, may work up slopes and on level surfaces.

8. Glaciers striate the sides and bottoms of their ravines very unequally, acting with great force and effect only on those places where their weight impinges most heavily. Icebergs, on the contrary, being carried by constant currents, and over comparatively flat surfaces, must striate and grind more regularly over large areas, and with less reference to local inequalities of surface.

9. The direction of the striæ and grooves produced by glaciers depends on the direction of the valleys. That of icebergs, on the contrary, depends upon the direction of marine currents, which is not determined by the outline of surface, but is influenced by the large and wide depressions of the sea bottom.

10. When subsidence of the land is in progress, floating ice may carry bowlders from lower to higher levels. Glaciers can not do this under any circumstances, though in their progress they may leave blocks perched on the tops of peaks and ridges.

He further said, that, in all these points of difference, the bowlder clay and drift of Canada, and other parts of North America, correspond rather with the action of floating ice than of land ice. More especially is this the case in the character of the striated surfaces, the bedded distribution of the deposits, the transport of material up the natural slope, the presence of marine shells, and the mechanical and chemical character of the bowlder clay.

He also enumerated the following Post-pliocene plants as occurring, in nodules, at Green's Creek, and other places in Canada, to-wit: *Drosera rotundifolia, Acer spicatum, Potentilla canadensis, Gaylussaccia resinosa, Populus balsamifera, Thuja occidentalis, Potamogeton perfoliatus, P. pusillus, Equisetum scirpoides.* None of the plants are properly Arctic in their distribution, and the assemblage may be characterized as a selection from the present Canadian flora of some of the more hardy species having the most northern range. At Green's Creek (near Ottawa) the plant-bearing nodules occur in the lower part of the Leda clay, which contains a few bowlders, and is apparently, in places, overlaid by large bowlders, while no distinct bowlder clay underlies it. The circumstances which accumulated the thick bed of bowlder clay near Montreal, were probably absent in the Ottawa valley. In any case, we must regard the deposits of Green's Creek as coeval with the Leda clay of Montreal, and with the period of the greatest abundance of *Leda truncata*, the most exclusively Arctic shell of these deposits. In other words, he regarded the plants above mentioned as probably belonging to the period of greatest refrigeration of which we have any evidence—of course, not including that mythical period of universal incasement in ice, of which, in so far as Canada is concerned, there is no evidence whatever.

The Tertiary formation * exists in the southern part of the State of Illinois. It is best developed in Pulaski and Massac counties. It is represented by a series of stratified sands and clays of various colors, with beds of silicious gravel, often cemented into a ferruginous conglomerate by the infiltration of a hydroxyd of iron. In some places it contains green, marly sand, with casts of fossils, and along the edge of the Ohio, at extreme low water, at Caledonia, there is a thin bed of lignite. At Fort Massac, just above Metropolis, the ferruginous conglomerate is from forty to fifty feet in thickness. Near Caledonia, a section gave a thickness of $56\frac{1}{2}$ feet.

T. A. Conrad† described, from the Miocene of the Eastern and Southern States, *Nassa subcylindrica*, *Volutifusus typus*, *Cancellaria scalarina*, *Saxicava parilis*, *Spisula capillaria*, *Tellina peracuta*, *T. capillifera*, *Astarte compsonema*, *Lithophaga subalveata*, *Macoma virginiana*, *Mercenaria obtusa*, and *Cumingia medialis*.

Philip P. Carpenter‡ described, from the Pliocene of Santa Barbara, California, *Turritella jewetti*, *Bittium armillatum*, *Opalia insculpta*, *Trophon tenuisculptus*, and *Pisania fortis*.

In 1867, Prof. E. W. Hilgard§ said that nowhere has the geologist more need of divesting himself of reliance upon lithological characters, than in the study of the Mississippi Eocene. Not only do the materials of the different groups often bear a most extraordinary resemblance to each other, but their character varies incessantly, *in one and the same stratum*, within short distances. Hale remarks that in Mississippi, the Orbitoides limestone seems to be represented by blue marlstone, and so it is, sometimes. But while on the one hand we see the hard limestone of the Vicksburg bluff passing into blue marl (Byram, Marshall's quarry), we on the other hand find it passing equally into a rock undistinguishable from that of St. Stephens (Brandon, Wayne county) ; the varied fossils described by Conrad disappearing almost

---

* Geo. Sur. of Ill., vol. i.
† Am. Jour. Conch., vol. ii.
‡ Ann. & Mag. Nat. Hist., 3d ser., vol. xvii.
§ Am. Jour. Sci. & Arts, 2d ser., vol. xliii.

entirely, to be replaced by millions of Orbitoides imbedded in a semi-indurate mass of carbonate of lime, interspersed at times with similarly constituted conglomeratic masses of *Pecten poulsoni.* He could not, therefore, agree to the propriety of distinguishing as separate divisions the Orbitoides limestone, and the Vicksburg Group. The occurrence of a different species of Orbitoides (*O. nupera*) at Vicksburg, does not alter the case, for the undoubted *O. mantelli* occurs there also, in the solid rock. And there are few of the characteristic fossils of the Vicksburg profile, which do not, on some occasions, occur side by side with the *O. mantelli*, and its companions, *Pecten poulsoni*, and *Ostrea vicksburgensis.* Of course, the coral had its favorite haunts—the mollusks theirs. There is nothing surprising in the fact, that where one abounds, the others are usually scarce, or *vice versa.* He regarded the Shell Bluff Group of Conrad, or the Red Bluff Group—No. 4 of the Vicksburg section—which is characterized by the occurrence of *Ostrea georgiana*, as more or less co-extensive with the Vicksburg Group, and regularly associated with it, as a subordinate feature. Its inconsiderable thickness readily explains its entire absence at many points, where, stratigraphically, it ought to appear.

Prof. E. D. Cope* described, from the Miocene of Charles county, Maryland, *Eschrichtius cephalus*, *Rhabdosteus latiradix*, *Squalodon mento*, *Aetobatis profundus*, *Myliobatis gigas*, *M. pachyodon*, *M. vicomicanus*, *Raja dux*, *Notidanus plectrodon*, *Galeocerdo lævissimus*, *Sphyrna magna*, *Trionyx cellulosus*, *Thecachampsa contusor*, *T. sericodon*, *Orycterocetus crocodilinus*, *Priscodelphinus acutidens*, *Eschrichtius leptocentrus*, *Squalodon protervus*, and *Galera macrodon.*

T. A. Conrad† described, from the Eocene of Texas, *Venericardia mooreana;* from the Miocene of the Eastern and Southern States, *Pleuromeris decemcostata*, *Mactra contracta*, *M. virginiana*, *Lucina densata*, *Cardium emmonsi*, *Mercenaria percrassa*, *Mulinia parilis*, *Semele carolinensis*, *Abra nuculiformis*, *Corbula curta*, *Pecten tricarinatus*, *P. yorkensis*, *Sycotypus pyriformis*, *Cylichna*, *virginica*, *Zizyphinus briani*, *Z. punctatus*, *Neverita densata*, *N. emmonsi*, *Ptychosalpinx*, *scalaspira*, *Paranassa granifera*, *Bursa centrosa*, and *Busycon dumosum.* Prof. Gill described, from North Carolina, *Sycotypus elongatus.*

In 1868, Prof. J. W. Dawson‡ offered the following reasons, to show,

* Proc. Acad. Nat. Sci.
† Am. Jour. Conch., vol. iii.
‡ Acadian Geology.

that the drift deposits of eastern America are not to be accounted for upon the theory of a terrestrial origin or a supposed glacial period.

1. It requires a series of suppositions unlikely in themselves, and not warranted by facts. The most important of these is the coincidence of a wide-spread continent, and a universal covering of ice in a temperate latitude. In the existing state of the world, it is well known that the ordinary conditions required by glaciers in temperate latitudes are elevated chains and peaks extending above the snow-line; and that cases, in which, in such latitudes, glaciers extend nearly to the sea level, occur only where the mean temperature is reduced by cold ocean currents approaching to high land, as for instance, in Terra del Fuego, and the southern extremity of South America. But the temperate regions of North America could not be covered with a permanent mantle of ice under the existing conditions of solar radiation; for, even if the whole were elevated into a table-land, its breadth would secure a sufficient summer heat to melt away the ice, except from high mountain peaks. Either, then, there must have been immense mountain-chains which have disappeared, or there must have been some unexampled astronomical cause of refrigeration, as, for example, the earth passing into a colder portion of space, or the amount of solar heat being diminished. But the former supposition has no warrant from geology, and astronomy affords no evidence for the latter view, which, beside, would imply a diminution of evaporation, militating as much against the glacier theory as would an excess of heat. An attempt has recently been made by Professor Frankland to account for such a state of things, by the supposition of a higher temperature of the sea, along with a colder temperature of the land; but this inversion of the usual state of things is unwarranted by the doctrine of secular cooling of the earth; it is contradicted by the fossils of the period, which show that the seas were colder than at present; and if it existed, it could not produce the effects required, unless a preter-natural arrest were at the same time laid on the winds, which spread the temperature of the sea over the land. The alleged facts observed in Norway, and stated to support this view, are evidently nothing but the results ordinarily observed in ranges of hills, one side of which fronts cold sea-water, and the other land warmed in summer by the sun.

The supposed effects of the varying eccentricity of the earth's orbit, so ably expounded by Mr. Croll, are no doubt deserving of consideration in this connection; but I agree with Sir Charles Lyell in regarding them as insufficient to produce any effect so great as that refrigeration supposed by the theory now before us, even if aided by what Sir Charles truly regards as a more important cause of cold—namely, a different distribution of land and water, in such a manner as to give a great excess of land in high latitudes.

2. It seems physically impossible that a sheet of ice, such as that supposed, could move over an uneven surface, striating it in directions uniform over vast areas, and often different from the present inclinations of the surface. Glacier ice may move on very slight slopes, but it must follow these; and the only result of the immense accumulation of ice supposed, would be to prevent motion altogether by the want of slope or the counter-action of opposing slopes, or to induce a slight and irregular motion toward the margins, or outward from the more prominent protuberances.

It is to be observed, also, that, as Hopkins has shown, it is only the sliding motion of glaciers that can polish or erode surfaces, and that any internal changes, resulting from the mere weight of a thick mass of ice resting on a level surface, could have little or no influence in this way.

3. The transport of bowlders to great distances, and the lodgment of them on hill-tops, could not have been occasioned by glaciers. These carry downward the blocks that fall on them from wasting cliffs. But the universal glacier supposed could have no such cliffs from which to collect; and it must have carried bowlders for hundreds of miles, and left them on points as high as those they were taken from. On the Montreal Mountain, at a height of 600 feet above the sea, are huge bowlders of feldspar from the Laurentide Hills, which must have been carried 50 to 100 miles from points of scarcely greater elevation, and over a valley in which the striæ are in a direction nearly at right angles with that of the probable driftage of the bowlders. Quite as striking examples occur in many parts of the country. It is also to be observed that bowlders, often of large size, occur scattered through the marine stratified clays and sands containing sea-shells; and whatever views may be entertained as to other bowlders, it can not be denied that these have been borne by floating ice. Nor is it true, as has been often affirmed, that the bowlder clay is destitute of marine fossils. At Isle Verte, Riviere du Loup, Murray Bay, and St. Nicholas on the St. Lawrence, and also at Cape Elizabeth, near Portland, there are tough stony clays of the nature of true "till," and in the lower part of the drift, which contain numerous marine shells of the usual Post-pliocene species.

4. The Post-pliocene deposits of Canada, in their fossil remains and general character, indicate a gradual elevation from a state of depression, which on the evidence of fossils must have extended to at least 500 feet, and on that of far-traveled bowlders to several times that amount; while there is nothing but the bowlder clay to represent the previous subsidence, and nothing whatever to represent the supposed previous ice-clad state of the land, except the scratches on the rock surfaces. which must have been caused by the same agency which deposited the bowlder clay.

5. The peat deposits, with fir roots, found below the bowlder clay in Cape Breton, the remains of plants and land snails in the marine clays of the Ottawa, and the shells of the St. Lawrence clays and sands, show that the sea at the period in question had nearly the temperature of the present Arctic currents of our coasts, and that the land was not covered with ice, but supported a vegetation similar to that of Labrador and the north shore of the St. Lawrence at present. This evidence refers not to the later period of the Mammoth and the Mastodon, when the re-elevation was perhaps nearly complete, but to the earlier period contemporaneous with. or immediately following the supposed glacier period. In my former papers on the Post-pliocene of the St. Lawrence, I have shown that the change of climate involved is not greater than that which may have been due to the subsidence of land, and to the change of the course of the Arctic current, actually proved by the deposits themselves.

It has long been known to geologists, that in northeastern America, two main directions of striation of rock surfaces occur, from northeast to southwest, and from northwest to southeast; and that locally the directions vary from these to north and south, and east and west. It would seem that the dominant direction in the valley of the St. Lawrence, along the high lands to the north of it, and across western New York, is northeast and southwest; and that there is another series of scratches running nearly at right angles to the former, across the neck of land between Georgian Bay and Lake Ontario, down the valley of the Ottawa, and across parts of the eastern townships, connecting with the prevalent south and southeast striation, which occurs in the valleys of the Connecticut and Lake Champlain, and elsewhere in New England, as well as in Nova Scotia and New Brunswick. What were the determining conditions of these two courses, and were they contemporaneous or distinct in time? The first point to be settled in answering these questions is the direction of the force which caused the striæ. Now, I have no hesitation in asserting, from my own observations, as well as from those of others, that for the south-west striation the direction was *from the ocean toward the interior, against the slope of the St. Lawrence valley.* The crag-and-tail forms of all-our isolated hills, and the direction of transport of bowlders carried from them, show that throughout Canada the movement was from northeast to southwest. This at once disposes of the glacier theory for the prevailing set of striæ; for we can not suppose a glacier moving from the Atlantic up into the interior. On the other hand, it is eminently favorable to the idea of ocean drift. A subsidence of America, such as would at present convert all the plains of Canada and New York and New England into sea, would determine the course of the Arctic current over this submerged land from northeast to southwest; and as the current would move *up a slope,* the ice which it bore would tend to ground, and to grind the bottom as it passed into shallower water; for it must be observed that the character of slope which enables a glacier to grind the surface may prevent ice borne by a current from doing so, and *vice versa.*

Now, we know that in the Post-pliocene period, eastern America was submerged, and, consequently, the striation at once comes into harmony with other geological facts. We have, of course, to suppose that the striation took place during submergence, and that the process was slow and gradual, beginning near the sea and at the lower levels, and carried upward to the higher ground in successive centuries, while the portions previously striated were covered with deposits swept down from the sinking land or dropped from melting ice.

The predominant southwest striation, and the cutting of the upper lakes, demand an outlet to the west for the Arctic current. But both during depression and elevation of the land, there must have been a time when this outlet was obstructed, and when the lower levels of New York, New England and Canada were still under water. Then the valley of the Ottawa, that of the Mohawk, and the low country between Lakes Ontario and Huron, and the valleys of Lake Champlain and the Connecticut, would be straits or arms of the sea, and the current, obstructed in its direct flow, would set principally along these, and act on the rocks in north and south and northwest and southeast directions. To this portion of the process, I would attribute the northwest and southeast striation. It is true, that this view does not account for the southeast striæ observed on some high peaks in New England; but it must be observed that even at the time of greatest depression, the Arctic current would cling to the Northern land, or be thrown so rapidly to the west that its direct action might not reach such summits.

Nor would I exclude altogether the action of glaciers in eastern America, though I must dissent from any view which would assign to them the principal agency in our glacial phenomena. Under a condition of the continent in which only its higher peaks were above the water, the air would be so moist, and the temperature so low, that permanent ice may have clung about mountains in the temperate latitudes. The striation itself shows that there must have been extensive glaciers, as now, in the extreme Arctic regions. Yet I think, that most of the alleged instances must be founded on error, and that old sea-beaches have been mistaken for moraines. I have failed to find even in our higher mountains any distinct sign of glacier action, though the action of the ocean breakers is visible almost to their summits; and though I have observed in Canada and Nova Scotia many old sea-beaches, gravel-ridges, and lake-margins, I have seen nothing that could fairly be regarded as the work of glaciers. The so-called moraines, in so far as my observation extends, are more probably shingle beaches and bars, old coast-lines loaded with bowlders, trains of bowlders or " ozars." Most of them convey to my mind the impression of ice-action along a slowly subsiding coast, forming successive deposits of stones in the shallow water, and burying them in clay and smaller stones as the depth increased. These deposits were again modified during emergence, when the old ridges were sometimes bared by denudation, and new ones heaped up.

We now have, in all, exclusive of doubtful forms, about one hundred species of marine invertebrates, from the Post-pliocene clays of the St. Lawrence valley. All, except four or five species, belonging to the older or deep water part of the deposit, are known as living shells of the Arctic or boreal regions of the Atlantic. About half of the species are fossil in the Post-pliocene of Great Britain. The great majority are now living in the Gulf of St. Lawrence, and on the neighboring coasts; and more especially on the north side of the gulf and the coast of Labrador. In so far, then, as marine life is concerned, the modern period in this country is connected with that of the bowlder clay by an unbroken chain of animal existence. These deposits in Lower Canada afford no indications of the terrestrial fauna ; but the remains of *Elephas primigenius*, in beds of similar age in Upper Canada, show that during the period in question, great changes occurred among the animals of the land ; and we may hope to find similar evidences elsewhere, especially in localities where, as on the Ottawa, the debris of land-plants and land-shells occur in the marine deposits.

The Eocene of New Jersey* is known as the Upper marl bed, and has a thickness of 37 feet. Fossils are abundant wherever marl pits have been opened, between Deal on the sea shore and Clementon in Camden county.

The Miocene is recognized by its fossils in many localities in New Jersey. It is not always conformable with the Eocene below, and its thickness is variable.

In 1868, Prof. E. D. Cope† described the Miocene deposit of the western shore of Maryland, as consisting of a dark, sandy clay, varying from a leaden to a blackish color, through which water does not penetrate. Its upper horizon may be traced along the high shores and cliffs of the Chesapeake by the line of trickling springs which follow its upper surface. A great bed of shells occurs at from fourteen to twenty-two feet below its upper horizon.

He described, *Cetophis heteroclitus, Ixacanthus cœlospondylus, Priscodelphinus spinosus,* now *Belosphys spinosus, P. atropius,* now *B. atropius. P. stenus,* now *B. stenus. Zarhachis flagellator, Delphinapterus ruschenbergeri,* now *Tretosphys ruschenbergeri. D. lacertosus,* now *T. lacertosus. D. gabbi,* now *T. gabbi, D. hawkinsi,* now *T. hawkinsi, D. tyrannus,* now *Eschrictius tyrannus, E. pusillus, Megaptera expansa,* now *E. expansus;* from the Eocene green sand of Monmouth county, New Jersey, *Palœophis halidanus,* and *P. littoralis.*

Isaac Lea described, from a Miocene deposit, six miles northeast of Camden, New Jersey, *Unio alatoides, U. carriosoides, U. humerosoides, U. nasutoides, U. radiatoides, U. subrotundoides, U. roanokoides. U. ligamentinoides, U. grandioides,* and *U. corpulentoides.*

Dr. Joseph Leidy described, from blue clay and sand beneath a bed of bitumen of Pliocene age, in Hardin county, Texas, *Megalonyx validus, Trucifelis fatalis,* and *Emys petrolei;* from Douglas Flat, Calaveras county, California, *Elotherium superbum;* from Martinez, *Equus pacificus,* the largest known fossil horse tooth; from Ashley river, South Carolina, *Hoplocetus obesus;* from Gibson county, Indiana, *Dicotyles nasutus,* found when digging a well between 30 and 40 feet below the surface; from the Miocene of the Bad Lands of White river, Dakota, *Leptictis haydeni, Ictops dakotensis;* from Half-moon Bay

---

* Geo. of N. Jersey, 1868.
† Proc. Acad. Nat. Sci.

California, *Delphinus occiduus;* from Washington county, Texas, *Anchippus texanus;* from the Bad Lands of Nebraska, *Lophiodon occidentale,* and from Shark river, Monmouth county, New Jersey, *Anchippodus riparius.*

T. A. Conrad* described, from the Miocene of the Atlantic coast, *Volutella oviformis, Prunum virginiana,* now *Marginella virginiana, Mercenaria cuneata, Caryatis plionema, Carditamera recta* ; and from Wyoming, *Goniobasis carteri.*

Prof. O. C. Marsh† described, from the Tertiary at Antelope station, on the Union Pacific Railroad, 450 miles west of Omaha, in Nebraska, *Equus parvulus,* now *Protohippus parvulus.*

The Tertiary underlies a wide central belt in West Tennessee, and was subdivided by Prof. Safford,‡ in 1869, in ascending order, into 1, Porters' Creek Group ; 2, Orange Sand ; 3, Bluff Lignite ; 4, Postpliocene beds, on the Mississippi Bluff, consisting of Bluff gravel and Bluff loam ; and superficial gravel beds, in other parts of the State, consisting of ore-region gravel, eastern gravel, and lastly of bottoms, and alluvial beds.

The Bluff lignite consists, especially in the middle and southern parts of the State, of a series of stratified sands, with more or less sandy, slaty clay, characterized by the presence of well-marked beds of lignite; though, in the northern part of the State, its upper portion is frequently more or less indurated, presenting layers of soft sandstone with less lignite. The upper part of the series is generally well exposed below the gravel of the Mississippi Bluffs. At Memphis, however, it scarcely appears above low-water. About one hundred feet of the series has been seen. In this thickness it contains from one to three beds of lignite, which are from half a foot to four feet in thickness.

The outcrop of the Orange sand or Lagrange Group, forms more than a third of the entire surface of West Tennessee. It occupies a belt, about 40 miles wide, which runs in a northeasterly direction, through nearly the central portion of this division of the State. As seen in bluffs, railroad cuts, gullies, and in nearly all exposures, it is generally a great stratified mass of yellow, orange, red or brown, and white sands, presenting occasionally an interstratified bed of white,

---

* Am. Jour. Conch., vol. iv.
† Am. Jour. Sci. & Arts, 2d series, vol. xlvi.
‡ Geo. of Tenn.

grey, or variegated clay. The sand beds are usually more or less argillaceous; sometimes but little, or not at all so. Like the Ripley Group, it contains, occasionally, patches, plates, and thin layers of ferruginous, sometimes argillaceous sandstone, and as in that group, presents, locally, massive blocks of sandstone on high points. At La Grange, a fine section of the group, more than a hundred feet in thickness, is exposed. It includes within its outcrop, nearly all of Fayette, Haywood, Madison, Gibson, and Weakley counties; the larger parts of Hardeman, Carroll, and Henry; and small parts of Shelby, Tipton, Henderson, Dyer, and Obion. He supposed this group to be of Eocene age, and to have a thickness of about 600 feet. This group must not be confounded with the Post-pliocene Orange sand of Hilgard, which occurs in Mississippi and Louisiana.

The Porter's Creek Group contains proportionally more laminated or slaty clay than the Orange Sand or Lagrange Group. Along the Memphis and Charleston railroad, the belt of surface occupied by the group is about eight miles wide. It becomes narrower in its northward extension, and appears to be the northern extension of the lower part of Hilgard's Northern Lignitic Group. The thickness is from 200 to 300 feet, and in this are usually several beds of slaty clay from five to fifty feet in thickness. It is well exposed on Porter's creek, in Hardeman county, and on the road from Bolivar to Purdy, commencing about seven miles from the former place, and extending to or beyond Wade's creek.

Prof. E. W. Hilgard[*] described the Grand Gulf Group, Orange Sand and Loess at Port Hudson, Miss., and gave a descending section midway between Port Hudson and Fontania as follows: 1st, Yellow loam, sandy below, 8 to 10 feet. 2d, White and yellow hard pan, 18 feet. 3d, Orange and yellow sand, sometimes ferruginous sandstone, irregularly stratified, 8 to 15 feet. 4th, Heavy, greenish or bluish clay, 7 feet. 5th, White, indurate silt or hard pan, 18 feet. 6th, Heavy, green clay, with porous, calcareous concretions above, ferruginous ones below; some sticks and impressions of leaves, 30 feet. 7th, Brown muck and white or blue clay with cypress stumps, 3 to 4 feet.

At the stage of extreme low water the stump stratum is visible to the thickness of 10 feet at its highest point; showing several generations of stumps, one above another, also the remnants of many successive falls of leaves and overflows. The wood is in a good state of

---

[*] Am. Jour. Sci. & Arts, 2d series, vol. xlvii.

preservation. The stump stratum exists, at about the same level, over all the Delta plain of the Mississippi and along the Gulf coast from Mobile, on the east, to the Sabine river.

Dr. Joseph Leidy* described, from the White River Group of Dakota, *Oreodon affinis, O. bullatus, O. hybridus, Leptauchenia nitida, Homocamelus caninus, Cosoryx furcatus, Nanohyus porcinus, Protohippus placidus, Hipparion affine,* and *H. gratum.* He described from the Eocene near Fort Bridger, Wyoming,† *Omomys carteri, Trionyx guttatus, Emys wyomingensis,* and from South Bitter creek, near where it crosses the stage route, 70 miles west of the summit of the Rocky mountains, in western Wyoming, *Crocodilus aptus.*

Prof. E. D. Cope‡ described, from the Miocene of Shiloh, Cumberland county, New Jersey, *Tretosphys uræus, Zarhachis velox,* and *Trionyx lima;* from the mouth of the Patuxent river, Maryland, *Zarhachis tysoni.*

He described,§ from the Eocene marl pits, at Shark river, Monmouth county, N. J., *Hemicaulodon effodiens;* from Farmingdale, *Myliobates glottoides,* and *Cœlorhynchus acus;* from the Green River Group, on the upper waters of Green river, Wyoming, *Asineops squamifrons, Clupea pusilla, Cyprinodon levatus;* from the Miocene in Wayne county, North Carolina, *Pneumatosteus nahunticus;* from Duplin county, *Pristis attenuatus;* from Edgecombe county, *Eschrichtius polyporus;* from Quanky creek, Halifax county, *Mesoteras kerranus;* from Stafford county, Va., *Thinotherium annulatum.*

He described, from the Post-pliocene, at Savannah, Georgia, *Anoplonassa forcipata;* from cave Breccia, in Wythe county, Virginia, *Tamias lævidens, Sciurus panolius,* and *Galera perdicida.*

Prof. O. C. Marsh‖ described, from the Eocene, near Shark river, Monmouth county, New Jersey, *Dinophis grandis.*

T. A. Conrad¶ described, from the same locality, *Pecten kneiskerni, Crassatella littoralis, Crassina veta, Bucardia veta, Caryatis delawarensis, Protocardia curta, Onustus annosus,* and *Terebratula glossa.* And from the Miocene of St. Charles county, Maryland, and from Petersburg, Va., *Pecten cerinus, Callista virginiana, Saxicava insita, Scapharca tenuicardo, Mercenaria plena,* and *Capsa parilis.*

* Jour. Acad. Nat. Sci., vol. vii.
† Proc. Acad. Nat. Sci.
‡ Proc. Acad. Nat. Sci.
§ Proc. Am. Phil. Soc., vol. xi.
‖ Am. Jour. Sci. & Arts, 2d series, vol. xlviii.
¶ Am. Jour. Conchology, vol. v.

W. M. Gabb* described, from the Miocene in Contra Costa county, near Tomales bay, near Martinez, Walnut creek, Monterey county, San Emidio, Cerros island, and other places in California, *Dosinia mathewsoni, Pecten packhami, Triptera clavata, Trophon ponderosum, Neptunea recurva, Metula remondi, Agasoma gravida, A. sinnata, Ranella mathewsoni,* now *Bursa mathewsoni, Cuma biplicata, Ancillaria fishi, Neverita callosa, Cancellaria vetusta, Turritella hoffmanni, Trochita filosa, T. inornata, Pachypoma biangulata, Pandora scapha, Hemimactra lenticularis, H. occidentalis, Schizodesma abscissa, Chione mathewsoni,* now *Callista mathewsoni, C. whitneyi,* now *C. whitneyi, Callista voyi, Dosinia conradi, Tapes truncata, Cardium meekanum, Conchocele disjuncta, Mytilus mathewsoni, Modiola multiradiata,* now *Volsella multiradiata, Pecten cerrocensis, P. veatchi, Ostrea atwoodi, O. taylorana, O. veatchi, O. cerrocensis, Asterias remondi, Ficus pyriformis, F. nodiferus, Venus pertenuis.* From the fresh water Tertiary, or Pliocene, on Snake river, in Idaho Territory, *Melania taylori,* and *Lithasia antiqua;* from the Pliocene, near Santa Barbara, Humboldt bay, San Francisco county, Kirker's Pass, Sonoma county, and other places in California, *Cancer breweri, Surcula carpenterana, Pleurosoma voyi, Columbella richthofeni, Littorina remondi, Zirphaea dentata, Gari alata, Dosinia staleyi,* now *Tapes staleyi, Cyrena californica, Lucina richthofeni, Neptunea altispira, N. humerosa, Sigaretus planicostum, Cancellaria altispira, Acmaea rudis, Siliquaria edentula, Caryatis barbarensis, Saxidomus gibbosus.* And from the Post-pliocene, near Santa Barbara, and San Pedro, *Surcula tryonana, S. perversa, Clathurella conradana, Muricidea paucivaricata, Trophon squamulifer, Cancellaria gracilior,* and *C. tritonidea.*

Prof. Leo. Lesquereux† described,from the Lower Eocene or Northern Lignitic Group of Tippah, Miss., and La Grange, and Sommerville, Tennessee, *Calamopsis danai, Sabal grayana,* now *Sabalites grayanus, Salisburia binervata, Populus monodon, Salix wortheni, S. tabellaris, Quercus moori, Q. retracta, Celtis brevifolia, Ficus schimperi, F. cinnamomoides, Laurus pedatus, Cinnamomum mississippiense, Persea lancifolia, Ceanothus meigsi, Juglans appressa, J. saffordana, Magnolia laurifolia, M. lesleyana, M. oralis, M. cordifolia, Asimina leiocarpa,* and *Phyllites truncatus.*

---

* Pal. of Cal., vol. ii.
† Trans. Am. Phil. Soc., vol. xiii.

Oswald Heer* described, from the Tertiary of Alaska, *Pteris sithen-sis*, *Taxodium tinajorum*, *Taxites microphyllus*, *Phragmites alaskanus*, *Poacites tenuistriatus. Carex servata. Sagittaria pulchella, Vaccinium friesi, Diospyros stenosepala, Viburnum nordenskioldi,* *Hedera auriculata, Vitis crenata, Tilia alaskana, celastrus borealis, Ilex insignis. Trapa borealis, Juglans nigella, J. picroides, Spirœa andersoni,* and identified numerous plants with those described from the Miocene of Europe. He described the insect *Chrysomelites alaskanus*, and Dr. Carolus Mayer described, *Unio onariotis, U. athlios, Paludina abavia,* and *Melania furuhjelmi.*

The Jackson Group, in Louisiana,† consists of marine strata; of lignitic beds that tell of swamps; and of nonfossiliferous beds of laminated sands and clays. It spreads over the State north of the Vicksburg outcrop and west of the Bastrop Hills. The marine strata contain massive clays, often full of selenite. At Grand View there is a stratum of such clay 85 feet thick.

The Vicksburg Group, in Louisiana, consists of smooth, yellow and red clays, with a very small proportion of sand. Limestone nodules occur, generally, soft and yellow, but sometimes hard and white, and always full of casts of shells. It is exposed from Godwin's shoals to about six miles south of Natchitoches, and from a point below Montgomery to the Washita, below Grand View, but it never occupies an area more than about twelve miles wide.

In 1870, Dr. Joseph Leidy‡ described, from the Fort Bridger Eocene, of Wyoming, *Baptemys wyomingensis*, now *Dermatemys wyomingensis, Emys stevensonanus, Patriofelis ulta, Lophiodon modestus, Hyopsodus paulus, Emys jeansi. E. haydeni, Baena arenosa, Saniva ensidens;* from near the junction of the Big Sandy and Green rivers, *Palœosyops paludosus, Crocodilus elliotti;* from Black's Fork, *Microsus cuspidatus, Notharctus tenebrosus;* from the Tertiary of Colorado, *Megacerops coloradoensis;* from the Tertiary of the Rocky mountain region, *Oncobatis pentagonus, Mylocyprinus robustus;* from Henry's Fork of Green river, *Lophiotherium sylvaticum;* from the Miocene in the valley of Bridge creek, a tributary of John Day's river. Oregon, *Oreodon superbus, Anchitherium condoni;* from Gay Head, Martha's Vineyard, *Graphiodon vinearius;* from the

* Flora Fossilis Alaskana.
† Geo. of Louisiana, 1870.
‡ Proc. Acad. Nat. Sci.

Pliocene of the Niobrara river, *Merychochœrus rusticus;* from Dry creek, Stanislaus county, California, *Mastodon shepardi;* and from Tuolumne county, *Auchenia californica.*

Prof. O. C. Marsh* described, from the Eocene of New Jersey, *Thecachampsa minor;* from the Miocene of Edgecombe county, North Carolina,† *Catarractes antiquus;* from Maryland, *Puffinus conradi;* from the Niobrara river, *Grus haydeni, Graculus idahensis;* from Squankum, New Jersey,‡ *Rhinoceros matutinus;* from Shark river, *Dicotyles antiquus;* and from the Pliocene at Monmouth, *Meleagris altus.*

Prof. F. B. Meek described, from the Miocene, at Fossil Hill, Hot Spring mountains, Idaho, *Sphærium rugosum, S. idahoense, Ancylus undulatus, Goniobasis sculptilis, G. subsculptilis, Carinifex binneyi, C. concava* and *C. tryoni.*

T. A. Conrad§ described, from the Miocene of Virginia and South Carolina, *Artena undulata, Crepidula rostrata, C. recurvirostra, C. virginica, Persicula ovula,* and *Axinæa bella.*

The Grand Gulf Group of Louisiana‖ consists of nonfossiliferous clays and sandstones pretty regularly stratified, varied, occasionally, by clayey sand and beds containing twigs and leaves. The sandstone occurs in ledges from six inches to 20 feet in thickness. It is cut into four parts by the bluff and the alluvion of Red river and the Mississippi. One reaches the Vicksburg area and extends into Mississippi; another is southwest of Red river and extends into Texas; another is northeast of Red river as far as Sicily Island on the Ouachita; and the other is at the western part of the Avoyelles prairies.

In 1871, T. A. Conrad¶ described, from the Eocene at Claiborne, Alabama, *Caryatis exigua;* and from the Oligocene at Vicksburg, Mississippi, *Macoma sublintea,* and *Abra protexta.*

F. B. Meek** described, from the Bridger Eocene at Henry's Fork, Black's Fork, and Church buttes, Wyoming, *Viviparus wyomingensis.*

Brady and Crosskey†† described, from the Post-pliocene of Portland and Saco, Maine, and from Montreal, Canada, *Cythere macchesneyi, C. logani, C. cuspidata, Cytherura cristata, C. granulosa,* and *Cytheropteron complanatum.*

---

* Am. Jour. Sci. and Arts, 2d series, vol. 50.
† *Ibid*, vol. xlix.
‡ Proc. Acad. Nat. Sci.
§ Am. Jour. Conch., vol. vi.
‖ Geo. of Lou., 1871.
¶ Am. Jour. Conch., vol. vi.
** Proc. Acad. Nat. Sci.
†† Lond. Geo. Mag., vol. viii.

Dr. Joseph Leidy* described, from the Bridger Eocene of Wyoming, *Anosteira ornata, Hybemys arenarius, Testudo corsoni, Emys carteri, Baena undata, Trogosus vetulus,* now *Anchippodus vetulus, Sinopa rapax, Palæosyops major, Hyrachyus eximius, Paramys delicatus, P. delicatior,* and *P. delicatissimus,* all now *Plesiarctomys,* and *Mysops minimus.* He described from the Miocene of Alkali flats, Oregon, *Rhinoceras pacificus,* and from Crooked river, *Stylemys oregonensis,* now *Testudo oregonensis.*

Prof. E. D. Cope† described, from the Post-pliocene occurring in a limestone fissure in Chester county, Pennsylvania, *Megalonyx loxodon, M. sphenodon, M. tortulus, M. wheatleyi, Sciurus calycinus, Arvicola speothen, A. tetradelta, A. didelta, A. involuta, A. sigmodus, A. hiatidens, Erithizon cloacinum,* and *Praotherium palatinum.* He described from the Miocene near Tuxtla, Chiapas, Mexico, *Prymnetes longiventer.*

Prof. O. C. Marsh‡ described, from the Green river basin west of the Rocky Mountains, *Boavus agilis, B. brevis,* and *B. occidentalis;* from the Bridger Eocene of Wyoming, *Limnophis crassus, Lithophis sargenti, Crocodilus affinis, C. brevicollis, C. grinnelli, C. liodon, C. ziphodon,* now *Limnosaurus ziphodon, Glyptosaurus anceps, G. nodosus, G. ocellatus, G. sylvestris, Titanotherium* (?) *anceps, Lophiodon affinis, L. bairdianus, L. nanus, L. pumilis, Anchitherium gracile,* now (?) *Orohippus gracilis, Lophiotherium ballardi, Elotherium lentum, Platygonus zieglteri, Hyopsodus gracilis. Limnotherium elegans, L. tyrannus, Sciuravus nitidus, S. parvidens, S. undans, Triacodon fallax, Canis montanus, Vulpavus palustris,* and *Bubo leptosteus.*

He described from the Miocene at Scott's Bluff, on North Platte river, Nebraska, *Amphicyon angustidens;* from Northern Colorado, *Meleagris antiquus;* from Cumberland county, New Jersey,§ *Lophiodon validus,* now *Tapiravus validus;* and named, but did not describe, from Wyoming, *Amia depressa, A. newberryana, Lepidosteus glaber,* and *L. whitneyi.* Also from the Pliocene sands, near the headwaters of the Loup Fork river, Nebraska, (‖) *Platygonus striatus, Arctomys vetus, Geomys bisulcatus, Aquila dananus;* and from Oregon, *Platygonus condoni,* and *Dicotyles hesperius.*

* Proc. Acad. Nat. Sci.
† Proc. Am. Phil. Soc., vol. xii.
‡ Am. Jour. Sci. and Arts, 3d series, vol. i. & ii.
§ Proc. Acad. Nat. Sci.
(‖) Am. Jour. Sci. and Arts, 3d series, vol. ii.

In 1872, Dr. Dawson* said, that the Bowlder clay of Canada consists of hard, gray clay, filled with stones, and thickly packed with bowlders, and usually rests directly on striated rock surfaces ; though in Cape Breton, a peaty or brown coal deposit, with branches of trees, has been found to underlie it, and in some places there are deposits of rolled gravel beneath it. The stones are often scratched and ground into wedge-shapes, as if by the action of ice. At Isle Verte, Riviere du Loup, Murray Bay, Quebec, and St. Nicholas, on the St. Lawrence it is fossilferous, containing, *Leda truncata, Balanus hameri*, and *Bryozoa.*

In some localities the stones in the Bowlder clay, are almost exclusively those of the neighboring rock formations, in others those having traveled from a distance predominate ; occasional instances occur where bowlders have been transported to the northward. Though the Bowlder clay often presents a somewhat widely extended and uniform sheet, yet it may be stated to fill up small valleys or depressions, and to be thin or absent on ridges and rising grounds.

Beneath the Bowlder clay on the St. Lawrence and the Ottawa, there are two sets of striæ, a southeast set, and a southwest set. In Nova Scotia and New Brunswick, as in New England, the prevailing direction is southeastward, though there are also southwest and south striation, and a few cases where the direction is nearly east and west. At the Mile end quarries, near Montreal, the polished and grooved surface of the limestone, shows four sets of striæ. The principal ones have the direction of S. 68° W. and S. 60° W. respectively, and the second of these sets is the stronger and coarser, and sometimes obliterates the first. The two other sets are comparatively few and feeble striæ, one set running nearly north and south, and the other northwest and southeast. These last are probably newer than the first two sets. The locality is to the northeast of the mass of trap constituting the Montreal mountain, and evinces that the movement must have been up the St. Lawrence, which is the dominant direction of the striæ in this valley. It is the Bowlder clay connected with this S. W. striation, that is rich in marine fossils.

At the mouth of the Saguenay, near Moulin Bode, are striæ and grooves on a magnificent scale, some of the latter being ten feet wide, and four feet deep, cut into hard gneiss. Their course is N. 10° W. to N. 20° W. magnetic, or N. 30° to 40° W. when referred to the true

---

* Post-pliocene Geol.

meridian. In the same region, on hills 300 feet high, are roches moutonnees with their smoothest faces pointing in the same direction, or to the northwest. This direction is that of the valley or gorge of the Saguenay, which enters nearly at right angles the valley of the St. Lawrence.

In like manner at Murray Bay, there are striæ on the Silurian limestones near Point au Pique, which run about N. 45° W., but these are crossed by another set having a course S. 30° W., so that we have two sets of markings, the one pointing upward along the deep valley of Murray Bay river to the Laurentide hills inland, the other following the general trend of the St. Lawrence valley. The Bowlder clay which rests on these striated surfaces, is a dark-colored till, full of Laurentian bowlders, and holding *Leda truncata*, and also Bryozoa clinging to some of the bowlders. In ascending the Murray Bay river, we find these bowlder beds surmounted by very thick, stratified clays, with marine shells, which extend upward to an elevation of about 800 feet, when they give place to loose bowlders and unstratified drift.

The Bowlder clay over a large portion of the plain of Lower Canada is succeeded by the Leda clay, which varies in thickness from a few feet to 50 or perhaps 100 feet. The material of the Leda clay is of the same nature as the finer portion of the paste of the Bowlder clay, and the latter seems to graduate into the former. It sometimes holds hard, calcareous concretions, which, as at Green's creek, on the Ottawa, are occasionally richly fossiliferous. When dried, the Leda clay becomes of stony hardness, and when burned, it assumes a brick red color. When dried and levigated, it nearly always affords some foraminifera and shells of ostracoids; and in this, as well as in its color and texture, it closely resembles the blue mud now in process of deposition in the deeper parts of the Gulf of St. Lawrence. It extends west to where the Laurentian ridge of the Thousand Islands crosses the St. Lawrence, and where the same rocks cross the Ottawa, and in general may be said to be limited to the Lower Silurian plain, and not to mount up the Laurentian and metamorphic hills bounding it.

The Saxicava sand sometimes rests upon the Leda clay, sometimes upon Bowlder clay, and often on the older rocks. In some instances the surface of the Leda clay has been denuded and cut into deep trenches, and the sand rests abruptly upon it; in other cases there is a transition from one deposit to the other, the clay becoming sandy and gradually passing upward into pure sand. It must have been originally a marginal and bank deposit, depending much for its distribution on the movement of tides and currents. In some instances, as at Cote des Neiges, near Montreal, and on the terraces on the Lower St. Lawrence, it is obviously merely a shore sand and gravel, like that of the modern beach.

The terraces and inland sea cliffs have been formed by the same recession of the sea which produced the Saxicava sand. At Montreal, where the isolated mass of trap, flanked with Lower Silurian beds, constituting Mount Royal, forms a great tide-gauge for the recession of the Post-pliocene sea, there are four principal sea margins, with several others less distinctly marked. The lowest of these, at a level of 120 feet above the sea, corresponds, in general, with the level of the great plain of Leda clay in this part of Canada. On this terrace, in many places, the Saxicava sand forms the surface, and the Leda clay and Bowlder clay may be seen beneath it. Another at 220 feet in height furnishes Saxicava sand resting on Bowlder clay. Three other terraces occur at heights of 386, 440 and 470 feet, and the latter has, at one place, above the village of Cote des Neiges, a beach of sand and gravel, with *Saxicava* and other shells. Even on the top of the mountain, at a height of about 700 feet, large traveled Laurentian bowlders occur.

The prevalent Post-pliocene deposit on Prince Edward Island is a Bowlder clay, or in some places bowlder loam, composed of red sandstones. This is filled with more or less rounded and striated bowlders of red sandstone, derived from the harder beds of the island. At Campbellton, however, in the western part of the island, a bed of Bowlder clay is found filled with bowlders of metamorphic rocks, similar to those of the mainland of New Brunswick. Striæ on the northeastern coast of the island have a direction S.W. and N.E. ; and on the southwestern coast S. 70° E.

At Campbellton, in the sand and gravel above the Bowlder clay, *Tellina greenlandica* occurs, at an elevation of about 50 feet above the sea. On the surface of the country, there are numerous traveled bowlders. Those of granite, syenite, diorite, felsite, porphry, quartzite and coarse slates are identical, in mineral character, with those which occur in the metamorphic districts of Nova Scotia and New Brunswick, at distances from 50 to 200 miles to the south and southwest; though some of them may have been derived from Cape Breton on the East. Those of gneiss, hornblende schist, anorthosite and labradorite rock must have been derived from the Laurentian rocks of Labrador and Canada, distant 250 miles or more to the northward.

In Nova Scotia and New Brunswick the Bowlder clay or unstratified drift varies from a stiff clay to loose sand, and its composition and color generally depend upon those of the underlying and neighboring rocks. Thus over sandstone it is arenaceous; over shales, argillaceous; and over conglomerates and hard slates, pebbly or shingly. The greater part of the stones contained in the drift are, like the paste containing them, derived from the neighboring formations; though, in some instances, they have been transported from a distance. The transported bowlders have generally been drifted southward, though some have been carried northward, and others in different directions. They have especially been drifted from the more elevated and rocky districts to the lower grounds in their vicinity. The striæ upon the rocks vary from north and south to east and west, though there is a general tendency to a southern and southeastern course.

Alfred R. C. Selwyn* found many fine examples of ice-grooves and scratches on the rocky shores of Vancouver's Island, where they occur in different directions, and sometimes nearly at right angles to each other. He quoted, with approval, the statement of Prof. J. D. Whitney, that northern drift does not occur in California, and that no evidence of its occurrence has yet been detected on the Pacific coast, as far north as British Columbia and Alaska. This conclusion having been arrived at on the authority of Mr. W. D. Dall, naturalist, attached to the Collin's Overland Telegraph Company, and who states that though he had carefully examined the country over which he had passed, in Alaska, for glacial indications, he had not found any effects attributable to such agencies ; and that no bowlders, no scratches, or other marks of ice action had been observed by any of his party, though carefully sought for. And that inland, neither Mr. Selwyn nor his assistant Mr. Richardson observed any.

That the superficial deposits of British Columbia are chiefly developed in the ancient terraces or benches, which, throughout the country, are wonderfully regular and persistent, occuring from the coast up to elevations of nearly 4,000 feet, in the passes of the Rocky mountains. They give a marked and peculiar character to the scenery of the river valleys, rising like gigantic stairs, to elevations of sometimes more than four hundred feet above the adjoining river or lake. In some places two, three, four and five distinct steps can be seen ; while often they have either become merged into one by subsequent denuding agencies,

---

* Geo. Sur. of Canada.

or else the precipitous character of the side of the valley has altogether prevented their formation. The steps vary greatly in height, the greatest height observed being as much as one hundred feet ; in width, from one to five chains is not uncommon.

Nearly all the lakes in British Columbia occupy long, narrow depressions in the river valleys, and are, in fact, lake-like expansions of the rivers. There is no doubt that such lakes were at one time much more extended and more numerous than they now are ; and that, in many places, as, for instance, at Lytton, and on the north bend of the Thompson, and at Canoe river crossing, the terraces mark the old margins of these lakes, while in others they doubtless represent only the ordinary flood-flats of the rivers. The removal of the rocky barriers by which these inland waters were confined would result in the formation of such gorges and canons as we now find on the Fraser at Gale, and below Lytton, as well as on the North Thompson at Murchison's Rapids, and on Canoe river below the wide flats at the crossing, and would, without any general movement of elevation, drain off the waters of the lakes, leaving the old shore lines exactly as we now see them, at corresponding heights on both sides of the valleys. Ordinary alluvial river flats do not commonly occur in that manner, but where a flat occurs on one side there is usually a steep bank on the other, and especially is this so along rapid rivers which traverse a mountainous country.

Dr. F. V. Hayden* said, that Fort Bridger is located in what appears to the eye a sort of basin, inclosed by high, arid table lands, but really in a central portion of the drainage of Black's Fork. The beautiful valleys, Smith's, Black's, and Muddy, have been carved out of the horizontal strata, and between the streams are terraces and flat table lands, which give a singular outline to the surface of the country. No forces now in operation, in this vicinity, could have given the existing features to the surface of the country, and the cause must have been local, proceeding from the northern slope of the Uintas. The beautiful table-top divides between the valleys, and streams are extensions into the plains of the radiating ridges of the mountain slope, and are literally paved, in many places, with the water-worn bowlders of the purplish sandstones and quartzites, and with the carboniferous limestones that compose the nucleus of the Uinta range. Here and there we can see a flat-topped butte cut off by erosion from some of the intervening ridges, and rising above the surrounding country as

---

* U. S. Geo. Sur. of Wyoming.

a partial witness to the extent of the denudation. A little south or west of Fort Bridger, is an isolated butte called Bridger's Butte, which forms a prominent land mark to the traveler, and according to the barometer, rises 750 feet above the valley of Black's Fork, at the fort. The summit appears perfectly level, and was estimated to be about two miles in length, from north to south, and about a fourth of a mile in width, from east to west. The upper portion of the butte is composed of the somber, brown, indurated, arenaceous clays, gray and rusty brown sandstones of the Bridger Group, passing down into limestones and marls of the Green river beds. In the brown clays are abundant remains of turtles, with a few fragments of other vertebrate remains. The terraces along the valley of Black's Fork, are composed of yellowish and whiteish gray marls, and chalky limestones, some of the layers mostly formed of *Unio*, and other fresh-water shells. A few plants were found in the valley of Smith's Fork, in thin black, flinty layers, mostly ferns and leaves of deciduous trees. Between Fort Bridger and Henry's Fork, the indurated, arenaceous clays, of the Bridger Group, are weathered into remarkably unique forms. The absence of harder layers of sandstone did not admit of the weathering into pinnacles, turrets. steeples, domes, etc., as observed near Church Buttes. The surface, though very rugged and almost impassable, except along the valleys of the streams, is much more rounded ; the hills are more dome or pyramid shaped, and entirely destitute of vegetation, except the sage, and several varieties of chenopodiaceous shrubs. Passing up the Cottonwood Fork, the marls and limestones make their appearance, for a short distance, in the bluffs. The divide between the drainage of Smith's Fork and Henry's Fork, is a high ridge of the leaden-brown clays of the Bridger Group, which extends up and juts against the base of the Uinta mountains.

From this ridge to Green river, the valley of Henry's Fork forms a remarkable line of separation between the Bridger Group and the lower beds. This line of separation is somewhat of a surface one, yet it is so marked as to attract the attention of the commonest observer. The valley is quite broad, and on the south side the surface of the country to the summits of the mountains appear smoothed downward, in part grassed over. A close examination will detect some thin remnants of the Bridger Group underlaid by lower Tertiary beds, which have a tendency to weather into rounded. gently-sloping hills. On the north side, the arid, rugged, " bad lands" are very conspicuous, and rise up somewhat abruptly like a high wall. On the north side of the creek, there is a great thickness of the indurated clays of the Bridger Group.

There seems to be no unconformability of the beds included in this Group, and the different beds pass from one to the other gradually ; but to the leaden-gray, somber, indurated, arenaceous clays, which cover a large area east of Fort Bridger, and weather into such unique architectural forms, and contain a large variety of vertebrate remains, Dr. Hayden gave the provisional name of the "Bridger Group." The calcareous layers which underlie the Bridger Group, and are so well displayed lower down on Henry's Fork, he referred to the "Green river Group." Intercalated with the clays of the Bridger Group are beds of rusty-brown and gray sandstones, all tending to a concretionary structure, and disintegrating by exfoliation in thin concentric layers. Sometimes there are beds of sandstone which form an aggregate of concretions. In the whole mass, arenaceous materials predominate. As we descend, the calcareous sediments prevail, until chalky limestones and marl are greatly in excess.

The Green River Group is seen to the best advantage along the valley of Green river, where the sides of the bluff blanks rise to a perpendicular height of 500 feet or more. Ten miles east of Green river Station, the Green River Group disappears abruptly on the south side of Bitter creek, and the coal formations come up to view. On the north side, the eastern limit of the Green River Group is most sharply marked by a long, high, white bluff. that extends off, far to the northeast toward the South Pass.

The dip varies from 3° to 5°, and the laminated calcareous shales gradually pass down into yellow, gray, and brown indurated arenaceous clays, sands, and sandstones, until the well-defined coal strata are exposed, without the least appearance of discordancy.

In traveling from Bear river to Great Salt Lake valley, soon after leaving Carter station, toward the west, pinkish Tertiary beds are observed. They seem to rise from beneath the Bridger Group. Their dip is about northeast 3° to 5°, and they have evidently been disturbed slightly by the later movements which elevated the Uinta range. They are composed of red, indurated, arenaceous clays, with beds of grayish and reddish-gray sandstones alternating; and for this series of strata Dr. Hayden proposed the name of the "Wasatch Group." Pinkish and purplish clays are the dominant features, and give the lithological character of the group as far west as Echo cañon, when the conglomerates prevail. The latter is full of beds of sandstone, largely concretionary, but the sandstones or harder layers are seldom of a reddish color.

Before reaching Bridger station the strata on either side of the road are horizontal. or nearly so. A long, flat ridge extends down a little east of north from the Uinta mountains. between Black's Fork and the Muddy. This may be regarded as the geological divide between the waters of the Great Salt Lake Basin and the drainage of Green river. The Muddy is one of the branches of Black's Fork. which flows into Green river, and west of this stream we have what is called the eastern rim of the Great Basin of Salt Lake. If we were to travel southward to the foot of the Uinta mountains, from the railroad along this divide, we should be able to detect no well-marked line of separation between the Green River Group and the Wasatch Group. Bridger's Butte, as well as the entire eastern portion of this divide fronting the valley of Black's Fork, exhibits a large thickness of the somber, indurated sands, clays, and sandstones of the Bridger Group, passing down into light buff, chalky layers, with *Planorbis. Unio, Helix, Goniobasis*, etc. Within a distance of ten miles to the west of this butte the little streams cut through the pinkish beds of the Wasatch Group. then pass up into whiter. indurated. marly clays, with numerous concretionary layers, differing from the chalky beds of the Bridger and Green river basin. This divide probably forms the junction of two great fresh-water lake basins, that may have existed contemporaneously. The two great basins may have been connected with each other at different points at some stages of their growth, but there is an abrupt, persistent, very marked difference in the character of the sediments of the two basins. While the Green River and Bridger Groups abound with fossils, the Wasatch Group, like all the rocks of the west that are characterized by brick-red coloring matter, is comparatively quite barren. At Bridger station, and from Bridger to Aspen. which is about 24 miles. the ochreous beds of the Wasatch Group are well exposed on both sides of the road, and the valley through which the road passes from Piedmont to Aspen is carved out of this Group

The tunnel at the head of Echo canon is cut through the reddish and purplish indurated sands and clays of the Wasatch Group. It is 770 feet in length. The valley of Echo canon is one of erosion, and on either side the rocks rise wall-like 500 to 1,000 feet, or have been weathered into curiously castellated forms, and bear such names as

Witches' Rock, Eagle Rock, Hanging Rock, Conglomerate Peak, Sentinel Rock, Monument Rock, etc. Monument Rock is a regular obelisk of conglomerate, standing at the junction of the Echo with the Weber valley, and being about 250 feet high. Descending the Echo canon, the more rugged picturesque scenery is exhibited on the right hand, and descending the Weber the same lofty perpendicular walls-weathered here and there into all sorts of fantastic forms, continue to the Narrows, where the Weber river makes a bend to the left, and the conglomerates disappear. The whole series of these beds is referred to the Wasatch Group, and the thickness estimated at from 3,000 to 5,000 feet, the conglomerate portion being from 1,500 to 2,000 feet.

He proposed the name of the "Sweetwater Group," for a lake deposit found in the Sweetwater valley. There is a high ridge or divide, between the drainage of Wind river, North Platte, and Sweetwater, 300 to 400 feet above the channels of these streams, which is composed of the Tertiary beds. The Sweetwater forms a distinct concavity, with this high divide on the north and east, and the valley has been scooped out so that until we reach the Sweetwater Canon, near the South Pass, only the massive granite ridges rise up among the modern Tertiary beds, which jut close up against their base. This is a valley of denudation, over a space of at least 30 to 50 miles in width. All the unchanged formations, from the lignite Tertiary down to the massive feldspathic granites, have been worn away, leaving the granites scattered over the valley in the isolated ridges. At that time there was a fresh-water lake which occupied the entire valley, much as Salt Lake once occupied the great basin, concealing most of the granite ridges, while others rose above the waters like islands. Then was deposited what he called the Sweetwater Group, or perhaps a series of beds identical with the upper portion of the Wind river deposits. These were scooped out again in time, and the Pliocene marls and sands were deposited; and then again there was another scooping out of the valley, and finally a covering of the hills with drift.

The mountainous portions of Northern Utah* are full of beautiful park-like areas, which contain the evidences of an ancient lake. At Copenhagen there is a considerable drift or bowlder deposit with fine white or yellow marly sands and clays, in regular layers, showing the deposit to be Post-pliocene, and that the waters of the lake were comparatively quiet. Near Box Elder Canon are two kinds of terraces,

---

the usual lake terraces, of which there are two well-defined lines at least, and the river terraces, which are confined to the streams, and do not seem to have any direct connection with the former. The lowest plain valley opposite the canon, near the water's edge, is 4,344 feet above sea level; 1st terrace, 4,683 feet; 2d terrace, 4,776 feet; and 3d terrace, 4,858 feet. These terraces show the gradual decrease, step by step, of the waters of the ancient lake, and the operations of the little streams pouring into it from the mountains on either side. The amount of local drift that has been swept down through the gorges or canons and lodged at the opening is very great. At the immediate mouth of the canon, the bowlders are quite large, varying in diameter from a few inches to several feet. Westward toward the shore of the lake the bowlders diminish in size and quantity, and the finer sediments, as sands and marls, increase, showing a constant decrease in the power of the currents of the water after leaving the mouth of the canon.

The local drift is conspicuous in Logan Canon. It is composed of rounded bowlders, with clays and marls, reaching a thickness of 100 to 150 feet in regular and horizontal strata, attached to the sides of the gorge, and showing that, however turbulent the waters, the materials were deposited in a lake. At the entrance of the canon are some remarkable terraces, composed of sands, clays, marls and rounded bowlders.

A large portion of Utah is made up of nearly parallel ranges of mountains, trending nearly north and south, with intervening valleys of greater or less width, which, after their elevation, formed shore lines for detached lakes or bays. It would appear that the last lake-period of this portion of the west commenced in the Pliocene epoch, and continued on up to the present time; that the waters once filled all these valleys, so that they rested high upon the sides of the mountains, depositing what Prof. Hayden called the Salt Lake Group, gradually passing into the Post-pliocene deposits which verge upon our present period. It is quite possible that there have been oscillations of level in these modern lake-waters; but so far as the proofs go, this great inland lake may have continued quite uniform until the terrace epoch, and that then the waters gradually receded to their present position.

The immediate valley of Bear river, near the crossing, is interesting on account of the fine development of the lake-deposit, which is composed of clay, sand, and marl, yellow and rusty-drab color, and attains a thickness of 200 to 300 feet. The elevation of Bear river valley, at the bridge, is 4,542 feet, and the highest terrace on the east side is 4,737 feet, and the highest on the west side is 4,779 feet. The immediate valley of Bear river may be said to have been worn out of the Pliocene or lake deposit.

Among the lower ranges of hills that border the east side of the Great Snake river basin, especially from Port Neuf Cañon northward, the Pliocene deposits are well shown, and lie beneath the basaltic floor. In the Port Neuf Cañon this fact is illustrated by the wearing away of the cap or floor of basalt, in a number of localities, but on the sides of the hills this is shown with equal clearness by the elevations of the basalt. The dip of the beds is not great, usually not more than 5° or 10,° and in all cases in the direction of the great basin. This would indicate that there had been a moderate elevation of the mountain ranges, or a depression of the basin at a very modern date, even approaching very close to our present period. The effusion of such a vast amount of igneous matter from the interior of the earth, might suggest the possibility, or even probability, that the cause of the subsequent changes in the hills around the borders, was either contemporaneous or subsequent to the effusion of the melted material. If the elevation began with the eruption, it certainly continued long after it ceased, inasmuch as the basalt is lifted up in thick beds, at the same angle with the underlying strata. Not only in the valley of the Port Neuf and Snake river is the basalt found in conjunction with lake deposits, but in numerous localities all over the northwest, it seems to rest upon these Pliocene beds, readily adapting itself by the form of the under surface to the irregularities of the surface of the lake deposits.

Prof. Eng. W. Hilgard* divided the Eocene of Alabama and Mississippi in descending order, into, 1st, Vicksburg Group, 120 feet; 2d, Red Bluff Group, 12 feet; 3d, Jackson Group, 80 feet; 4th, Claiborne Group, 60 feet; 5th, Buhrstone Group, 150 feet; 6th, Flatwoods and Lagrange Lignitic Group, 450 feet, making a total thickness of 872 feet. The Lagrange and Porter's Creek Group of Safford is the same as the Flatwoods and Lagrange Lignitic. The Buhrstone Group of Tuomey is the same as the Siliceous Claiborne Group of Hilgard.

The Eocene is followed by the Grand Gulf Group, probably a deposit in brackish water, almost non-fossiliferous, and having a thickness of 250 feet.

Prof. Leo Lesquereux† described, from the Green River Group of

---

* Proc. Am. Ass., Ad. Sci.
† 1872, U. S. Geo. Sur. of Montana, etc.

Wyoming, high on hills from the river. *Ceanothus cinnamomoides*, now *Zizyphus cinnamomoides;* from the Bridger Group at Washakie station, near Bridger's Pass, *Rhamnus intermedius, Liquidambar gracile,* now *Aralia gracilis,* and *Quercus æmulans;* and from Barrell's Springs, *Equisetum haydeni.*

After reviewing the state of the knowledge of the Tertiary and Cretaceous flora of this country, he arrived at the following conclusions, to-wit:

1. The Tertiary flora of North America is, by its types, intimately related to the Cretaceous flora of the same country.

2. All the essential types of our present arborescent flora are already marked in the Cretaceous of our continent, and become more distinct and more numerous in the Tertiary; therefore the origin of our actual flora is, like its *facies,* truly North American.

3. Some types of the North American Tertiary and Cretaceous flora appear already in the same formations of Greenland, Spitzbergen, and Iceland; the derivation of these types is, therefore, apparently, from the arctic regions.

4. The relation of the North American Tertiary flora with that of the same formation of Europe, is marked only for North American types, but does not exist at all for those which are not represented in the living flora of this continent. Therefore, the European Tertiary flora partly originates from North American types, either directly from our continent, or derived from the arctic regions.

5. The relation of the Tertiary flora of Greenland and Spitzbergen with ours indicates, at the Tertiary and Cretaceous epochs, land connection of the northern islands with our continent.

6. The species of plants common to the Cretaceous and Tertiary formations of the arctic regions, and of our continent, indicate, in the mean temperature, influencing geographical distribution of vegetation, a difference, in $+$, equal to about 5° of latitude for the Tertiary and Cretaceous epochs.

7. The same kind of observation on the geographical distribution of vegetable species, shows at the Tertiary and Cretaceous times, differences of temperature according to latitude, analagous to what is remarked at our time, by the characters of the southern and northern vegetation.

Prof. E. D. Cope* referred the Bridger Group to the Eocene, and described, from Cottonwood creek, Wyoming, *Mesonyx obtusidens, Triaco-*

---

* Pal. Bull., No. 1, and Proc. Am. Phil. Soc., vol. xii.

don aculeatus, Lophiotherium pygmœum, Anostira œdemia, now *Plastomenus œdemius, A. molopina,* now *P. molopinus, A. trionychoides,* now *P. trionychoides, Trionyx concentricus, T. thomasi,* now *Plastomenus thomasi, Axestus byssinus, Bœna hebraica, Testudo hadriana,* now *Hadrianus corsoni, Emys polycyphus, E. terrestris, Helotherium procyoninum,** *Stypolophus pungens, Pantolestes longicaudus, Pseudotomus hians, Hadrianus octonarius, Hadrianus allabiatus,*† *Protagras lacustris;* from the Bad Lands of Black's Fork of Green river, Wyoming, *Stypolophus brevicalcaratus, S. insectivorus. Miacis parvivorus, Tomitherium rostratum,* and *Emys latilabiatus.*

He described,‡ from the Eocene of the upper waters of Bitter creek, Wyoming, *Synoplotherium lanius, Crocodilus clavis, Rhineastes peltatus. R. smithi, Loxolophodon cornutus,*§ *L. furcatus, L. pressicornis,* and *Palœosyops vallidens.* From the northern part of the Eocene basin of Green river, *Anaptomorphus œmulus,*‖ *Crocodilus sublatus,*¶ *C. sulciferus,* and *Anostira radulina.* From the lower beds of the Green River Group, near Black Buttes, *Alligator heterodon.* From the Wasatch Group, near Evanston, Utah, *Bathmodon radians, B. semicinctus, Notharctus* (now *Hyracotherium*) *vasacciensis, Notomorpha gravis, N. testudinea.* From the Eocene, at Osino, 25 miles northeast of Elko, Nevada, *Trichophanes hians* and *Amyzon mentale.* From the Green River Group of Wyoming,** *Erismatopterus ricksecke i,* and *Osteoglossum,* now *Dapedoglossus encaustum.*

He described, from the Eocene of New Jersey.†† *Lembonax propylœus, L. insularis,* and *Thecachampsa serrata.* And from the Miocene near San Diego, California, *Eschrichtius davidsoni.*

Prof. O. C. Marsh described,‡‡ from the Eocene near Fort Bridger. and near Henry's Fork, Wyoming, *Palœosyops laticeps, Telmatherium validus. Hyrachyus princeps, Homacodon vagans, Limnocyon verus, Viverravus gracilis, Nyctitherium velox, N. priscus, Talpavus nitidus, Limnofelis ferox, L. latidens, Limnocyon riparius, L. agilis, Thinocyon velox, Viverravus* (?) *nitidus, Thinolestes anceps. Telmalestes crassus, Limnotherium affine, Orohippus*

---

* Pal. Bull. No. 2, and Proc. Am. Phil. Soc.
† Pal. Bull. No. 3, and Proc. Am. Phil. Soc.
‡ Pal. Bull. No. 6, and Proc. Am. Phil. Soc.
§ Pal. Bull. No. 7, and Proc. Am. Phil. Soc.
‖ Pal. Bull. No. 8, and Proc. Am. Phil. Soc.
¶ Pal. Bull. No. 9, and Proc. Am. Phil. Soc., vol. xii.
** U. S. Geo. Sur. of Wyoming.
†† Proc. Acad. Nat. Sci., Phil.
‡‡ Am. Jour. Sci. and Arts, 3d ser., vol. iv.

pumilus, Helohyus plicodon, Thinotherium validum, Passalacodon litoralis, Anisacodon elegans, Centetodon pulcher, Stenacodon rarus, Antiacodon venustus, Bathrodon annectens, B. typus, Mesacodon speciosus, Hemiacodon gracilis, H. nanus, H. pucillus, Centetodon altidens, Entomodon comptus, Entomacodon minutus, Centracodon delicatus, Nyctilestes serotinus, Ziphacodon rugatus, Harpalodon sylvestris, H. vulpinus, Orotherium uintanum, Helaletes boops, Paramys robustus, Tillomys senex, T. parvus, T. lucaris, Sciuravus parvidens, Colonymys celer, Apatemys bellus, A. bellulus, Entomacodon angustidens, Triacodon grandis, T. nanus, Euryacodon lepidus, Palæacodon ragus, Aletornis nobilis, A. pernix, A. venustus, A. bellus, A. gracilis, Uintornis lucaris, Thinosaurus agilis, T. crassus, T. grandis, T. leptodus, T. paucidens, Glyptosaurus princeps, Oreosaurus ragans, Tinosaurus stenodon, Glyptosaurus brevidens, G. rugosus, G. sphenodon, Oreosaurus lentus, O. gracilis, O. microdus, O. minutus, Tinosaurus lepidus, Iguanavus exilis, Tinoceras grandis, Dinoceras lacustris, and *Oreocyon latidens.* He described, from the Postpliocene, near Bangor, Maine, *Catarractes affinis,* and from Monmouth county, New Jersey, *Meleagris celer,* and *Grus proavus.* Of the above list, it is stated by Cope that the new generic names are not generally defined.

Dr. Joseph Leidy[*] described, from the Bridger Group of Wyoming, *Uintacyon edax, U. vorax, Chameleo pristinus, Lepidosteus atrox, L. notabilis, L. simplex, Amia gracilis, A. media, A. uintensis, Hypamia elegans, Pimelodus antiquus, Phareodus acutus, Hyrachyus nanus, Microsyops gracilis, Palæacodon verus, Hipposyus formosus, Palæosyops junior, P. humilis,* and *Uintatherium robustum.* From the Niobrara Group, on the Niobrara river, in Nebraska, *Felis angustus;* from Green river, *Oligosomus grandævus;* from the Black Foot country at the head of the Missouri, *Tylosteus ornatus;* and from the Pliocene of Oregon, *Hadrohyus supremus, Rhinoceros pacificus,* and *Stylemys oregonensis.*

Prof. F. B. Meek [†] described, from the Green River Group at Washakie, Wyoming, *Unio washakiensis,* and from Pacific Springs, *Bythinella gregaria.*

T. A. Conrad[‡] described, from the Eocene of North Carolina, *Ostrenomia carolinensis;* and from the Miocene of the same state, *Donax idoneus.*

---

* Proc. Acad. Nat. Sci.
† Geo. Sur. of Wyoming.
‡ Proc. Acad. Nat. Sci.

In 1873, Prof. E. D. Cope* described, from the Bridger Group of Bitter creek, and Cottonwood creek, *Limnohyus lævidens;* from a bluff on Green river, near the mouth of the Big Sandy, Wyoming, *Palæosyops fontinalis;* from the summit of Church Butte, *Trionyx heteroglyptus;*† from the Bad Lands of Cottonwood creek, *T. scutumantiquum, Pappichthys plicatus, P. sclerops, P. lævis, P. symphysis, Rhineastes radulus ;* from Ham's Fork, *Bœna ponderosa, Clastes anax ;* from the Green River Group, near Evanston, Utah, *Bathmodon latipes;* from near Black Buttes, *Emys euthnetus, E. megaulas, E. pachylomus;* from Upper Green river, *Pappichthys corsoni, Rhineastes calvus, R. arcuatus;* from Green river basin,‡ *Antiacodon furcatus,* now *Sarcolemur furcatus, Orotherium* (now *Hyracotherium*) *index;* from Cottonwood creek, *Microsyops vicarius, Oligotomus cinctus ;* from South Bitter creek, *Paramys leptodus, Eobasileus galeatus, Achænodon insolens,* from Henry's Fork, *Palæosyops diaconus, Hyrachyus implicatus;* from near Evanston, *Phenacodus primævus.*

He described, from the Miocene of Colorado,§ *Hyopsodus minimus, Hypertragulus calcaratus, H. tricostatus,* and *Menotherium lemurinum;* from the Miocene of the Western plains,‖ *Aelurodon mustelinus,* now *Mustela parviloba, Aphelops megalodus. Palæolagus agapetillus,¶ Colotaxis cristatus, Hyracodon quadriplicatus,* now *Anchisodon quadriplicatus, H. arcidens. Symborodon torvus. Miobasileus ophryas, Megaceratops acer, M. helocerus, Peltosaurus granulosus, Testudo amphithorax, T. cultratus, T. laticuneus, T. ligonius, Domnina gradata,** Herpetotherium fugax, Daptophilus squalidens, Tomarctus brevirostris, Stibarus obtusilobus, Canis gregarius. Isacis* (now *Miodectes) caniculus, Palæolagus triplex. P. turgidus. Tricium avunculus, T. leporinum, T. paniense, Gymnoptychus minutus, G. nasutus, G. trilophus, Anchitherium cuneatum,* and *Trimerodus cedrensis.*

Prof. O. C. Marsh†† described, from the Eocene deposits of Wyoming and Oregon. *Dinoceras mirabilis, Orohippus agilis, Colonoceras agrestis, Dinoceras lucaris, Oreodon occidentalis, Rhinoceras annectens R. oregonensis. Tillotherium hyracoides ;* from the Miocene of Colora-

* Proc. Am. Phil. Soc., vol. xiii.
† U. S. Geo. Sur., Wyoming, etc.
‡ Pal. Bull. vol. xii.
§ Proc. Acad. Nat. Sci
‖ Pal. Bull. vol. xiv.
¶ Pal. Bull. No. xv.
** Pal. Bull. No. xvi.
†† Am. Jour Sci. and Arts, 3d ser., vol. v.

do, *Brontotherium gigas,* and *Elotherium crassum;* and from the Upper Eocene of Wyoming,* *Dinoceras laticeps.*

Dr. Joseph Leidy† described, from the Bridger Group in the Buttes of Dry creek, *Hyopsodus minusculus, Mysops fraternus, Washakius insignis, Saniva major ;* from the Grizzly Buttes, *Sinopa eximia ;* from the Buttes, ten miles from Dry Creek Canon, *Amia uintaensis;* from the junction of Sand and Green rivers, *A. media ;* from Henry's Fork, *A. gracilis ;* from Dry creek, *Hypamia elegans ;* from the junction of Big Sandy and Green rivers, *Lepidosteus atrox,* now *Clastes atrox ;* from Washakie station, *L. simplex, L. notabilis,* now *Clastes notabilis ;* from Big Sandy and Green rivers, *Pimelodus antiquus, Phareodus acutus, Clupea alta,* now *Diplomystus altus ;* from the Miocene of Bridger creek, a tributary of John Day's river, one of the branches of the Columbia, in Oregon, *Dicotyles pristinus, Elotherium imperator ;* from Washington county, Texas, *Anchitherium australe ;* from Red Rock creek, a tributary of Jefferson Fork of the Missouri, *Anchitherium agreste ;* from Richmond, Virginia, *Procamelus virginiensis, Tautoya conidens,‡ Acipenser ornatus ;* from the Post-pliocene of California, *Felis imperialis,* and *Auchenia hesterna.*

Prof. F. B. Meek§ described, from Church Buttes, *Physa bridgerensis;* from twelve miles south of Fort Bridger, *Pupa leidyi;* and from the upper beds exposed at Separation, on the U. P. R. R., *Limnæa compactilis.*

Prof. Lesquereux described, from South Park, near Castello's Ranch, *Ophioglossum alleni,* and *Planera longifolia;* from Elko station, *Sequoia angustifolia, Thuya garmani,* and *Abies nevadensis.*

In 1874, Prof. E. D. Cope‖ described, from the Bridger Group of South Bitter creek, *Eobasileus galeatus,* and *Achænodon insolens ;* from the Miocene of Colorado, *Symborodon hypoceras, Anchitherium exoletum,* and *Hippotherium paniense.* He described from the Eocene of the Middle and South Parks, Colorado,¶ *Amyzon commune,* and *Clupea theta,* now *Diplomystus thetus ;* from the White River Group, *Hypertragulus tricostatus, Elotherium ramosum,* now *Pelonax ramosus,* and *Menotherium lemurinum ;* from the Loup Fork Group,

* Am. Jour. Sci. and Arts, 3d ser., vol. vi.
† Cont. to Ext Vert. Fauna, W. Terr.
‡ Proc. Acad. Nat. Sci.
§ 6th Ann. Rep. U. S. Geo. Sur. Terr.
‖ 7th Ann. Rep. U. S. Geo. Sur. Terr.
¶ Bull. U. S. Geo. Sur. Terr.

Protohippus sejunctus, Procamelus angustidens, P. heterodontus, and *Merycodus gemmifer,* now *Blastomeryx gemmifer.*

He determined that the lacustrine deposit in the valley of the Rio Grande, called the Santa Fe marls, is of Pliocene age, and described[*] *Martes nambianus,* now *Putorius nambianus, Cosoryx ramosus,* now *Dicrocerus ramosus, C. teres,* now *D. teres, Hesperomys loxodon,* now *Eumys loxodon, Panolax sanctæfidei, Cathartes umbrosus,* now *Vultur umbrosus, Mastodon productus,* and *Steneofiber pansus.*

Prof. O. C. Marsh[†] described, from the Eocene of Wyoming, *Orohippus major, Stylinodon mirus,* and *Tillotherium latidens;* from the Miocene of Colorado, *Brontotherium ingens;* from Nebraska, Dakota and Oregon, *Miohippus annectens, Anchitherium anceps, A. celer, Anchippus brevidens,* and *Elotherium bathrodon;* and from Pliocene strata of the west, *Pliohippus pernix, P. robustus, Protohippus avus, Morotherium gigas,* and *M. leptonyx.*

Prof. Leo Lesquereux described,[‡] from Elko, Nevada, *Lycopodium prominens, Myrica partita, Quercus elkoana, Diospyros copeana, Sapindus coriaceus;* from Middle Park, *Salvinia cyclophylla, Ulmus tenuinervis, Sapindus angustifolius, Staphylea acuminata, Rhus drymeja, R. haydeni, Pterocarya americana;* from Green river, *Equisetum wyomingense;* from Florissant, South Park, *Acorus affinis, Myrica copiana, Weinmannia rosæfolia, Ilex subdenticulata, I. undulata, Paliurus florissanti, Cæsalpinia linearis, Acacia septentrionalis.*

The Eocene[§] is found in North Carolina, between the Neuse and the Cape Fear, and in limited outcrops throughout the triangular region between Newbern, Goldsboro and Wilmington. It consists of a light colored, consolidated marlite, as in the steep bluffs on the Neuse, 10 miles below Goldsboro, or of a shell conglomerate as seen about Newbern, and 8 or 10 miles up Trent river, or of a white calcareous sandstone, more or less compacted, as on the Neuse near Goldsboro; or of a gray and hard limestone, as about Richlands in Onslow; or of a coarse conglomerate of worn shells, sharks' teeth, and fragments of bones and stony pebbles, as in the upper part of Wilmington and at Rocky Point; or of a fine shaly infusorial clay, light gray to ash colored, as in Sampson county near Faison's depot. The outliers show that the formation, though limited in thickness, had a great horizontal extent, and once extended quite into the hill country of the State, and

---

* Proc. Acad. Nat. Sci. Phil.
† Am. Jour. Sci. and Arts, 3d ser., vol. vii.
‡ 7th Ann. Rep. U. S. Geo. Sur. Terr.
§ Geo. of N. Carolina, 1875.

nearly 150 miles from the present coast line, and to an elevation of nearly 400 feet.

The Miocene occurs in disconnected patches, in river bluffs and in ravines over the seaboard region, and extending from the shore and the western margins of the sounds 50 to 75 miles inland. It consists of beds of clay, sand and marl, which are locally filled with shells from 2 to 8 feet, and occasionally 10 to 20 feet.

Prof. Theo. B. Comstock* said the Green River Group is used to designate that portion of the fresh-water Tertiary strata which lies directly above the coal group, and which is the present surface formation over a large portion of the Green river basin, north of Fort Bridger. The upper limit is not readily definable at present, the transition between the beds of this and the overlying group being rather gradual, but the general character of the two formations, both lithologically and palæontologically, differs greatly. The Green river beds are mainly composed of a series of shales, marls, and harder calcareous strata, the latter especially containing quantities of the remains of fresh-water forms of life, with laminated layers, literally filled with the remains of land plants of the Phænogamous series. The outline of the ancient lake-basin, in which these strata were deposited, is not fully determined, but there are indications that its eastern boundary was outside of the present limits of the Green river basin, and there is no room for doubt that the Uinta mountains, and the Wahsatch chain, then, as now, towered above its surface. Northward it is equally clear that the Wind River Range formed the shore of the great lake, with probably more or less of gently sloping border during a portion of the era of Lower Eocene deposition. The excessive erosion has exposed the beds over the route from Fort Bridger to near South Pass, and generally speaking, the rock contains a considerable portion of calcic carbonate, with an abundance of ferric oxide produced by decomposition and oxidation. Gypsum and calcite of different varieties are abundant, frequently occurring as thin, papery seams between the rock-layers, at other times forming masses of considerable extent. Some of the layers are little more than a pure clay shale, while there are a few quite arenaceous beds and some compact limestones. The texture of the different beds is quite variable, but, in general, the streams which have cut their channels through them are walled by nearly vertical cliffs, and the buttes and benches for the most part have

---

* Rep. of Reconnaissance of Yellowstone river and N. W. Wyoming

quite precipitous sides. Numerous joints occur in many of the strata, particularly in the more compact kinds, and fine examples of concretionary structure or weathering are not rare. The tendency of the thick beds of marly sandstone on the banks of Green river, at the crossing, to weather spheroidally, is very noticeable, and this is repeated in various degrees in the argillaceous and calcareous rocks as well.

The Bridger Group, though succeeding the Green River Group, is closely related to it, for the transition from one to the other is not abrupt, either in the structure of the beds or their contents. The Group is exposed at the surface over a considerable extent of country, northward and eastward from Fort Bridger as far as Little Sandy river and beyond, forming the top layers of numerous isolated buttes, During this epoch it is probable that the land was covered with fresh water in a lake as large as in the previous era, if not more extensive. The beds are mainly composed of dull-colored, indurated clays, and arenaceous layers of considerable thickness, the latter usually brownish, or dull yellow or gray, often with more or less of a concretionary structure. The clays are generally compacted, but they become disintegrated upon exposure to the atmosphere, and readily yield to the eroding forces. Some thinner layers of more calcareous material, with silicious seams, often affording interesting concretions, are interspersed, but they are rather exceptional than otherwise. The Green river and Bridger Groups are readily distinguished by the effects produced by erosion. The former presenting nearly vertical cliffs, so that the impression in crossing the country where it forms the surface rocks is that of traveling over an ordinary plain with occasional descents, by a succession of terraces, to the narrow valleys of the streams. On the contrary, where it is concealed, or only occasionally capped by the Bridger Group, the country is very irregular, often simulating the " Bad Lands;" the beds of the latter being eroded without complete denudation, so that they stand out in buttes, or rude architectural forms.

The deposits in the Yellowstone Lake basin, and in the valley of the main river and its tributaries, which may be regarded as Pliocene, are mainly the sediments of an ancient lake, of which the present body of water is the representative on a much reduced scale. Beautiful and highly instructive sections of the old beach formations are exposed in the valleys of the streams, particularly in the lower valley of Pelican creek, and far down the Yellowstone river, where they become more complicated and more interesting. An examination of these shows that the lake formerly extended over a much larger area, and that it has held its place amid changes of great importance. It was during the latter portion of the Tertiary age that much of the volcanic activity took place which was so general over this portion of the country, though probably only the closing stages of the lava flows are represented by the eruptive deposits of the Pliocene epoch. A section on the present lake shore, between Bluff Point and Steam Point, in descending order, is as follows:

1. Grass-covered soil passing gradually to loose sand, 2 feet.

2. Various sand, gravel, and spring deposits with scattered irony concretions, 6 feet.

3. White and dark lake sand, very thinly laminated with beach structure, and occasional irony layers, 5 feet.

4. About 15 feet of thinly laminated, blue-black clay, locally contorted and beautifully cut by a small rill, emanating as a spring from one of the irony layers in No. 3. The water is slightly chalybeate.

Other sections show the same general features with more or less variation. They represent the upper portion of the Pliocene series, deposited toward the close of volcanic activity, hence the occasional beds of volcanic ejectamenta which were poured out into the lake, are mainly composed of volcanic sand and the finer textured conglomerates, as may well be seen near Steamboat springs. As we descend the valley of the Yellowstone river, we find the lower members of the group well exposed, and the beds of unmodified non-molten material becoming more common, with increasing proportions of the molten or lava series, until the latter are almost universal, and doubtless represent an earlier period, though frequently largely concealed by the subsequent spring deposits. Near the close of the Pliocene epoch, the internal fires had so far died out that the igneous ejections were of fitful occurrence, and geysers, solfataras, fumaroles, etc., abounded to an almost incredible extent, giving rise to enormous deposits of siliceous and calcareous material, which has continued to be deposited with decreasing vigor until the present day.

Prof. G. K. Gilbert[*] found a section of Tertiary on the east face of San Pitch Plateau, at Wales, Utah, 1.292 feet in thickness, another near the head of the main Sevier river, in Utah, 560 feet, and another on the north fork of Virgin river, between Mountain Lakelet and Rockville, in Southern Utah, estimated at 3,000 feet.

Prof. E. D. Cope[†] described the Puerco marls as in all probability

---

[*] Geo. Sur. W. 100th Meridian, vol. iii.

[†] Ann. Rep. Explr and Sur., W. 100th Meridian App. L. L.

a lacustrine formation of Eocene age, though having examined an out-crop for forty miles, he discovered no fossil remains except fossil wood. He said the material is so easily transported that the drainage chan-nels are cut to a great depth, and the Puerco river becomes the recep-tacle of great quantities of slimy-looking mud. Its unctious appear-ance resembles, strongly, soft soap, hence the name *Puerco*, greasy. These soft marls cover a belt some miles in width, and continue at the foot of another line of sandstone bluffs, which bound the immediate valley of the Puerco to a point eighteen miles below Nacimiento.

This section of the Eocene strata in the region west of the Sierra Madre Range in New Mexico consists of green and black marls, which he named the Puerco Group, 500 feet; sandstone of the Wasatch Group 1,000 feet, and red and gray marls of the same group, 1,500 feet; mak-ing a total thickness of 3,000 feet.

He described,* from the Eocene of New Mexico, *Ambloctonus sin-osus, Protolomus secundarius, P. multicuspis, P. strenuus, Diacodon alticuspis, D. cœlatus, Pelycodus frugivorus, Pantolestes chdcensis, Opisthotomus astutus, O. flagrans, Antiacodon mentalis, A. crassus, Hyrachyus singularis, Hyracotherium tapirinum, H. angustidens, H. cuspidatum, Bathmodon latidens. B. cuspidatus, Diplocynodus sphenops, Crocodilus grypus, C. wheeleri,* and *Dermatemys (?) costilatus.*

He described,† from the Miocene of Cumberland county, New Jersey, *Phasganodus gentryi, Sphyrænodus silovianus,* and *Agabelus porca-tus;* from Flower's marl pit, Duplin county, North Carolina,‡ *Pristis attenuatus;* from Edgerton's plantation, in Wayne county, *Pneuma-tosteus nahunticus;* from Halifax county, *Mesoteras kerrianus,* and *Delphinapterus orcinus.* From the Loup Fork Group of New Mexico,§ *Pliauchenia humphreysana, P. vulcanorum, Hippotheri-um calamarium,* and *Aphelops jenezanus;* and from the Pliocene of the West, *Canis ursinus.*

Prof O. C. Marsh‖ described, from the Eocene of Wyoming, *Lemu-ravus distans, Tillotherium fodiens;* from Utah, *Diceratherium ad-venum, Diplacodon elatus, Orohippus uintensis,* and *Agriochœrus pumilus.* From the Miocene bad lands of Nebraska, *Laopithecus ro-bustus, Anisacodon montanus;* from the John Day river in Oregon,

---

* Geo. Sur. W. 100th Meridian, Syst. Catal. of Vertebrata.
† Proc. Am. Phil. Sci., vol. xiv.
‡ Geo. of N. Carolina.
§ Proc. Acad. Nat. Sci.
‖ Am. Jour. Sci. and Arts, 3d ser., vol. ix.

where the beds have an estimated thickness of 5,000 feet, *Diceratherium armatum. D. nanum, Thinohyus lentus,* and *T. socialis.*

T. A. Conrad* described, from the Eocene at Wilmington, North Carolina, *Terebratula demissirostra;* and from Beaufort, *Pecten anisopleura* and *P. carolinensis.*

From the Miocene near Wilmington, and other places in North Carolina, *Liropecten carolinensis, Ostrea perlirata, Placunomia fragosa, Raeta alta, R. erecta, Abra bella, A. holmesi, Noetia protexta, N. filosa, Mercenaria carolinensis. Leptothyris parilis, Trachycardium bellum, Mysia carolinensis, Saxicava protecta, Turritella perexilis, T. carolinensis, Fissurella carolinensis, Littorina carolinensis, Busycon kerri, B. amoenum,* and *B. concinnum ;* from Suffolk, Va., *Zizyphus virginicus.*

W. H. Dall† described, from the Miocene at Cerros Island, California, *Waldheimia kennedyi;* from the Pliocene at San Diego, *Chrysodomus diegoensis.* And R. C. Stearns‡ described, from the same strata, *Opalia anomala,* and *O. varicostata.*

In 1876, Prof. J. W. Powell§ subdivided the Tertiary rocks of the plateau province of the west in ascending order, into the " Bitter Creek Group," which is synonymous with the Wasatch Group, and has a thickness of 5.000 feet. It is succeeded by the Lower Green River Group, consisting of shales, often bituminous; sandstones; carbonaceous shales and lignitic coal near the base. Thickness, 800 feet.

This group is well exposed along Green river, from Green River station southward for 10 miles; in many of the escarpments of the Quien Hornet mountain, and a few miles northeast from the head of Vermilion canon; on Snake river, six miles above the northern foot of Junction mountain; and on the elevated ledges known as Pine Bluffs, near the sources of the eastern tributaries of Vermilion creek. The beds are all fresh water.

The Upper Green River Group consists of sandstones, sometimes argillaceous limestones, carbonaceous shales and lignitic coal, near the middle and in the lower part massive or irregularly bedded sandstone, ferruginous. Unconformable by erosion with lower Group. Thickness, 500 feet.

The plant beds of this group are well exposed to the north of Green River station, and between that point and Alkali stage station, in many gulches and canons; in the cuts of the Union Pacific Railroad

---

* Geo. of N. Carolina.
† Proc. Cal. Acad. Sci., vol. v.
‡ Proc. Acad. Nat. Sci. Phil.
§ Geo. of Uinta Mountains.

between Green River station and Bryan and in the escarpments on either side of Henry's Fork at many places. The Tower sandstone is well shown in the cliffs at Green River station, and in that vicinity and below the mouth of Currant creek. The Tower sandstone is laid down unconformably on the Lower Green River Group, the unconformity being represented by gentle valleys of erosion.

The Bridger Group consists of Bad Land sandstones (chiefly green sands) limestones, shells, marls, and concretionary and stratified flints. Thickness, 2,000 feet.

It is well exposed in the vicinity of Fort Bridger, at Church Buttes, at Haystack mountain and in the Cameo mountains. An outlying patch is found north of the Dry mountains between Vermilion creek and Snake river. Unconformity with the beds of the Lower Green River Group may be seen in the vicinity of Carter station, but unconformity with the Upper Green River Group has not been observed. The two are separated, however, upon lithological grounds, though the plane of demarkation is obscure. The moss agates for which the region about Fort Bridger has been noted are from irregular beds and aggregations of chalcedony in this Group.

The Brown's Park Group consists of sandstones, gravels, limestones. concretionary and stratified flints. Unconformable with all underlying rocks. Thickness, 1,800 feet.

It is well represented at Brown's Park, in northeastern Utah, and in northwestern Colorado. About five miles above the confluence of Snake river with the Yampa, the beds may be seen resting unconformably against Carboniferous strata, and on going north they may be observed to rest unconformably with the Bridger Group.

In Brown's Park, it lies in a deep basin of erosion, the bottom and sides of which are composed of Uinta sandstone. This basin is in the very axis of the Uinta uplift. Its sandstones are Bad Land rocks of exceedingly fine texture. In some places there are extensive and irregular aggregations of chalcedony.

The Bishop's Mt. Conglomerate, which is unconformable by plication and erosion with underlying rocks. Thickness, 300 feet. It is found on the summits of Bishop and Quien Hornet mountains, and upon various tables in the Uinta mountains. On the north side of Connor basin, at the head of Sheep creek, this conglomerate has a thickness of more than 1,000 feet. It is neither a marine nor lacustrine deposit. but a subærial one.

Prof. Powell says, witnessing on every hand the accumulation of such gravels in valleys and over plains where mountains rise to higher altitudes on either side, and having in many cases actually seen the cliffs breaking down, and the gravels rolling out on the floods of a storm, I am not willing to disregard explanations so obvious, and so certain, for an extraordinary and more violent hypothesis. Irregular accumulations of clay, accumulations of sand, of gravels, and bowlders, having, in a general way, all the lithologic characteristics of "drift," are very common in the Rocky mountain region, and in many cases their origin can be traced to ordinary atmospheric agencies acting on the adjacent hills and mountains; and no glaciers or icebergs are needed for their explanation.

We learn from Dr. Hayden,* that on the high divide between the drainage of the Arkansas and South Platte rivers, there occur fresh-water lake deposits, having a thickness of 1,000 or 1,500 feet, and covering an area of about 40 miles from north to south, and 50 miles from east to west, or about 2,000 square miles, called by Dr. Hayden, in 1869, the "Monument Creek Group," from the fact that the atmospheric agents have carved out of the beds peculiar monuments or columns. He referred the deposits to Miocene or Pliocene age ; later, in 1873, Prof. Cope, from the evidence of the hind leg and foot of an *Artiodactyle*, and a fragment of *Megaceratops coloradoensis*, referred the deposits on the Colorado divide, perhaps the same, to the age of the Miocene. The texture of the rocks is quite varied.

The lower portion is composed of rather massive beds of sandstone, varying from a pudding-stone to a fine-grained sandstone, usually of a light color, sometimes of a yellow or iron-rust, with their intercalations of arenaceous clay. In the distance, the whole group, in many localities, presents a chalky-white appearance. At the immediate base of the mountains, just south of the small lake on the divide, the rocks are variegated sandstones, brick-red, white and yellow, varying in texture from a fine sandstone to a pudding-stone, with all the signs of deposition in moving waters. Still farther north, on the divide proper, the beds jut against the granites, inclining not more than 3°, and are made up of a coarse aggregate of feldspar and quartz crystals, so that it resembles a very coarse granite. It is plain that the sediments of this group were derived very largely from the granitoid rocks. The sediments become finer and finer as they recede eastward from the foot of the mountains into the plains.

---

* U. S. Geo. and Geogr. Sur. of Colorado and Adjacent Territory.

To the eastward of the line of the Denver and Rio Grande Railroad, the surface is cut up into more or less rectangular masses, with rather broad table-shaped summits, varying from 400 to 800 feet in height. The sides are often very steep, almost inaccessible. At a remote period in the past, the erosion has been very great, carving out by an almost inappreciably slow process, these broad valleys, leaving these buttes here and there, composed of horizontal beds, .to aid in forming some conception of the amount of denudation which has taken place. It is not possible at the present time to estimate the original thickness of this group, but believe it to have been very much greater than the highest beds now existing would indicate. The summits of many of the buttes are capped with a greater or less thickness of a beautiful purplish trachyte, which must have ascended in the form of dikes from beneath, and flowed over the surface. Much of the trachyte is a sort of breccia, composed of rather coarse sandstones, which must have been caught in the melted material. It is quite evident that these outflows occurred during the existence of the lake, though at a late period. Dr. Hayden synchronized the age of this group with the upper portion of the White River Group far to the northward, and probably with the fresh-water deposits in the South Park.

Lake basins have occupied a large part of the country from the Isthmus of Darien to the Arctic Circle. In many instances they were merely expansions of river valleys, like the greater number of the lake basins of the present time. During the later Cretaceous and early Tertiary periods, the western portion of the continent was covered with immense lakes, but during the Pliocene and the interval to modern time, thousands of small lakes, with a few of large size, were distributed over the great area west of the Mississippi, and the basins with their peculiar deposits are found in the parks, among the mountains, and along every important valley.

Dr. Hayden believed there are evidences of glacial action and morainal deposits in the valley of the Upper Arkansas river, at elevations of 9,000 feet and upward, and along both flanks of the Sawatch mountains; but, he said that he observed no proof of any wide extended drift-action, like that of the New England States, in the Rocky mountains, as the superficial deposits are all of local origin, and the source is limited to the drainage of the streams in which the deposits are found. For example, all the marls and coarser deposits in the valley of the Upper Arkansas, have the same origin, and the forces that produced them were limited geographically to the drainage of that stream. That not a fragment of rock had been transported even from so short a distance as beyond the drainage west of the Sawatch, or east of the Park ranges. He placed the superficial deposits in one great period, extending from the Pliocene up to the present time, because in the aggregate they afford no proof of any break in the order of time. In the valley of Roaring Fork in the Elk mountains, the morainal deposits are remarkable for their thickness. The surface is covered with huge bowlders, some angular, and others partially rounded. The terraces are very conspicuous, rising, in some instances, to 1,000 feet or more above the bed of the stream, and strewed over with huge bowlders. None of the stray materials in any of the valleys or gorges seem to have been transported a very great distance, and never, under any circumstances, is there any drift or glacial deposits from a neighboring drainage; in other words, the loose material does not pass from one independent valley to another. So it is all over the Rocky mountain region. All the drift or Postpliocene deposits are local.

Prof. E. D. Cope* described, from the Eocene of New Mexico, the giant bird *Diatryma gigantea;* and from the Pliocene, phosphate beds of South Carolina, *Cyclotomodon vagrans.*

Prof. O. C. Marsh† described, from the Eocene of the Rocky mountain region, *Eohippus validus, E. pernix, Parahyus vagus, Dromocyon vorax, Dryptodon crassus,* and *Coryphodon hamatus.*

Dr. Joseph Leidy‡ described, from the Eocene of New Jersey, *Myliobates fastigiatus,* and *M. jugosus;* from the Pliocene beds of Ashley river, South Carolina, *Belemnoziphius prorops, Choneziphius liops. C. trachops. Eboroziphius coelops, Proroziphius macrops, Myliobates magister, M. mordax,* and *Proroziphius chonops.*

Prof. C. A. White§ described, from the Eocene at Bijou basin, 40 miles east of Denver, Colorado, *Corbicula powelli, Mesodesma. bishopi. Phorus exoneratus;* from Crow creek, *Melania laranda:* from the West, *Tulotoma thompsoni;* from the Lower Green River Group, 8 miles below Green River station, Wyoming, *Helix riparia;* from the Upper Green River Group, at Henry's Fork and Alkali station, *Unio shoshonensis. Succinea papillispira, Pupa incolata,* and *P. arenula.*

Prof. F. B. Meek‖ described, from the White River Group, on Pinot's creek, *Limnœa shumardi.*

---

* Proc. Acad. Nat. Sci.

† Am. Jour. Sci. and Arts. 3d ser., vols. xi and xii.

‡ Proc. Acad. Nat. Sci.

§ Geo. of Uinta Mountains.

‖ Hayden's U. S. Geo. Sur. Terr.

G. T. Bettany[*] described, from the Miocene of John Day's river, Oregon, *Merycochœrus leidyi*, and *M. temporalis*.

J. A. Allen[†] described, from the lead crevices and superficial strata of the lead region of Wisconsin, Iowa, and Illinois, of supposed Pliocene age, *Canis mississippiensis*, and *Cervus whitneyi*. Charles M. Wallace found flint implements in the stratified drift, near Richmond, Virginia, which he referred to Post-pliocene age.

In 1877, Dr. F. M. Endlich[‡] found the Puerco Group forming the lowest member of the Wasatch, and well developed in southern Colorado. It was best observed on the Lower Animas, where it consists of 1,000 to 1,200 feet of variegated shales and marls. At the base, they are a muddy green, changing into yellow or almost blue. Farther up, pink, pale orange, lilac, and reddish colors predominate, varied by interstrata of white or light yellow. Thin beds of sandstone merely of local occurrence, however, separate these beds; not forming definite recognizable horizons. Farther east, these variegated marls gradually change into shales and sandstones, so that they are no longer characteristic. Above them there occur 1,000 feet of yellow to brown sandstones and shales. As a rule the beds of sandstone are heavy, weathering massively, but they frequently show but small thickness, and are interstratified with yellow and grayish shales. In some of the shales, indications of coal may be observed, but nowhere throughout the San Juan region was any vein found that would have been sufficiently large, or of good quality to be worked.

All the lower cañons of the San Juan drainage, and that of the river itself, are formed by this series of sandstones, and others superincumbent. Over the entire region which they cover, they are uniform, both in occurrence and in lithological character. Their very small dip to the south, 2° to 4°, and their total thickness of 3,000 feet, enables them to extend over a large area of country.

Dr. B. F. Mudge found the Pliocene strata of Kansas resting directly upon the Cretaceous. The material of the Pliocene deposits consists of sandstone of various shades of gray and brown, occasionally whitened by a small admixture of lime. The lower strata are usually composed of finer sand than the upper, and much more loose and friable in their texture. The overlying beds are of coarser ingredients, consisting of water-worn pebbles of metamorphic rocks, quartz, green-

---

[*] Quar. Jour. Geo. Soc. Lond., vol. xxxii.
[†] Am. Jour. Sci. and Arts, 3d ser., vol. xi.
[‡] 9th Ann. Rep. U. S. Geo. Sur. Terr.

stone, granite, syenite, and sometimes fragments of fossil wood from an older formation. These portions of the deposit, when crumbled, and the finer parts washed away, have much the appearance of drift, and have been mistaken for it.

At Breadbowl Mound, Phillips county, it is about 200 feet above Deer creek, and at Sugarloaf Mound, in the western part of Rooks county, it is about 300 feet above the Solomon river. On Prairie Dog creek, in Norton county, it is 400 feet in thickness, and in the extreme northwestern part of the State it is still thicker. The formation like all the rest in the State, appears to dip slightly to the northwest.

In the southern portion of the Pliocene, in the vicinity of Fort Wallace and Sheridan, the hill-tops are covered with a stratum about eight feet in thickness, very hard and siliceous. The material varies from coarse flint-quartz to chalcedony. The latter mineral shades from milk white to transparent, sometimes presenting a semiopal appearance. The so-called moss agate is found in the upper few inches of the stratum. This cap rock is interesting to the mineralogist by showing the moss agate in its various stages of formation. The lower portion of the eight feet indicates an imperfect chemical solution of the silica and black oxide of manganese, therefore the crystalization of the latter is imperfect. As we examine the strata from the bottom to the top, we find the chemical conditions more favorable and complete, so that the distinct quartz, chalcedony, and manganese of the bottom become more commingled toward the upper inch or half inch, where the silica must have been sufficiently fluid to allow the manganese to assume the form of sprig crystals. This peculiar deposit is common on all the high hill-tops of Wallace county.

In King's Geo. Sur.,* the Tertiary is divided into Eocene, Miocene, and Pliocene, each of which is again sub-divided in ascending order as follows. Eocene—1. Vermillion Creek Group; 2. Green River Group; 3. Bridger Group; 4. Uinta Group. Miocene—1. White River Group; 2. Truckee Group. Pliocene—1. North Park Group; 2. Humboldt Group; 3. Niobrara Group; 4. Wyoming Conglomerate. The "Vermillion Creek Group," is a synonym of the Wasatch, and the "Uinta Group," of the Brown's Park Group, and worse than all, the "Niobrara Group" was a pre-occupied name for a Cretaceous Group.

S. F. Emmons estimated the Eocene of the Green river basin at 7,500 feet in thickness. The beds of the Wasatch series, which are

---

* Geo. Sur., 40th Parallel.

chiefly arenaceous, were deposited, in greater thickness than either of the other groups, and extended from the base of the Park range to the flanks of the Wasatch mountains. The beds of the Green river series contrast with those of the other two groups by the relative prevalence of calcareous material, and the fineness of their sediments. They consist of a lower series of calcareous sandstones and impure limestones, containing some lignite seams, overlaid by a great thickness of remarkably fissile calcareous shales, abounding in remains of fish and insects, which reach an aggregate thickness of about 2,000 feet, and are characterized throughout by their prevailing white color. The Bridger Group consists of a thickness of about 2,500 feet of arenaceous beds, with a small development of calcareous material, of a prevailing dull, greenish-gray color, characterized by the great quantity of vertebrate remains which have been buried in them. Its greatest development is in the southern portion of the Bridger basin. In the Washakie basin, on the western borders of the Little Muddy creek, and at Washakie mountain and Cathedral bluffs, the Wasatch series are exposed, weathering in castellated forms, and recognizable from great distances by their bright pinkish and reddish coloring. Washakie mountain and the line of bluffs which extend to Cathedral bluffs, are formed of beds of the Green river series in the upper portion, and with the red Wasatch beds at the base, the line of division can be distinctly traced, descending somewhat in horizon toward Barrel springs, and ascending again beyond toward Cathedral bluffs. A section taken at Sunny Point, near Little Snake river, gave a thickness from the river to the summit of the cliff of about 2,000 feet. The upper 950 feet belonging to the Green river series, and the remaining 1,050 feet to the Wasatch Group. The Green River Group is exposed in the valley of Brown's Park, which is a bay-like depression, from 6 to 8 miles in width, occupying the geological axis of the eastern end of the Uinta mountains, from 1,000 to 1,200 feet in thickness. Throughout the valleys of the Little Snake and Yampa rivers, these groups have been worn into rounded ridges, where, generally, only disintegrated material is found.

In the basin of Vermillion creek, the beds of the Wasatch Group have their greatest development. It was on one of the broad benches, between the branches of this creek, to the east of Ruby Gulch, that the originators of the famous diamond fraud, of the summer of 1872, located their pretended discovery. An exposure of coarse, iron-stained sandstone, on the surface of the mesa, at the foot of Diamond Peak, was strewn by them with rough diamonds and rubies, which were ingeniously mixed with the soil around, so as to make it appear that they came from the decomposition of the sandstone.

Along Bear river, in Utah, from Bear River City to Evanston, the hills on either side are occupied by the nearly horizontal beds of the Wasatch Group. The greater part of Bear river plateau is covered. with a considerable thickness of these beds, which are in general rather coarser and more conglomeratic than those of the Aspen plateau. Its summit varies in width from 2 to 4 miles, beyond which to the eastward these beds are exposed in the deep canons of Woodruff, Randolph and Saleratus creeks, from 2,000 to 2,800 feet in thickness.

He found the Savory plateau region covered, principally, by horizontal beds of the North Park Tertiary, which he referred to the Pliocene, and which, as proved by exposures in the deeper cuts, on its northern edge, overlie the upturned edges of Cretaceous and earlier beds, while the higher portions of the ridges are capped by remnants of the Wyoming conglomerate. The best exposures are found in the open valleys at the heads of Savory and Jack's creeks, and on the pass between the Archaean body of the Grand Encampment mountains and the Savory plateau. A thickness of not less than 1,000 feet of these beds is here exposed, which is made up in the upper portion of a thickness of about 300 feet of a drab, earthy, somewhat porous, limestone, sometimes inclosing small pebbles, underlaid by beds, which grade off insensibly from limy sandstones into coarse gravel beds.

They occupy the valley of the North Platte to the South of Jack's creek, forming long, gentle slopes, extending up from the river to the flanks of the Grand Encampment mountain, which, though so covered by recent deposits that only few exposures of the underlying Tertiary are found, sufficiently show the continuity of their original deposition. Their beds may be traced along the line of bluffs bordering the valley of Sage creek on the south and west. Here the upper member is a hard silicious shale, more like an older rock, under which are seen the white limy sandstones ; the lower beds being concealed beneath debris accumulations.

Arnold Hague found the White River Group along the south and east face of Chalk bluffs, in Wyoming, resting unconformably upon the Laramie Group, and protruding from beneath the Pliocene beds. The strata are exposed near Carr's station, on the Denver Pacific Railroad, eastward across Owl creek, the tributaries of Crow creek, and beyond. The thickness of the group is estimated at 300 feet, and is of Miocene age.

He estimated the thickness of this Pliocene lake strata, which he called the Niobrara Pliocene exposed in Wyoming, at from 1,200 to 1,500 feet.

The beds are found lying unconformably upon the older uplifted strata, and overlapping the area of the Miocene basin. South of the Union Pacific Railroad, they occur abutting against Mesozoic formations; just north of Granite Canon, they lie next the Archaean mass ; and a short distance beyond, at the mouth of Crow Creek canon, are found essentially horizontal against nearly vertical Palæozoic limestones. From Crow creek, northward, they may be seen resting directly upon every formation, from the Archæan to the Fox Hills Group.

The strata consists of marls, clays, coarse and fine sandstones, conglomerates, with some nearly pure limestones. Fine, marly sandstones are the predominant beds.

Overlying the Pliocene lake deposits on Sybille creek and its tributaries, and in the region of Chugwater and Pebble creeks, there occur beds of coarse and fine conglomerate, having a thickness of 300 or 400 feet. These beds have been called the Wyoming Conglomerate.

In North Park, Pliocene beds lie unconformably upon the older rocks, resting in places against every formation from Archæan to the top of the Cretaceous, and are seen in undisturbed condition resting against the basalts. They extend over the entire Park basin, giving it the level, prairie-like aspect, which it presents from all the higher elevations.

He referred the Tertiary beds in the eroded basins and valleys worn out in the rhyolite in the Toyabe range of the Nevada basin, and noticeable on Silver and Boone creeks to the Truckee Miocene.

S. F. Emmons found the same formation in the valley of Reese river, near Ravenswood Peak, along the foot hills, both to the east and west of the Soldier's Spring Valley basin, in the low depression of Indian valley, and in the re-entering bay north of Black Canon, with a thickness of over 700 feet.

The Truckee Miocene is so named from Truckee range, Nevada, which extends in a north and south line for 72 miles, and consists, for the greater part of the distance, of a single narrow ridge, barely more than 5 miles from base to base, but widening considerably at the southern end, where it is made up of broad fields of Tertiary eruptive rocks.

Alfred R. C. Selwyn said* that between Blackwater and Stewart's

---

* Geo. Sur. of Canada.

Lake, and thence to the Finlay Rapids, on Peace river, the country, with some exceptions, is more or less overspread with drift material ; much of this has been derived from the abrasion of the Tertiary formations, through which many of the principal valleys of the country have been cut, exposing alternating beds of clay, lignite, sand and rounded gravel, capped by vast sheets of volcanic products, chiefly porous and compact lavas—columnar and concretionary—and dense dolerite, forming high hills or undulating stony table-lands, such as that which is crossed by the wagon road between Clinton and Bridge creek, at an elevation of 4,000 or 5,000 feet. From Mr. Horetzky's description of the abrupt character of the country on the Susqua river, and in the vicinity of Fort Stager on the Skeena, these Tertiary volcanic products are supposed to be extensively developed in that region. The lignite Tertiary strata which are assumed to have preceded the latest of these volcanic outbursts, occupy undefined, but extensive areas between Fort George and McLeod's lake ; and probably continue thence to the valley of Nation river, with only such interruptions as are the result, partly, of the original unevenness of the surface upon which they were laid down, and partly of the subsequent denuding agencies to which they have been subjected, giving rise to outcroppings of the older rocks, either as hills or ridges rising above the general level of the country, or appearing as rocky bars or canons in the deep-cut channels of the rivers. The general similarity of some of the sands and gravels of the drift period to those of Tertiary age, makes it difficult, without close and critical examination of each exposure, to determine to which period they should be referred, and the distribution of the drift upon the Tertiary deposits is so irregular as to make it quite impracticable to define their respective limits.

At about three miles below Nation river, a steep cliff rises on the right bank of Parsnip river, from the water's edge to 70 or 80 feet. At the base, stiff blue clays are seen, and these are overlaid by layers of sand and fine gravel, passing at the top into coarse rounded gravel. This is, probably, near the northern limit of the Parsnip river lignite-Tertiary basin, as a short distance further a rocky ridge crosses the river and crops out in both banks, the country then rising rapidly, on one side to the Rocky mountains, and on the other to the watershed between the Omineca and the Parsnip rivers. On the eastern side of the mountains there do not appear to be any deposits which can be referred with certainity to the lignite-Tertiary series. At intervals along the river, on both sides, deposits of stratified sand and gravel, cut into benches and terraces, extend from the water to elevations of seven or eight hundred feet. Somewhat similar sands and gravels are thinly spread over many parts of the great prairie plateau, which stretches eastward from the base of the mountains. A section of these, about thirty feet thick, consisting of brown sand, and reddish rusty-looking gravel in thin bands, is seen capping the steep hill of horizontal Cretaceous shales and sandstones, which rises to an elevation of 550 feet above the river, immediately in rear of the Hudson bay post at Dunvegan. In these high gravels the pebbles are small and pretty uniform in size, in which respect they seem to differ from those of the lower benches, which are much coarser; the small and large pebbles being irregularly distributed through them. These upper gravels can not well be distinguished from those which, near Quesnel, occupy a position immediately beneath the basaltic lava flows, and perhaps they belong to the same epoch.

George M. Dawson said that along the foot of the bank of the Fraser river, in front of the town of Quesnel, a considerable thickness of the lignite-bearing formation is shown. The lowest seen is situated about a mile above the confluence of the Quesnel with the Fraser river, and consists of impure lignites and clays, with layers of soft sandstone and ironstone concretions. These are followed in ascending order by clays and arenaceous clays of pale-grayish, greenish and yellowish tints, with a general southward or southwestward dip at low angles. These fill the trough of a shallow synclinal over which the town of Quesnel stands. On the south bank of the Quesnel river, the impure lignites and associated beds rise again to the surface, and in some sections of 15 or 20 feet, the lignite may constitute 1-6th of the whole. It is not, however, in well-defined beds, but interstratified throughout with clays and appears to have been deposited in the form of drift-wood by somewhat rapidly flowing water, and is not so pure as to be of any economic importance. Small spots and drops of amber are abundant in some layers. Half a mile below the mouth of the Quesnel river, on the east bank of the Fraser, a cliff exposes about 100 feet in thickness of this lignitiferous group. The plants, from the Quesnel beds, and also from the lignitiferous beds on the Blackwater, are to a great extent identical with those described by Prof. Heer from the "Miocene" of Alaska, though the age of these beds may be and probably is older than the Miocene.

The basaltic series, consisting of several or many horizontal or overlapping flows, with the exception of those areas of older rocks protruding through them, or exposed in the river valleys where they have been cut away, extend from the lower portion of the Chilcotin river westward to that part of the Chilanco due south of Puntz lake; on the Chilco, to a point a few miles west of the 104th meridian, and on the Chilcotin itself, may stretch to Chizicut lake, and thence extend northeastward, their boundary nearly following the Clusco river for some distance. They characterize the greater part of the Nazco valley, and the plateau extending east and west from it, and probably reach the western slope of the range of hills crossing the Blackwater at the upper canon. The rocks exhibited in these flows are usually true basalts or dolerites of various textures, and from iron-gray to dark greenish and nearly black colors, and often contain much olivine. The vesicles are comparatively seldom filled with infiltrated minerals, though near the sources of the Nazco they are almost invariably so, the material being pale chalcedony, passing over in some instances to chrysopraze. In this vicinity, and near Cinderella mountain, some beds are wacke-like and scoriaceous; and the soil of the water shed region between the Nazco and Bac-zac-coh, on the Cluscus trail, seems to be almost entirely composed of fine rusty pumiceus fragments.

Samuel H. Scudder described, from a very fine grayish and greenish-white fire-clay, in thin layers, with coniferous and angiospermous leaves and seeds, $8\frac{1}{2}$ inches thick, which is superimposed upon a two-inch layer of carbonaceous clay, or impure lignite or matted leaves, mingled with clay, and succeeded by 36 feet of sands and clays, at Quesnel, the following insect remains, to-wit: *Formica arcana, Hypoclinia obliterata, Aphænogaster longæva, Pimpla decessa, P. saxea, P. senecta, Calyptites antediluvianum, Boletina sepulta, Brachypeza abita, B. procera, Trichonta dawsoni, Anthomyia inanimata, A. burgessi, Heteromyza senilis, Sciomyza revelata, Lithortalis picta, Lonchæa senescens, Palloptera morticina, Prometopia depilis* and *Lachnus petrorum.*

Robert Bell, in his report on an exploration between James bay and lakes Superior and Huron, says, that in the region about the height of land, at the head of the east branch of the Montreal river, the lower levels are filled with great mounds and steep ridges of gravel and cobble-stones. The valley of this river, for some miles before it joins the main stream, is also covered with similar materials. The first limestone pebbles were observed on the Mattagami, 24 miles below Kenogamisse Lake. Along the Missinibi river, for many miles above its junction with the Mattagami, a blue clay, only occasionally holding pebbles, underlies the gray and drab bowlder clay, which is overlaid by gravel, sand, and gravelly earth. Marine shells were collected along this river from the Grand Rapids, and along the Missinibi from near Round bay, all the way to Moose factory. They appear to be derived from a pebbly, drab clay, associated with the bowlder drift. The greatest elevation above the sea at which they were collected, is about 300 feet, but they were found along the Kenogami, a branch of Albany river, at the height of 450 feet. Among the fossils collected are, *Rhynchonella psittacea, Leda truncata, L. pernula, Cardium islandicum, Tellina grœnlandica, Macoma sabulosa, Saxicava arctica, Balanus crenatus, Mya arenaria, M. truncata, Mytilus edulis*, and *Buccinum undatum.*

Milne and Murray* found the drift on Newfoundland containing shells similar to those still living, in the surrounding sea, as well as striated angular stones, terraces, and raised beaches tending to show that Newfoundland was at no very remote period below the present level of the sea. The surface of the rocks is often roundly smoothed and striated as if produced by coast-ice acting in a rising area.

Prof. E. D. Cope† described, from the shales of the Green River Group, Wyoming, near the main line of the Wasatch mountains, *Dapedoglossus testis, Diplomystus dentatus, D. analis, D. pectorosus, Erismatopterus endlichi, Amphiplaga brachyptera, Asineops pauciradiatus, Mioplosus abbreviatus, M. labracoides, M. longus, M. beani, Priscacara serrata, P. cypha*, and *P. liops*; from the Eocene of the Rocky mountains,‡ *Clastes aganus, Trionyx radulus, T. ventricosus, Plastomenus serialis, Stypolophus hians, Tomitherium tutum, Plesiarctomys buccatus, Coryphodon obliquus, C. lobatus, Orotherium loewi*; from the Loup Fork Group, *Canis wheeleranus, Dicrocerus trilateralis*, and *D. lehuanus*; from the Eocene§ in Macon Co., Ga., *Amphiemys oxysternum*; from the Upper Miocene of Montana, *Pithecistes brevifacies, Brachymeryx feliceps, Cyclopidius simus, C. heterodon, Blastomeryx borealis*; from the Loup Fork Group of Northwestern Kansas, *Dicotyles serus, Tetralophodon campester*; from the Pliocene of Oregon, *Cervus fortis, Anchybopsis altarcus, A. angustarcus, A. gibbarcus*; from Washington Territory, *Taxidea sulcata*; from Southwestern Texas, *Pseudemys bisornatus, Cistudo marnochi, Anchybopsis breviarcus*, and *Cœnobasileus tremontigerus.*

* Geo. Mag. 2d ser., vol. iv.
† Bull U. S. Geo. Sur. Terr., vol. iii.
‡ Wheeler's Sur. W. 100th Mer., vol. iv.
§ Proc. Am. Phil. Soc.

Prof. O. C. Marsh* described, from the Miocene of the Rocky mountain region, *Moropus distans, M. senex,* and *Allomys nitens;* from the Green River Group of Wyoming, *Heliobatus radians;* and from the Pliocene of the Rocky mountains, *Moropus elatus, Tapiravus rarus, Bison ferox, B. alleni,* and *Crocodilus solaris.*

Prof. F. B. Meek† described, from the Miocene at Cache valley, Utah, *Limnæa kingi.*

Dr. C. A. White‡ described, from the Wasatch Group at Black Buttes Station, Wyoming, *Unio provitus, U. holmesanus, U. endlichi, U. couesi;* and from Wales, in Utah, and the Canon of Desolation, of Green river, *Unio mendax;* from the Tertiary,§ at Last Bluff, Utah, *Physa pleuromatis;* and from Joe's Valley, *Viviparus ionicus.*

In 1878, Prof. C. A. White‖ said the Wasatch Group is the lowest of a series of three fresh-water Tertiary Groups, all of which are intimately connected, not only by an evident continuity of sedimentation throughout, but also by the passage of a portion of the molluscan species from one group up into the next above. Not only were these three groups, aggregating more than a mile in thickness, evidently produced by uninterrupted sedimentation, but it seems equally evident that it was likewise uninterrupted between the Laramie and Wasatch epochs, although there was then a change from brackish to fresh waters, and a consequent change of all the species of invertebrates then inhabiting those waters.

The Wasatch Group, for which " Vermilion Creek Group" and " Bitter Creek Group" are uncalled-for synonyms, in the Green river region, consists very largely of soft, variegated bad-land sandstones, that reach a thickness of about 1,500 feet, together with from 100 to 300 feet of the ordinary indurated sandstones, alternating with badland material at the base, and a similar amount of similar material at top, the estimated aggregate thickness being about 2,000 feet.

Resting immediately and conformably upon the Wasatch are the strata of the Green River Group. Although intimately connected with the former by continuous sedimentation and specific identity of molluscan species, they differ considerably from those of that group in general aspect, and in composition also. The group is lithologically

---

* Am. Jour. Sci. and Arts, 3d ser., vol. xiv.
† U. S. Geo. Expl., 40th Parallel.
‡ Bull. U. S. Sur., vol. iii., No. 3.
§ Wheeler's Sur. W. 100th Mer., vol. iv.
‖ 10th Ann. Rep. Hayden's U. S. Geo. Sur. Terr.

separable into a lower division, having a thickness of about 900 feet, and an upper division having a thickness of about 500 feet.

The Bridger Group in the typical localities rests conformably upon the Green River Group, into which it passes without a distinct plane of demarkation among the strata. Its molluscan fossil remains correspond closely with those of the Green River Group, some of the species being common to both, all indicating a purely fresh condition of the waters in which the strata of both groups were deposited. In the valley of Red Bluff Wash, between Raven ridge and White river, where they are covered by the Brown's Park Group, the thickness is only about 100 feet.

The Brown's Park Group is unconformable with the Bridger Group, but it can not be of later date than Pliocene, for the following reasons: In many places the strata still remain in a nearly horizontal position, but in others they have been considerably displaced, as, for example, by being flexed up against the flanks of the Uinta mountains, and also, in a similar manner, against the Dry mountains, northeastward from Brown's Park. This shows that, although much movement of displacement took place before the deposition of the Brown's Park strata, as shown by their unconformity with those of the older groups, a considerable amount of movement, even of mountain elevation, has taken place since their deposition. Beside this, a large proportion of the immense denudation which the strata of that region have suffered, is known to have taken place since the deposition and partial displacement of the Brown's Park Group, because these strata are involved with the others in that denudation. Furthermore, a remarkably extensive outflow of basaltic trap, covering a large region which lies mainly to the eastward, but which formerly extended much within the limits of this district, took place after the deposition of this Group, and also after it had suffered displacement and erosion to some extent, at least. This is known to be the case, because the trap is found resting upon the unevenly eroded surface of a portion of this group, at Fortification Butte. That portion occupies a higher level than does the principal portion of the group, and the trap rests unconformably upon the Laramie and Cretaceous strata, in the immediate vicinity, as well as upon the Brown's Park strata, in such a manner as to show that little, if any, movement has taken place since the trap outflow. The denudation of the rocks of that region has been so great since the trap outflow, that the latter rock has been removed from a large part of the surface it once occupied, leaving only here and there mere shreds of the once massive and extensive sheet upon the higher hills.

The Brown's Park Group occupies that expansion of Green river valley which is known as Brown's Park. From there it extends eastward and around the eastern end of the Uinta uplift, except a few miles interruption of its continuity there, and thence extends westward along the southern base of the Uinta mountains a large part of the length of that range. It extends northward from the eastern portion of the Uinta mountains, as far as Dry mountains and Godiva ridge. Remaining patches of it show that the formation formerly extended eastward as far as the foot hills of the Park range. It occupies nearly the whole surface of the western portion of Axial basin, comparatively small areas immediately east and immediately north of Yampa mountain, and a considerable portion of the space between Junction mountain and the eastern end of the Uinta uplift, all of which spaces are in unbroken continuity. It, also, occupies a large space from Raven ridge and Red Bluff Wash extending far westward.

F. M. Endlich observed the Wasatch and Green River Groups spread over an area in the White river region of western Colorado, of more than 3,000 square miles. A section on Douglas creek, a branch of White river, showed a thicknesss of 1,500 feet for the Wasatch Group. A stratum of brick-red sandstone, 160 feet in thickness, and placed immediately below the middle of the Group, served as a landmark for identification. Inferior beds of coal occur in the upper part of the Group. Groups of columnar monuments, and monuments composed of shales with cappings of sandstones are not uncommon.

Fine exposures of the Green River Group occur in the Book Cliffs, just north of the Grand river. Geognostically and lithologically speaking, it is separable into an upper and lower division. The lower arenaceous division having a thickness of 2,400 feet, as obtained from the southern bold escarpment of the plateau, and corroborated by observations elsewhere; and succeeded by laminated shales, having a thickness of 1,000 to 1,200 feet; the upper division consisting of yellow and brown sandstones, with thin interstrata of dark shales, and having a thickness of 1,100 to 1,200 feet. These sandstones, by erosion and weathering, have assumed many fantastic shapes, some imitating the ruins of some ancient building, and others rising in spires for several hundred feet above their gently sloping surroundings. A group of three of these weathered monuments near Asphalt Wash, in White river valley, one of which is 80 feet high, received the name of the "Happy Family."

On the White river drainage he observed no evidence pointing to the former existence of glaciers. The numerous canons cut through the soft shales, marls, and sandstones, are formed so regularly, and agree so thoroughly with the pronounced stratigraphical conditions, that they admit of no other agency having shaped them than water. Ascending any one of them toward the main divide, the upward slope is found very even, its valley widening wherever other creeks or streams enter, and its entire character in conformity with the view regarding it as the result of the action of flowing water.

Dr. A. C. Peale made a section of the Roan cliffs, at White Mountain, on Grand river, where he found the thickness of the Wasatch Group, measured by angles taken with the gradienter, to be 1,650 feet, and the Green River Group, 2,282 feet.

George M. Dawson[*] referred the lignite and basaltic series in the basins of the Blackwater, Salmon, and Nechacco rivers, and on Francois lake, in British Columbia, to one group, which, on the evidence of the fossil plants, corresponds with the Miocene of Alaska and Greenland. The basaltic and other igneous flows form the latter part of the group, but blend with the underlying sedimentary beds, and form an integral part of the whole. No trace, however, is found of rocks due to volcanic action since the period of the drift. The sources of the immense flows of molten matter have been numerous ; for, beside the many dykes found traversing the older rocks, which may, at one time, have been fissures giving exit to lava streams, beds characterized by a roughly brecciated character appear in many places, and can scarcely have been formed far from the mouths of larger or smaller vents, capable of ejecting fragments. Between the region of the upper waters of the Blackwater and Salmon rivers, and the Bella Coola, three masses of broken mountains represent as many centers of former very great volcanic activity.

Samuel H. Scudder described, from the Tertiary at Quesnel, British Columbia, *Sciara deperdita, Euschistus antiquus, Lachnus quesneli, Bothromicromus lachlani,* and *Aranea columbiæ.*

The striæ upon the rocks of New Hampshire[†] are extremely variable in their course. A few extremes are as follows : S. 2° E. ; S. 83° E. ; S. 58° W.; N. 40° W. ; N. 83° E.

Bible hill, in Claremont, rises about 350 feet above the plain of the village, at its northern base. What is supposed to be the normal direction of the striæ is about S. 12° W., which occurs commonly west of

---

* Geo. Sur. of Canada.
† Geo. of N. Hampshire, vol. iii.

the summit of the hill for two or three miles, reaching beyond the Connecticut. North of the village, it is S. 15° E. ; among the houses, S. 41° E. ; and on the east side of the hill, S. 23°–25° E., in a valley leading to Unity. On the south slope of Green mountain, east of the village, are intersections of the almost east course with that of about S. 12° E. On the westerly side of the top of Bible hill, the most common course is S. 6° E., with S. 25° E. This is a half mile east from Brown's, Clark's, and Stone's, where the westerly course has been noted. We now proceed three fourths of a mile northeast to the " Flat Top," a spur of the hill, with scarcely any depression between. At the commencement. where the northeast slope begins, are striæ S. 57° E., pointing back to Little Ascutney, and crossing others S. 1° W. Next are some S. 46° E. pointing to Ascutney, apparently marked on the lee side of striæ, pointing S. 1° W. to S. 1° E. Another ledge has striæ S. 46° E. crossed by others S. 1° E.; then S. 16° E. crossed by S. 41° E., and S. 51° E., the middle one the most common. Another ledge shows, in a narrow compass, the courses S. 21°, 36°, 41°, and 57° E. Where the courses are so numerous, there is a marked tendency to irregularity ; the striæ do not preserve their parallelism. A change of 10° or 15° degrees in direction will occur in a distance of less than a yard. Flat Top hill shows more of the irregularities than the highest summit to the southwest. Near the aqueduct, at the base of Flat Top, the course is S. 17° E.

Prof. E. D. Cope* described, from the Miocene of Oregon, *Steneofiber gradatus, Entoptychus cavifrons, E. planifrons, E. crassiramis, Pleurolicus sulcifrons, Meniscomys hippodus, M. multiplicatus, Temnocyon altigenis, Canis cuspigerus, C. geismarianus, Machærodus strigidens, M. brachyops, Anchitherium equiceps, A. brachylophum, A. longicristes, Stylonus seversus, Dæodon shoshonensis,* and *Hyopotamus guyotianus.*

He described,† from the Upper Miocene of Montana, *Ticholeptus zygomaticus;* from the Loup Fork beds of Kansas, *Aphelops fossiger, A. malacorhinus,* and *Mylagaulus sesquipedalis :* from the Green River Group. on Bear river, Wyoming. *Priscacara oxyprion, P. pealei, P. clivosa, Dapedoglossus æquipinnis ;* and from Florissant, Colorado, *Trichophanes foliarum;* from the Pliocene of Oregon, *Auchenia vitakeriana, Mylodon sodalis, Graculus macropus, Anser hypsibatus,* and *Cyg-*

nus paloregonus; and from a lacustrine, Post-pliocene deposit in Vandenburg county, Indiana, *Cariacus dolichopsis.*

Messrs. Henry F. Osborn, Wm. B. Scott, and Francis Speir, Jr.,[*] described, from the Eocene of Wyoming and Colorado, *Megencephalon primœvus, Leurocephalus cultridens, Hyrachyus imperialis, H. intermedius, H. crassidens, Helaletes latidens, Ithygrammodon cameloides, Uintatherium leidyanum, U. princeps, Paramys superbus, Crocodilus parvus,* and *Tricophanes copei.*

J. A. Allen[†] described, from the insect bearing shales of Florissant Colorado, *Palœospiza bella.*

Prof. Leo Lesquereux[‡] described, from the Pliocene Chalk bluffs of Nevada county, California, *Sabalites californicus, Betula aequalis, Fagus pseudo-ferruginea, Quercus nevadensis, Q. distincta, Q. gœpperti, Q. royana, Q. pseudolyrata, Castaneopsis chrysophylloides, Salix elliptica, Platanus appendiculata, P. dissecta, Liquidambar californicum, Ulmus californica, U. pseudofulva, Ficus sordida, F. tiliaefolia, Aralia whitneyi, A. angustiloba, Cornus kelloggi, Magnolia lanceolata, M. californica, Acer aequidentatum, Zizyphus microphyllus, Z. piperoides, Rhus mixta, R. myrciaefolia, Juglans californica, J. laurinea, J. egregia;* from Table mountain, Tuolumne county, *Quercus elœnoides, Q. convexa, Salix californica, Ulmus affinis, Ficus microphylla, Persea pseudo-carolinensis, Cornus ovalis, Acer bolanderi, Ilex prunifolia, Rhus typhinoides, R. bowenana, R. metopioides, R. dispersa, Cercocarpus antiquus;* and from Oregon, *Xanthoxylon diversifolium, Juglans oregoniana,* and *Quercus bowenana.*

Prof. W. H. Dall[§] described, from the Pliocene of California, *Axinea profunda, Pecten expansus, P. hemphilli, P. stearnsi, Anomia limatula,* and *Scalaria hemphilli.*

In 1879, Dr. A. C. Peale[‖] found the Wasatch Group in the Green river basin resting unconformably upon strata from the Silurian to the Laramie; along the southwestern slopes of the Wind River mountains, upon the granitic rocks; south of Thompson plateau upon the Jurassic rocks of Meridian ridge; and north of Thompson plateau on Jurassic, Cretaceous and Laramie beds. The strata along the southwestern slopes of the Wind river mountains were evidently derived

---

[*] Pal. Rep. Princeton Sci. Exped.
[†] Am. Jour. Sci. and Arts, 3d ser., vol. xv.
[‡] Rep. Foss. Plants Aurif. Grav. Deposits.
[§] Proc. U. S. Nat. Museum.
[‖] 11th Ann. Rep. U. S. Geo. Sur. Terr.

from the disintegration of the granitic rocks of the mountain range. They consist of yellow, gray, and pink sands and marls, which dip from 5° to 10° from the mountains. West of Green river the character of the beds is similar to those on the east. They are generally brick-red in color, and weather into picturesque bad-land forms. Along the edge of the basin they are composed of conglomerates which contain pebbles of limestone derived from the adjacent mountains. The red character of the strata is due to the wearing away of the red Mesozoic rocks. The thickness exposed along the western edge north of Thompson plateau is from 500 to 800 feet. On the Bear lake plateau the thickness is greater, especially toward the west, and on the eastern flanks of the Bear River range it is still greater; it increases also to the southward until it is several thousand feet in thickness. The line separating the Wasatch from the Green River Group is lithological. All the variegated beds that lie below the laminated, light-colored sandstones, are referred to the Wasatch Group, and all above to the Green River Group.

The area between Green river and the Big Sandy is covered with the Green River Group until the northern portion of the basin is reached. North of the New Fork it is present only as cappings of the mesas that stand between the streams. Along the east side of the Green, from New Fork southward, the Green river shales and sandstones form bluffs several hundred feet in height. On the west side of the river above La Barge creek, the group is present only in isolated mesas. South of that stream, however, it is the surface formation rising from Green river to the westward, and breaking off in bluffs that face Meridian ridge. It consists of a series of light colored sandstones which are succeeded by calcareous layers and fissile shales. In the Ham's Fork plateau the group forms the surface of a shallow synclinal; it is highly fossiliferous, and contains near the top a layer of bituminous shale. An excellent fossil locality may be found on Twin creek, at the South end of the Ham's Fork plateau. It was at Station 14, south of Horse creek and west of Green river, where beds of limestone were found completely covered with the petrified cases of caddis-flies described by Dr. Scudder, under the name of *Indusia calculosa.*

The Bridger Group may be observed extending northward from Ham's Fork toward Slate creek, breaking off in low bluffs, in which the sombre clays and sands of the group are exposed. Between the mouth of the Big Sandy and the Green, on the east side of the former, there are variegated sands and marls belonging to this group, which weather into bad lands.

The Pliocene or Salt Lake Group in Bear river valley consists of yellow sands and marls, white limestones and shales, and pea green shales and sands. The thickness is estimated at from 600 to 700 feet.

The Salt Lake Group or Pliocene of Cache valley is succeeded by a later Pliocene deposit, for which Dr. Peale proposed the name Cache Valley Group. The beds near the center of the valley deposited by the lakes, and still remaining in horizontal position, are those to which he applied this new name, but without more reasons than he adduces it would have been just as well not to have proposed it.

F. M. Endlich found east of the Wind River Range a series of variegated, arenaceous marls, resting upon the yellow sandstones of the Laramie Group. These are of the age of the Lower Wasatch or Lower Eocene. They are nearly horizontally stratified, and are carved into typical "Bad Lands" by fluviatile erosion. A variety of colors presents itself in these marls. Gray and reddish-brown predominate, interchanging, however, with yellow, white, greenish, and maroon. Without any apparent separation of strata, these colors and shades form bands resembling well-defined bedding. Rapidly denuded by erosion, the slopes presented by these marls are generally entirely bare of vegetation. Thin bands of highly argillaceous sandstones, occurring sporadically within the series, sometimes give rise to the formation of low regular bluffs. These marls southeast of Beaver creek, have a thickness of 450 to 500 feet. They are supposed to be parallel with the Puerco marls of New Mexico and Colorado. They are succeeded by the yellow sandstones and shales of the Upper Wasatch Group.

An extensive section of the Wasatch Group may be obtained north of Salt Wells. The lower marls reach a thickness of 600 to 700 feet, and the upper sandstones and shales attain a thickness of about 600 feet. A number of volcanic eruptions have taken place in this region. Several buttes occur, from one of which called Essex mountain, the extension of the Wasatch Group may be traced by its color. The Red Desert of this region is derived from the upper members of the Wasatch Group. Not a drop of water is to be found in this desert. The thickness of the Upper Wasatch diminishes in the direction of the Sweetwater.

The Wasatch Group is succeeded by the Green River Group. Packer's creek flows in a southerly direction into Bitter creek, a short distance east of Rock springs. West of it there is a high ridge composed of the light Green river shales. The strata have a gentle dip to the westward. Erosion has removed large masses of the strata, and exposed the Wasatch for a number of miles up stream. Eastward the Green River makes a sharp turn and passes north of Essex mountain. In this section the lower members of the Green River Group are composed of gray and bluish shales, more or less calcareous and arenaceous. Higher up the shales are yellow and light brown, mostly very sandy, but containing strata of impure argillaceous limestones. Above these follow concretionary sandstones and shales, of yellow and rusty brown color. The former contains one very prominent horizon of silicious material, appearing in the form of chalcedony and agate. Near the base a thin seam of oolite occurs. West of Packer's creek, the total thickness of this group is from 1,700 to 1,800 feet. Of this the upper sandstones with their shales, occupy about 800 to 900 feet, and the arenaceous beds near the base, about 150 to 200 feet, which leaves an average thickness for the shales of 700 to 800 feet. Both the shales and sandstones diminish in thickness in their northern extension.

The Green River Group is succeeded by the Bridger Group in this section, wherever the bluffs rise high above the general level as on Steamboat buttes. A thickness is preserved of about 500 feet, but most of the group has been eroded. Toward the south and southwest it becomes thicker. On the northern edge of the Sweetwater plateau the Wasatch Group is succeeded by a local deposit, called the Sweetwater Group. It consists of brown, yellow and white arenaceous marls and clays, and near the top some sandstones without clearly defined stratification. It is not conformable with the Wasatch. This group has suffered greatly by erosion, but retains a thickness in some places of 1,200 to 1,400 feet. The hills south and southwest of Saint Mary's ranch, the central butte in Elkhorn Gap, the Sweetwater hills and numerous bluffs are composed of the strata of this group. It is of Miocene age.

The Sweetwater Group is succeeded by the Pliocene. Near the base it consists of a very loosely aggregated sandstone, of a light gray or yellowish color. Above this there is a succession of light marls and indurated clays. Usually these are either very light yellow or white, but pink and greenish beds are not wanting. Toward the eastern termination of the group the strata become highly silicious. Thoroughly permeated by silica, the clays become very hard and brittle. The former occurs also in the shape of narrow veins, concretions and even strata. Innumerable moss agates are strewn over about six square miles, near Agate lakes, north of the Sweetwater. All of them are water-worn. It is an accepted fact that the "moss" in agates is but the result of impeded crystallization. The best exposures of the Plio-ocene series are low down on the Sweetwater, and along the northern edge of the plateau. The thickness is estimated at from 700 to 900 feet.

The Wyoming Conglomerate is structureless, and composed of the most varying material. It is the product of all formations existing within a given area. Along the entire northern slope of the Sweet-water and Seminole hills there are enormous deposits of it. The thick-ness is estimated at from 10 to 400 feet. It is also abundant along the southern slope of the Sweetwater mountains, in the Pliocene valley west of South Pass, and is scattered to a greater or less extent all over the country, which has been subjected to extensive erosion. The maximum accumulation occurs along the shores of the former Ter-tiary lakes, and was probably carried there by the waters draining in-to them, and it is, therefore, of the age of the younger Pliocene marls and shales.

George M. Dawson* said that in the plateau region in the southern part of British Columbia, lying east of the Coast Ranges, terraces are exhibited on a scale scarcely equaled elsewhere. They border the river valleys, are found attached to the flanks of the mountains to a great height, though none have been found in this region equal to the elevation of that on Ilgachuz mountain in the north—5,270 feet. The higher terraces can be due to nothing else than a general submergence of the country. Five of the best marked terraces on the southern slope of Iron mountain, at the mouth of the Coldwater, have the follow-ing elevations above the sea, viz: 2,386, 3,063, 3,392, 3,611, and 3,715 feet. The last mentioned is, the highest observed, and is quite narrow. Above this, the drift covering becomes thinner, but rolled stones, some of them certainly from a distance, occur to the very summit—5,280 feet above the sea. The elevation of the white silt terrace bordering Okanagan lake, is 200 feet above the lake, or 1.277 feet above the sea. Leaving this to ascend the Okanagan mountain, south of the Mission, a great series of high terraces is passed over. The heights of six of these are as follows: 1,862, 2.042, 2.141, 2,645, 2,800, and 2,839 feet. On the northern slope of the same mountain, six principal terraces have the following heights: 1,451, 1,579, 1,962, 2,452, 2,553, and 2,879 feet.

---

* Geo. Sur. of Canada.

A hill on the east side of McDonald's river, near Nicola lake, is terraced at different levels to the height of 800 feet above the lake, or 2,600 feet above the sea. On the Coldwater, near the first bridge, a terrace fringes the west side of the valley, at the height of 200 feet above the river, or 2,955 feet above the sea. On Whipsaw creek, Similkameen river, a terrace occurs 200 feet above the stream near Powder camp, or 3,845 feet above the sea. Between Powder camp and Nine-mile creek some of the more prominent benches have the following elevations above the sea: 2,956, 3,078, 3,237, and 3,252 feet. The trail, when some distance north of the South Similkameen, above its junction with the north fork, passes over several broad terrace flats, two of these are elevated 2,632 and 2,683 feet above the sea. Near the junction of the north and south forks a terrace-flat occurs 300 feet above the river, or 2,264 feet above the sea. Further down the Similkameen, in a grassy hill above Keremeoos, a terrace is seen 1,000 feet above the river, or 2,300 feet above the sea. In a wide valley between Okanagan and Vermilion forks, a rather irregular bowldery bench occurs with an elevation of 3,713 feet. It is on the rim of the valley and far above the stream.

In the valleys of streams draining westward from the mountains, there is a remarkable absence of detrital deposits, and though a few terraces occur, the valleys are much contracted, and in a region so mountainous that it is generally difficult to decide precisely what significance attaches to them. Not only may some of them be merely river-terraces, but others may simulate beach-terraces, but owe their origin solely to the damming up of valleys by glacier ice or moraines. At the summit between the Coldwater and Coquihalla, a terrace occurs at an elevation of 3,286 feet. On the Skagit another occurs at an elevation of 1,997 feet, and on the Uztlihoos, tributary to the Anderson river, narrow but well marked benches occur at 3,087, and 3,582 feet.

Robert Bell, in his report on an exploration of the east coast of Hudson's bay, says that the striæ in the southern part of the Eastmain coast have a southwesterly course, but in going northward the direction gradually changes till it has become nearly west at Cape Jones. From this point northward the course continues west and north of west, or toward the center of the bay. The grooving is remarkably well preserved on the bare hills and on the rocks generally from Great Whale river northward. In this region one can not help being struck by the more modern appearance of the glaciated surface than in the inhabited part of Canada to the south. The course of the striæ in sixty-six localities between Sherrick's Mount and Cape Dufferin vary from S. 45° W. to N. 35° W., many of them are S. 60,° 70,° or 80° W., while an equal number are N. 60,° 70,° or 80° W. The bowlder clays abound with marine shells. He found abundant evidence that the sea level is falling at a comparatively rapid rate in Hudson's bay. On the islands and shores all along the Eastmain coast the raised beaches are very conspicuous at all heights up to about three hundred feet, immediately near the sea, but, no doubt, higher ones will be found further inland. Driftwood (mostly spruce) is found almost everywhere, above the highest tides, in a more and more decayed state the higher above the sea, up to a height of at least thirty feet, and in some places up to forty and fifty feet, above which it has disappeared by the long exposure to the weather. Judging by the rate of decay of spruce-wood in this climate its preservation in large quantities, during an elevation of the land, or rather a fall in the water, to the extent of thirty feet, would indicate a change in the relative level of the sea, amounting to perhaps between five and ten feet in a century.

The striæ observed at eleven places on the east shore of Lake Winnipeg vary from S. 15° W to S. 45° W.; at thirty-four places along the boat route from Lake Winnipeg to Hudson's bay, they vary from S. 50° W. to S. 20° E.; and at twenty-one places along the Nelson river, from Great Playgreen lake downward, they vary from S. 25° W. to S. 80° W. The bearings refer to the magnetic meridian.

G. F. Mathew found the course of the grooves and scratches on the rocks in the southern counties of New Brunswick having both southeasterly and southwesterly bearings. A southeasterly course is most prevalent in the western part of Charlotte county, and a southwesterly course most prevalent in the valleys east and northeast of St. John. These two general courses, as well as the intermediate ones, are controlled by the contour of the surface of the land in the several districts where they occur; for, as a general rule, the furrows conform to the direction of the river valleys, or at least are influenced in their course by these depressions.

Prof. E. D. Cope* described, from the Truckee beds of the White River Group of Oregon, *Hesperomys nematodon, Sciurus vortmani, Paciculus insolitus, Canis lemur, Amphicyon entoptychi, Archælurus debilis, Hoplophoneus platycopis, Chænohyus decedens, Thinohyus trichænus, Palæochœrus subæquans, Merycopater guiotianus, Coloreo-*

---

* Proc. Am. Phil. Soc.

don ferox, and *C. macrocephalus; Enhydrocyon stenocephalus,** E. basilatus,* now *Hyænocyon basilatus, Pæbrotherium sternbergi, Boochærus humerosus;* and from the Loup Fork Group of Cottonwood creek, Oregon, *Lutrictis lycopotamicus,* and *Protolabis transmontanus;* from the Green river shales of Wyoming,† *Xiphotrygon acutidens;* from the White river beds of Colorado, *Anchisodon tubifer;* and from the Post Pliocene of Shasta county, California, *Arctotherium simum.*

Samuel H. Scudder‡ described, from the thinly bedded, almost paper-like, yellow, gray siliceous Tertiary shales, on the North Fork of the Similkameen river, three miles from its mouth, *Penthetria similkameena, Hygrotrechus stali, Cercopis selwyni, Planophlebia gigantea, Calidia columbiana;* from Nine-mile creek, which flows into Whipsaw creek, a tributary of the Similkameen, *Trox oustaleti, Gallerucella picea, Tenebrio primigenius;* and from the Nicola river, *Nebria paleomelas, Cercyon (?) terrigena, Buprestis tertiaria, B. sepulta,* and *Cryptohypnus (?) terrestris.*

Prof. J. W. Dawson described, from Nine-mile creek, *Equisetum similkamense.*

In 1880, Prof. E. D. Cope§ described, from the Truckee Miocene of the John Day river, Central Oregon, *Nimravus confertus, N. gomphodus, Pogonodon brachyops; P. platycopis, Hoplophoneus cerebralis, H. strigidens, Dinictis cyclops;* from the Loup Fork Group of Nebraska, *Peraceras superciliosus;* and from a cave on the Schuylkill river, in Pa., of Post-pliocene age, *Smilodon gracilis.*

Dr. C. A. White‖ described, from the Wasatch Group, near the head of Soldier's Fork, Utah, *Planorbis militaris;* from the Green River Group, on Henry's Fork of Green river, Wyoming, *Planorbis æqualis,* and from three miles east of Table Rock Railroad station, *Limnæa minuscula;* and from the Upper Green River Group of Henry's Fork, Wyoming,¶ *Pupa atavuncula.*

Angelo Heilprin** described, from the Eocene of Clarke county, Alabama, *Cytherea nuttalliopsis, Pseudoliva scalina, Lævibuccinum lineatum, Fusus subtenuis, F. interstriatus, F. engonatus, F. sub-*

---

* Bull. U. S. Geo. Sur. Terr., vol. v.
† Am. Nat., vol. xiii.
‡ Geo. Sur. of Canada.
§ Am. Nat., vol. xiv.
‖ Proc. U. S. Nat. Mus.
¶ 12th Rep. U. S. Geo. Sur. Terr.
** Proc. Acad. Nat. Sci.

scalarinus, Pleurotoma moniliata, Pyrula multangulata, Solarium cupola, S. delphinuloides, and *Dentalium microstria.*

In 1881, Prof. E. D. Cope* described, from beds supposed to belong to the Puerco Group, *Periptychus carinidens,* and *Deltatherium fundaminis.* From the Wind river Eocene,† *Calamodon 'cylindrifer, Esthonyx acutidens, Sciurus ballovianus, Pantolestes secans, Microsyops scottianus, Miacis canavus, M. brevirostris, Didymictis dawkinsianus, Ictops didelphoides, Bathyopsis fissidens, Lambdotherium brownianum, Hyracotherium venticolum,* and *Phenacodus trilobatus;* from the Miocene of the John Day river in Oregon, *Nimravus gomphodus, N. confertus, Coloreodon ryderanus, Palæochœrus platyops, Protolabis prehensilis,* and *Eumys lockingtonianus;* and from the Amyzon shales in the South Park of Colorado, of Upper Eocene or Lower Miocene age, *Charadrius sheppardanus.*

We have passed, in historical review, the Tertiary as it has been discovered, and is now known on the eastern, southern, and western parts of the continent of North America, leaving for further consideration, only the drift or fresh-water Pliocene and Post-pliocene of the central part. The reason for separating the rocks in this manner may be found in the fact that there is no connection between the marine drift of the New England States and northeastern shore of the continent, and the fresh-water drift or lake drift of the central part, and as to the western part or Rocky mountain region, it has never been subjected to any general drift action, though here and there the waters from the local lakes have left their drift in and about the streams that drained them.

It may be important here to remark, that in this historical review, geologically speaking, we have not found any Glacial Period or Glacial Epoch, nor palæontologically speaking, have we found any evidence whatever of such a period, nor have we found any phenomenon requiring the intervention of such a period to explain it; but, on the contrary, all the phenomena are to be accounted for, without change of climate, and without the violation of any of the laws of nature, which are now in operation, and form the basis, from which the geologist judges of the past. And when we come to a review of the drift of the central part of the continent, it will appear equally as clear that no part of it was the result of glaciers, and that so far as North America is concerned, the so called Glacial Period never had an existence.

---

* Am. Nat., vol. xv.
† Bull. U. S. Geo. Sur., vol. vi.

Before we proceed, however, with the fresh-water or lake drift of the central region, it may be proper to recapitulate some of the facts, which we have already considered, and to call further attention to the total absence of evidence to support the theory of a Glacial Period.

The marine Eocene, commencing in New Jersey, with a thickness of only 37 feet, and covering but a narrow surface area, crosses the State of Maryland at Fort Washington; Virginia, by the way of Fredericksburg, Richmond and Petersburg ; North Carolina, by way of Newbern and Wilmington; South Carolina, by way of Charleston and Shell Bluff, on the Savannah river; Georgia, by way of Milledgeville; Alabama, by way of Claiborne; and Mississippi, by way of Jackson and Vicksburg. In South Carolina, it covers a large area, and attains a thickness of 1,000 or 1,100 feet. In its surface expansion, it is exposed in Florida, and reaches up into Tennessee, where it is called the Porter's Creek Group and Orange Sand, and attains a thickness of between 800 and 900 feet. In Alabama and Mississippi, it is subdivided into the Vicksburg Group, Red Bluff Group, Jackson Group, Claiborne Group, Buhrstone Group, and Flat Woods and Lagrange Lignitic Group, and covers a large area, and attains a thickness of 872 feet. It crosses Louisiana, and offers numerous exposures in Texas. It also appears in limited exposures in California. But nowhere is it conformable with the underlying rocks. It is extremely fossiliferous in many of its exposures, and the general *facies* of the shells has a striking generic resemblance to the living mollusca of the same latitude, though none of the species are supposed to have survived.

The marine Miocene, beginning at Martha's Vineyard, though it may exist as far north as the State of Maine, crosses New Jersey through Cumberland county, and forms a border upon the east and south of the Eocene exposures, a large part of the way to the Mississippi river, and west across the States of Louisiana and Texas. It is not conformable with the Eocene, and in some parts does not, therefore, intervene between it and later deposits, as in South Carolina for instance its very existence has been doubted. But on the western coast, and especially in California, it is highly developed. Between Canada de las Uvas and Solidad Pass the thickness is 2,500 feet, and in other places the maximum is evidently much greater. The Coast range is composed in large part of strata of this age, and hence its elevation has been since the Miocene period. As far as we may be able to judge of the climate and temperature of this period, by the fossils obtained from this region, it was the same that it is now; and, indeed, we might go far anterior to this for the same climate except so far as the proportion of the land and water surface may have acted to change it. It, too, is highly fossiliferous in some of its exposures, and the shells, generally, belong to living genera and many of the species still survive in the waters bordering the adjacent coast.

The marine Pliocene strata are found in Maryland, superimposed upon the Miocene, and in South Carolina upon the Eocene, and, generally, forming a narrow border at the east of these outcrops on the Atlantic coast and a wider border on the south adjoining the Gulf coast. Fossil shells of species now living on the adjacent coast abound intermingled with those which have become extinct. The number of living species indicates, so far as one may be capable of judging, identically the same climate on the eastern coast of the United States that now prevails, substantially the same may be said of the Pliocene of the Pacific coast, and especially of the California strata of this age and the living and extinct species. Indeed, there is no palæontological evidence that the Pliocene climate was different from the present, on this continent, nor could we reasonably suppose it to have been different, because the outlines of the continent were nearly the same as they are now. The Pliocene so graduates into the Post-pliocene at many places that the separation of the two is very difficult, and in others it is wholly impracticable, and, in such cases, an arbitrary approximating line for separation is assumed.

The marine Post-pliocene of the eastern coast, south of the State of New York, and bordering the Atlantic and the Gulf, and also on the Pacific coast, is usually found conformable with the Pliocene below, and always graduating into the present or modern times without a break stratigraphically or palæontologically. In South Carolina it forms a belt along the coast 8 or 9 miles wide, and the fossils nearly all belong to living species now inhabiting the coast. There are, in layers of blue mud, and also in the sands which succeed them of this age, the bones of horses, hogs, dogs, rabbits, beavers, tapirs, and other mammals that flourished, as far as we can judge, throughout the period. Here rests the evidence that the climate of South Carolina, during the Post-pliocene, was substantially the same that it is at present, and it seems to be conclusive, in the absence of any geological evidence to the contrary. The stratigraphical indications of the Post-pliocene of Texas and California, and the palæontological evidences, without a single exception, are that there has been no change in the climate of these States since the Pliocene age. That man was an inhabitant of this continent during part or all of the Post-pliocene period, no reasonable man will doubt, for his bones and his stone implements have frequently been found in the Ashley beds of South Carolina, with the remains of the extinct Mastodon and Mammoth, and living mammals that are well known to have been contemporaneous. This mixture of the bones and implements of man, with the remains of living and extinct mammals, is also well known from the labors of Prof. Whitney, in California.

Sometime during the Pliocene or Post-pliocene era, and, most likely, commencing during the first and continuing into the second, a portion of the northeastern coast, about Hudson's Bay, and the Gulf of St. Lawrence, and that arm of land south of the Gulf, and east of Lake Champlain and the Hudson river, known as New Brunswick and the New England States, was submerged or overflowed by ocean currents, with the exception of a few mountain elevations. The depression in the Hudson Bay region has been fully set forth in the foregoing pages. It appears that the rocks are striated in nearly all directions, and that upon the striated surface there rest marine clays full of fossils belonging to the living species of that region, and numerous bowlders from the contiguous mountains and hills. The scratches are evidently the work of floating icebergs and shore ice, during the period of submergence. There is no general radiation of detritus from mountain ranges to evidence the existence of glaciers in this region, nor any other evidence tending to show that the climate was materially different from what it is now. The fossiliferous marine clays and sands prove the submergence, and all other phenomena, including the scratches, follow as the necessary results of submergence, in that latitude, without the intervention of glaciers; and, furthermore, there is nothing to warrant the supposition of a glacial period within this area. And as the Laurentian range of mountains is south and east of this submerged area, and rises to the height, in some places, of 3,000 to 4,000 feet, and generally has an elevation of 1,500 feet or more above the level of the ocean, no reasonable theorist will claim that a glacier would ascend this range of mountains for the mere purpose of going south, and yet how could we have a continental sheet of ice moving south unless it did. Modern ice has a tendency to move down an incline, rather than to ascend rugged elevations and mountain chains, and an ordinary philosopher would suggest, that if we must have a Post-pliocene glacier, on the northern side of the Laurentian mountains, that we let it slip down hill instead of up, even if the direction is to the north. Of course this would destroy much of the beauty and symmetry of the glacial theory, but there would be one thing in its favor—it would not be reversing the laws of nature.

South of the Laurentide mountains, the surface of the rocks beneath the bowlder clay is generally striated in the direction of the valleys. This pursuit of the valleys by the lines of striation may be observed from the mouth of the Gulf of St. Lawrence up the St. Lawrence, and down the Champlain and Hudson river valleys. No one who has read the description of these markings by Prof. Dawson can have any doubt that the bodies which produced them came from the Atlantic ocean at the eastern end of the Laurentian range of mountains, and following up the St. Lawrence were drifted to the south at various angles, some floating over New Brunswick, and others over Maine, and others passing up so far as to drift through Lake Champlain, and re-enter the Atlantic ocean by the Hudson river, while others drifted past Montreal, and were driven into the mouth of the Ottawa river valley, and the ancient valley of Ontario.

In New Brunswick, which is immediately south of the gulf, the striæ are related to the contour of the surface of the land, and conform to the direction of the river valleys. A southeasterly course prevails in the western part of Charlotte county, and a southwestern course in the valleys east and northeast of St. John. A map of the State of Maine, showing the course of the rivers will show the course of the striæ. The appearance of the surface geology of this State early suggested the fact that a great rush of waters poured over it from a northerly source, and transported, by its power, the surface debris which had accumulated in earlier ages by subærial forces, and large masses of rock from parent ledges, and deposited them in regions more or less distant from the several sources, and as they passed along they striated and grooved the rocks against which they impinged, or over which they rubbed in the traveled course. The course of the striæ is, therefore, in nearly all directions. If the rivers are flowing in valleys, bearing to the southeast, the striæ have that course, or if south or southwest, the striæ conform to the valley. Sometimes striæ have been found which ultimately varied at right angles from their original direction. The Katahdin mountains formed an obstruction around which the striating agency operated, but it did not cross the summit. The striæ are found upon the north side of the mountains, and not upon the south side, unless for a short distance where the slope is small. The striæ in the States of Vermont and New Hampshire are in all directions, and it is with difficulty that any two sets are found exactly agreeing in their course, though as in Maine they conform to the direction of the valleys. The greater part of Massachusetts and Connecticut is covered with the drift sand, gravel, bowlders or clay, and the grooves, furrows and scratches upon the surface of the rocks in place, have a general southerly direction, though varying with the contour of the valleys to a southeasterly or southwesterly course. At the Island of New York, there is abundant evidence that a current swept over it from the northwest to the southeast. The furrows are most strongly marked on the northwestern slopes of the hills, and least so on the southeastern. In many instances they are very distinct on the western and northwestern slopes, extending to the highest points of the rock, but no traces are to be seen on the eastern and southeastern slopes, although both slopes are equally exposed. The striæ are most numerous in the middle part of the island, somewhat less in the western, and least in the eastern. It appears that the current was deflected southward by some force, at an angle to its course in the middle part of the island. Throughout all this region south of the Gulf of St. Lawrence, and the St. Lawrence valley, we have in the course of the striæ, and the distribution of clay, sand, gravel, and bowlders, the evidence of an overflow of the whole country, except the higher hills and mountains, and the evidence that this overflow was by subsidence of the coast, and that the Arctic current, instead of leaving the coast on approaching the mouth of the Gulf, as it does now, flowed into the Gulf and across the depressed New England area, transporting its fields of ice, which grounded upon the northern slopes of hills and mountains, and rubbed the rocks in the valleys and plains wherever the surface soil and subærial accumulations were swept off by the grinding weight of a mass driven by a current through water too shallow to float it. However, the evidence of submergence does not rest alone upon these appearances, but stands upon the incontestible ground of palæontology.

Throughout nearly all this region the striated rocks are succeeded by fossiliferous, bowlder-bearing, marine clays and sands. In the Gaspe peninsula ocean terraces and stratified clay containing marine testacea occur at the height of 600 feet above the sea. In the St. Lawrence valley, the valley of the Ottawa, Champlain region of Vermont, and over the triangular area of 9,000 square miles extending from Ottawa to Lake Champlain, the marine fossils occur in the bowlder clay at all elevations as high as 500 feet or more above the level of the ocean. The fossiliferous marine clays and sands form a coating for a large part of the face of New Brunswick, and sea beaches, sea bottoms, and fossiliferous clays form almost a continuous belt on the coast of Maine, 150 feet above the ocean, and extending up the rivers to the same height. These facts prove the submergence of the country, beyond a doubt, to a depth much greater than 600 feet below the present level of the ocean; because the marine shells must have some depth of water as well as the clay, in which to encase them, in order to produce fossilization. Nor would we expect, on account of the ocean currents that swept over the region in question, to find marine remains, except in very deep water, where the shells or bones might receive a covering of drift materials sufficient to preserve them from the disintegrating and denuding agencies which have prevailed, during the long train of centuries that have elapsed since the deposit.

The nodules at Green's creek are in the lower part of the Leda clay, which contains bowlders, and is succeeded by very large bowlders, while no bowlder clay underlies it. The plants contained in these nodules are characterized as a selection from the present Canadian flora of some of the more hardy species, having the most northern range, and the animals such as may now be found in the Arctic current and the Gulf of St. Lawrence. It appears, that the Arctic current, that entered by the way of the Gulf of St. Lawrence, backed its waters up the Ottawa valley, and that the plants from the heights of the Laurentian range of mountains, on the border of the valley, found their way into an eddy, where the blue clay was precipitated, and the *Mallotus villosus,* molluscan shells and hardy vegetation were so beautifully coffined in enduring nodules of stone. Dr. Dawson collected and identified from the marine deposits ten species of plants, and 195 species of radiates, molluscs, articulates, and vertebrates, and the whole of these, with three or four exceptions, he affirmed to be living northern or Arctic species, belonging, in the case of the marine species, to moderate depths, or varying from the littoral zone to 200 fathoms. The assemblage is identical with that of the northern part of the Gulf of St. Lawrence and Labrador coast, at present, and there is nothing in it to indicate any change of climate, beyond that which would necessarily follow, by changing the Arctic current, so as to throw it into the gulf and across the New England States.

There is nothing in all this area that indicates the existence of even a local glacier with any degree of certainty, though it may not be considered impossible that a small glacier should have existed upon the top of some of the highest mountain peaks of New England, when the Arctic current was flowing across the lower land. There is nothing to indicate a glacial period, but, on the contrary, every known geological and palæontological fact tells us that it never existed. And in the face of all these evidences furnished by scientific investigation, without the intervention of any extraordinary or unusual exercise of the powers of nature, except the depression and elevation of a coast line, which is proven by the deposit of the shells and bones of marine animals, it is difficult to understand how any one can conceive of a continental sheet of ice rising up from Hudson's bay, crossing over the Laurentian mountains, going down to the depths of the Gulf of St. Lawrence, and then ascending the mountains of Maine, New Hampshire, and Vermont, for no other purpose than that of taking a trip south; and if the imagination extends that far, it is still more incomprehensible why any one should believe it.

The submergence and elevation of this margin may have included the whole of the Pliocene, and part of the Post-pliocene periods, for the vegetable remains, in the peat beds of Brandon, Vermont, and in Nova Scotia, and other places which were covered by the drift, and evidently mark the age next preceding it, have been doubtfully identified with both the Eocene and Miocene, and other palæontological evidence is wanting, except so far as furnished by the Post-pliocene, and probably Pliocene fossils enclosed within the drift itself.

The submergence and elevation of this coast, preceded the lake drift of the central part of the continent, or at least could not have been contemporaneous with it, as will be shown in the sequel. Lake Ontario is an old river channel with the adjacent low lands covered with water. It is about 245 feet above the ocean. It will be readily seen that with the coast submerged this lake would fall at the east end 245 feet, which would bring it within less than one third of its present dimensions, and leave the maximum depth of the channel less than 500 feet. And with the elevation of the coast, as there is no canon to the sea, the elevation of the lake would follow to its present level. The consideration of this subject, however, belongs to succeeding pages, and we will now pass to a brief summary of the Tertiary of the Rocky mountain region or western part of the continent.

The gradual elevation of the western ranges of mountains through the later Cretaceous and all of the Tertiary time, and the formation of bays and arms of the sea and lakes, which have drained themselves more or less completely, and yet in ever continuing succession, have made it possible for the geologist to link the Tertiary with the

Cretaceous, and to bind the Eocene, the Miocene and Pliocene with the present as one connected age. The lower Eocene lake deposits are found superimposed conformably upon the brackish deposits of the Fort Union Group. The Eocene is divided in ascending order, into the Wasatch, Green river and Bridger Groups, though these are found conformable with each other in some places and mark a continuing age. The Wasatch is again divided by having the lower marls called the Puerco Group, and the Green River Group is divided, for convenience, in some places, into an upper and lower Green River Group. It would seem that all other names proposed for the fresh-water Eocene deposits are synonyms, though the equivalency of strata has not, probably, in all cases, been determined. The Miocene is known in the lower part as the Wind River Group, and higher as the White River Group, and sometimes the latter name is used to the exclusion of the former. In some places the upper Miocene is called the Truckee Group. The Brown's Park Group, Sweetwater Group and Monument Creek Group are Miocene, but their exact position is not so fully determined. The two latter are supposed to be equivalent to part of the White River Group, and the former may be so too. The Pliocene is very properly called the Loup Fork Group. It has also, in part, received the name of the Salt Lake Group, and a conglomerate of the age of the upper part of the Pliocene is called the Wyoming Conglomerate. The distribution of these Groups and questions of synonymy, have been considered at some length, in preceding pages, and in the near future the nomenclature will no doubt be more definitely established.

The northern drift does not occur in California, nor on the Pacific coast as far north as British Columbia and Alaska. There are no indications throughout the Rocky mountain region of any general ice action. There are no such exhibitions of scratched and grooved rocks succeeded by fossiliferous marine clays and sands with bowlders, as occur in the New England States and St. Lawrence region, nor of scratched rocks and ancient soils succeeded by clay, sand and gravel with bowlders, as occur in the central part of the continent; but, on the contrary, the whole region may be regarded as an absolutely driftless area, except as to local drift produced upon the shores of the Tertiary lakes, and more or less distributed by the rivers, that in the course of time cut out the canons which drained them. On the borders of the ancient lakes, on the borders of the ancient lake-like expansions of the rivers, and on the borders of the ancient rivers, there are terraces which mark old shore lines at various places from Mexico to Alaska, and especially throughout British Columbia. These terraces show only the ordinary subaërial denudation since they constituted the shore lines of lakes and rivers; but they are standing monuments of evidence to disprove the existence of a glacial period on this continent, or the existence of a continental ice sheet; for no one can conceive of the movement of such a heavy body of ice across a valley, without disturbing the graveled terraces that border it, upon both sides, at different elevations. The natural towers that stand as an evidence of erosion from the Wasatch times to the present; from the Green River Eocene to the present; from the Bridger Eocene to the present; from the White River Miocene to the present; the columnar masses, irregular pyramids, sandstone towers, and turreted outliers of the Bad Lands of Colorado, Wyoming, Montana and British Columbia; the monuments on Monument creek; the Garden of the Gods; the buttes in all the mountain chains; the transverse ridges, lone mountains and exalted peaks; and the whole array of canons from Texas and Mexico to Alaska, all alike, tell us, in language unmistakable, that no glacial sheet ever moved south upon the western plains or mountain ranges. No geologist has ever found a rock or bowlder that had crossed the dividing ridge from one valley to another in all this western region of the United States and British America. No one has ever found any evidence of any general drift action, or general ice action in any part of the territory. Then, why talk about a continental ice sheet or glacial period ?

Many of the phenomena attributed by glacialists to a continental sheet of ice belong to the ordinary eroding atmospheric causes, others to drifting sand, others to land slides, others to land slips or avalanches, which have been precipitated into the bed of the river, producing a dam that backed the water up until a lake was formed, and the quantity of water became so great as to force its way through the barrier, and cast the increased volume with terrific force upon the valley below. Lyell notices the devastating effects of one of these land slips from the White mountains of New Hampshire, into the Saco river, in 1826, and points out its insignificance, when compared with those occasioned by earthquakes, when the boundary hills, for miles in length, are thrown down into the hollow of a valley. The effects of even extraordinary floods, in river valleys, seem to be overlooked by some glacialists; and, in this connection, it will not be without interest to call attention to one that happened in the Connecticut.

In the winter of 1780,* well known for being one of the severest ever

---

* Hayden's Geological Essays.

experienced in this country, the ice in the Connecticut river was increased to a great thickness and solidity. In many instances, the
water in the river was literally frozen to the bottom. In the month of
January, as usual, there came a great and sudden thaw, accompanied
with incessant torrents of rain, which appeared to spread over an immense extent of country. The consequences were such as might be expected; the snow which was over five feet deep, was quickly melted; every
stream as suddenly became a river; and every river threatened to become an ocean. The Connecticut river was very soon raised almost to
a level with its banks, and the ice, which was two and a half feet thick
was borne away by the current in the most terrific majesty; for where-
ever it was impeded in its progress, by an island, or the narrowing of
the shores of the river, it was broken up, and immense masses raised
into the air, until their elevated portions, preponderating over their
floating foundations, were left to fall on the surrounding ice with a report, equal in some instances to that of small pieces of ordnance.
This scene of awful grandeur was extended for miles to the north and
south, and while thousands were contemplating the frightful spectacle,
the ice, being very solid, and hurried on by a powerful current became
obstructed at the mouth of the straits twenty-five miles below, near
Middletown, and the whole force of the river for a short time was impeded: the water set back and upward, and enormous masses of ice
were hurried over the banks of the river, into the creeks and larger
streams to a considerable distance from the river, into the meadows
and low grounds: When on a sudden, from the pressure above, the obstruction at the straits gave way, and this threatening appearance almost in a minute vanished; the water fell to its natural state, and left
huge masses of transparent ice in the meadows and intervals, to be re-
removed only by the powerful influence of a summer's sun. When this
was accomplished, in the following season, large pieces of rocks and
heaps of rolled pebbles were left exposed to view on an alluvial surface,
on which before a stone could not be found for its weight in gold.
These rocks and stones, from their characters, were known to be the
same as those which composed the bed of the river many leagues
above.

We now come to the consideration of the sand, gravel and bowlders constituting the drift of the central part of the continent; the scratches and furrows upon the rocks; the ancient soil beneath the drift ; and the animal and vegetable remains which immediately preceded the drift, and also such as are found within it.

It is idle to talk of continental elevations or depressions, for the whole science of geology and palæontology teaches us of the gradual growth or formation of continents. The appearance of islands above water, until an archipelago is formed, followed by the slow filling up of the shallow places and the intermittent local elevation of mountain chains, through vast geological ages, until the islands are thoroughly united into one vast body or continent, is the history of all continental elevations, and science teaches us of none other, and if continents have been depressed they must now be beneath the ocean, for we know nothing of such phenomena.

We have already seen the vast deposits of the Triassic and Jurassic periods, followed by the marine and brackish water deposits of the Cretaceous age that so well nigh formed the outlines of this continent. The elevation of the mountain chains that caused the formation of vast internal lakes, which have slowly drained themselves through all Tertiary time, and the slight elevation of some parts of the coast during the same period has given us the present form of our continent.

As soon as an island appeared above the ocean the denudation of its surface, from atmospheric causes, began. The rains at once commenced the excavation of valleys and ravines, and when the islands began to assume a continental shape, the valleys must necessarily have been correspondingly increased in size. As the Appalachian range dates back, in part, as far as the close of palæozoic time, so the Ohio river and other streams from this mountain chain have the same age. Another drainage system existed from the Laurentian mountains by a way that has been interrupted and thrown into a series of lakes, but the ancient valley has been traced from Lake Huron through Lakes Erie and Ontario. To the west and north of this drainage system, vast internal lakes were formed by the elevation of the western mountain chains, which overflowed and drained themselves across the central part of the continent, and produced, as we will see, in the sequel, all the phenomena of the drift.

As heretofore, we will follow the historical and chronological order of discovery as far as practicable.

In 1817, Dr. Daniel Drake,[*] of Cincinnati, wrote an essay upon the alluvial and drift formations of Ohio and the surrounding country. The letter was not published, however, until 1825. He supposed that the gravel and sand which spreads itself over the western part of Ohio, and is not found over eastern Kentucky, is the result of an inundation, having its origin north of the lakes, and that the large bowlders and blocks of stone, distributed over the country, were transported by large fields of ice and icebergs, which floated from the arctic regions during this inundation. He said, the ice to which they were attached could not of course pass a certain latitude ; and from the great increase of these masses as we advance toward the north, it would seem that many of the icebergs suffered dissolution long before they arrived at this locality.

In 1820, Caleb Atwater[†] stated, that an arrow-head was found in the alluvium, when digging a well at Cincinnati, 90 feet below the surface ; that a human skeleton was found in the alluvium at Pickaway plains, 17½ feet below the surface, that could not have been interred by human hands in that position ; and he figured and described a human skull of a very low grade, which was found nine feet below the surface, in such a position as to suggest its contemporaneity with the drift era.

In 1825, Sayers Gazley[‡] found fossil wood in Hamilton county, Ohio, below the gravel, and intermixed with it and bluish earth, at depths from 10 to 40 feet below the surface, and apparently where the trees had originally grown.

In 1838, Prof. James Hall[§] observed the indications of diluvial action, in western New York, in the accumulations of gravel, sand, pebbles and bowlders of all dimensions strewn over the surface. In some places slight scratches were observed on the rocks, while in others they were numerous and deep, often extending for several feet, and in

---

[*] Trans. Am. Phil. Soc., vol. ii.
[†] Am. Jour. Sci. and Arts.
[‡] Ibid.
[§] Geo. Rep. N. Y.

one case a continuous furrow was found 100 feet in length. The general direction of these scratches is N.N.E. and S.S.W. though they vary a little. One of the remarkable features of the country is a "Lake ridge" passing through the four lake counties nearly parallel to the lake shore, and from four to eight miles distant from the lake. The width of the ridge at the base is from four to eight rods, and narrowing toward the top to only two or three rods in width. In many places it much exceeds this width. The elevation of this ridge above lake Ontario is from 160 to 200 feet, though it varies a little from this at some places. The whole of the ridge is superficial, being composed of sand, gravel and pebbles, in all respects similar to those forming the beaches along the present lake shore. South of the ridge there are numerous parallel ridges, composed of sand and gravel, rising about 25 to 35 feet above the general level, and having uniformly a north and south direction, but never crossing the lake ridge. The opinion expressed in relation to this ridge is that it once constituted part of the shore of the lake, and consequently that the water in the lake was once 160 or 200 feet higher than at present, and that the north and south ridges resulted from the overflow of the lake and the pouring out of its waters in a southerly direction.

Prof. J. W. Foster* separated the surface deposits of Central Ohio into : 1. Vegetable mold; 2. Loam, or a mixture of sand and clay; 3. Sand and pebbles; 4. Yellow clay; 5. Dark blue clay effervescing with acids. The whole of which has a thickness of from 50 to 150 feet. And also over the surface of the country there are scattered bowlders of granite, syenite, quartz, etc. In the region about Columbus, some of these erratic blocks contain 1,000 cubic feet. Not even a primitive pebble has been found on the highlands east of Zanesville, showing that the valley of the Muskingum formed a connection of the currents of water, that swept over the country, with the Ohio river. He described from an excavation for the canal at Nashport, Ohio, *Castoroides ohioensis.* It was taken from a layer of dark carbonaceous silt, below a yellowish clay bed 14 feet in thickness, but above a layer of pebbles of primitive rocks and the blue clay at the bottom of the canal.

Prof. John Locke found the surface of the rocks at Light's quarry seven miles above Dayton, about 448 feet above the Ohio river at Cincinnati, planed, scratched and grooved. The quarry had been stripped of soil, more or less, over ten acres. The natural surface of the stone is very rough, and in some places this roughness was un-

* Ohio Geo. Rep. 1838.

touched, in others the prominences were just touched by the grinding operation, partially worn down, or entirely obliterated, leaving a flat, but unpolished surface, and in many other places the surface is polished, and grooved. The grooves are, in width, from lines scarcely visible, to those three fourths of an inch wide, and from one fortieth to one eighth of an inch deep, and traverse the quarry from between N. 19° to N. 33° west, to the opposite points, in lines exactly straight, and in fassicles of sometimes 10 in number, exactly parallel, cleanly engraved in compact limestone, without seam or fault of any kind, and in a surface ground down to a perfect plane. The grooves appear as if they had been formed by icebergs floating over the terrace, which is the highest in the neighborhood, and dragging gravel and bowlders frozen into its lower surface, over the plane of the stone.

In 1842, Lardner Vanuxem* found the drift scratches in Central New York confined to no particular rock, and at no particular elevation, but not uncommon, and corresponding, in direction, with the course of the valley, or of the valleys in which they occur. One of the best localities for observing the phenomena is at a quarry two and a half miles northeast of Amsterdam. The surface of the rock is covered with soil and earth, which, when removed, shows a water-worn surface with two or three sets of scratches, exhibiting great regularity, and having a common direction toward the east, one set of which is about eight degrees south. The scratches, including furrows, are generally from a mere line to one fourth of an inch wide, and from one to two tenths of an inch and more in depth. Some of them show that the moving power which produced them, passed over the surface with a vibratory or tremulous motion.

In 1843, Prof. James Hall† said that the northern part of the fourth district of New York, and the low slopes and deeper valleys of the southern part, are covered to a greater or less depth by superficial materials of more northern origin, mingled with those of the rock on which the deposit rests. All the formations have suffered greatly from denudation, and the abraded fragments of each constitute a large proportion of the superficial detritus resting on its southern neighbor. The size of the fragments always bears a proportion to the distance they have been transported from the parent rock. Often, a huge mass of a northern rock rests upon the margin of the one next south of it, while at a distance of 10 or 20 miles farther south, only small

---

* Geo. 3d Dist., N. Y.
† Geo. Sur. 4th Dist., N. Y.

pebbles of the same occur. In some places the coarser and finer materials are intermingled, in the greatest confusion, and heaped up into conical hills thickly scattered over the surface. And again the same materials are accumulated in long hills or ridges having a determinate direction, and sloping down from a high northern elevation to the general level of the country south.

On one hand, we have comparatively an evenly distributed deposit, as if made by the retiring waters of an ocean; on the other, the long hills, with certain directions, show a determinate course and more powerful current in the ocean, while the irregular, conical and dome-shaped hills, with deep, bowl-shaped cavities, show the force of contending currents, or of other obstructions, in the course of the transported materials.

The great bulk of the deposit, whether evenly distributed or irregularly raised into hills and ridges, is composed of the rock but a short distance on the north, or perhaps of the one on which it rests, with a constantly decreasing proportion of rocks of northern origin. The materials of the primary rocks constitute but a comparatively small proportion of the superficial accumulations of western New York. The local origin of the drift is shown by the sections everywhere examined. A section on Irondequoit bay, is as follows: 1. Medina sandstone, shaly with bands of green. 2. Fragments and rolled masses of the sandstone below, with gravel and sand; this contains a few pebbles of the shaly, calcareous sandstone next on the north. 3. Bed of fine sand. 4. Stratum of sandstone pebbles, cemented into a conglomerate by oxide of iron and carbonate of lime. 5. Stratum of pebbles and sand. 6. A coarse deposit of pebbles of the Medina sandstone below, with gravel and sand. 7. The soil of sandy loam. Another section 70 miles farther west on the bank of lake Ontario, at the town of Wilson, in Niagara county, is as follows: 1. Red clay and gravel of the Medina sandstone. 2. Blue clay and gravel. The pebbles are principally of the rocks of the Hudson River Group. 3. Gravel, clay and sand, of the neighboring rocks, folding over and passing beneath No. 2. 4. The soil of clayey loam with clay below. The sections of the drift almost universally correspond with these, and their explanation, viz: a bed of broken fragments, with worn pebbles resting upon the rock from which they are derived. The granite and other materials of a far northern origin rarely constituting a part. And where they do form a part, the deposit may have undergone some subsequent change.

Grooves or striæ are found upon the surface of all the rocks beneath the drift in the fourth district, which are of sufficient hardness to receive and retain such impressions. From the Medina sandstone, at the level of lake Ontario, to the summit of the Carboniferous conglomerate, in the southern part of the State, some of the strata in every group bear upon their surface these markings of former abrasion, and evidence of moving force. The direction of these striæ vary but few degrees from N. 35° E. and S. 35° W. in their general course. Short and shallow striæ are abundant, which vary ten and fifteen degrees from this direction, but these have no continuous course, and apparently fall into the main direction after a few feet. These markings range from the slightest possible scratch, to grooves of half an inch in width and one fourth of an inch in depth. The grooves seem to have been made by a hard substance, moved with great force and under great pressure, for fragments are found broken out as the grooves approach a fissure in the strata, as if crushed out by some heavy body, and sometimes the grooves are observed following, somewhat obliquely, the fractured slope. The outcropping edges of strata, previously polished and grooved, are often found overturned, upon the rock, in place.

At Rochester, the surface of the limestone is finely striated, and almost perfectly polished by the abrading force. The material here resting upon the rock is fine sandy loam; in another locality a mile farther south, it is covered by coarse limestone gravel and sandstone pebbles, with bowlders of granite. The striæ here are N.N.E. and S.S.W. At Black Rock, the surface of the Corniferous limestone shows that the nodules of hornstone interrupted the progress of the striæ and stand above the surrounding polished surface. The direction here is N. 15° E. and S. 35° W. At the cliff of Lake Erie in Portland, Chautauqua county, the rocky strata below have been uplifted, broken and contorted; the fragments intermingled with clay and gravel, and the same pressed beneath the strata, which otherwise appear to be in place.

The terrace at Lewiston is formed by the upper part of the Medina sandstone, the Clinton Group and the Niagara shale, capped by about twenty feet of Niagara limestone. The top of this terrace is 350 feet above Lake Ontario, and more than 200 feet above the plain about Lewiston. The Niagara shale is carried away so as to leave the limestone of the Clinton Group forming a projecting shelf about 100 feet below the top of the terrace. The surface of this projecting shelf is deeply grooved and striated, the grooves having a general southern tendency, but more irregular than where they are seen upon the limestone on the top of the terrace; and at this place, the surfaces 200 feet lower, and 100 feet higher, are scored in like manner. What agency could produce this effect? Here is an abrupt elevation of 100 feet above the striated surface; and it seems hardly possible that an island of ice, loaded with granite bowlders, could have stranded upon this projecting shelf, and produced the scoring, and that, at the same time, others above and below could be made in like manner.

The fourth district, in its greatest elevation of about 2,000 feet above tide water, descends to the level of Lake Ontario, 240 feet above tide. for the most part, in a series of steps or terraces over the successive formations; the surfaces of these, from the highest to the lowest, are grooved and striated, and in the limestones often beautifully polished. There is no high land on the north. from which glaciers could originate to cover this entire surface. The relative levels, as well as the directions of the water courses, must also have been different, to have allowed of such effects from glaciers; for, under present circumstances, we should hardly expect to find a glacier advancing from the valley of Lake Ontario, toward the southern margin of the State, and ascending nearly 2,000 feet in 100 miles. Even admitting the glacial theory to be true, it is probable that the glaciers would originate among the mountains of Canada, or farther north among the primary rocks; and in this event, we might expect to meet, intermingled with the earliest drift, a considerable proportion of granite and other pebbles and bowlders of the older rocks, which is not the case.

There is another fact worthy of notice. The vertical faces of joints, when much separated and nearly coinciding with the direction of these grooves, are polished in the same manner as the surfaces. The chinks and fissures, in the harder rocks along the sea shore. are polished, in like manner, by the washing in of sand and pebbles by the advancing and retiring waves.

The first plateau above Lake Ontario is often plentifully covered with bowlders. These usually lie upon the surface, and always upon the top of the drift. They are not evenly distributed, but often appear in immense numbers, scattered over several acres; while beyond this, for a great distance, few are to be found. There appears to be no law regulating their distribution, though they are more abundant in the eastern than in the western part of the district. The bowlders are often in immense numbers on the low ground just north of the Ridge road from Wayne county to the Niagara river, and appear as if they had been brought there while the water was limited by this barrier, and spread over the bottom in shallow water near the shore. In higher situations, and just beneath the great limestone terrace they again appear in abundance, as if this elevation prevented their farther advance to the south. The bowlders are most abundant in Wayne and the eastern part of Monroe county; going westward from the Genesee they are less so, becoming extremely rare in Erie and Niagara counties. As we ascend the second limestone terrace formed by the Helderberg range of limestones extending westward, bowlders become perceptibly less numerous; they are irregularly scattered, and at few points present the thickly covered fields which are observed farther north. Very few ascend the slope formed by the passage of the Hamilton Group to the rocks above; and in all the previous cases, they seem to have been brought on, at intervals, in great numbers, and their limits bounded by the different elevations of the surface. As we pass southward over the higher groups, bowlders become exceedingly rare; and finally toward the southern margin of the State they are rarely seen.

Some of them bear evidence of much wearing, being actually striated upon the surface, and sometimes flattened on one side, as if held in that position while moved over a bottom of gravel or sand resting upon the strata beneath. For the most part, however, they bear no evidence of attrition beyond what similar masses do a few miles from their parent rock, and thus offer no argument for their mode of transportation. Many of them are angular, and with no appearance of attrition beyond what the weathering in their present situations would produce. The process by which fragments of granite become rounded bowlders, is illustrated by the desquamation which takes place in some granites, the weathering in place, and the attrition in mountain streams soon after leaving their native beds. A large proportion of the bowlders of western New York are of dark felspathic granite and red granites like those of the northern part of the State. Some other varieties occur, which are likewise referable to the same region. A few of crystalline limestone with serpentine, and a few of specular iron ore have been found which are like rock found in St. Lawrence county.

In many places, the drift hills have no definite direction, but those north of the great valleys of Seneca and Cayuga lakes are long elevated ridges, rising abruptly on the north, to a height of 50 or 60 feet, and sloping gradually down to their southern termination. The form of the hills is precisely such as would be made by a powerful current passing southward through these valleys, piling up the coarser materials at the northern extremity, and moving the finer ones farther on, until they were in some measure protected by this barrier before they were deposited.

One of the most interesting of the superficial deposits is the Lake ridge, which, from Sodus in Wayne county, with some trifling exceptions, is a traveled highway nearly as far as the Niagara river. Beyond this it can be traced to the head of Lake Ontario. It follows the general course of the Lake; being at its nearest point about three miles distant, and at its greatest about eight miles. In some places it is strongly defined, descending toward the lake twenty or thirty and even fifty feet in a moderate slope. It consists of sand and gravel, and contains fragments of wood and shells, and in every respect it resembles the sea beaches. It was undoubtedly the ancient beach of Lake Ontario, or a body of water which once stood at this elevation. The top of the lake ridge is 158 feet above Lake Ontario at Lockport; 185 feet at Middleport, and 188 feet at Albion and Brockport.

Beside this well-defined ridge or ancient beach there are a number of less distinctly defined terraces of gravel and sand at much higher elevations, on the hill sides, leading to the supposition that the water of the Lake stood more than 750 feet higher than at present, or that the country has been correspondingly depressed.

Prof. W. W. Mather* found that the drift scratches, grooves and furrows conform in their directions to those in which currents would flow, if the country were mostly covered by water. In some parts, they correspond in direction to the main water-sheds ; in others they do not, but where they do not, the deviation is owing to some topographical feature which disturbed the course of the currents of water.

In 1845, Alexander Murray† found the drift of western Canada, consisting of various beds of clay, sand and gravel, interspersed with large bowlders. The thickness frequently reaches 200 or 300 feet. The clay cliffs of Scarborough, are 320 feet. The ridges running parallel to the north shore of Lake Ontario, are 200 or 300 feet, and the highlands in Oxford, are 100 or 200 feet, and even more, and the banks of Grand river often expose a considerable amount of drift. The southern shores of lake Simcoe, are extensive sandy plains, which are in many places thickly strewed with bowlders, and bear proof of having once been the bottom of the lake. Wherever gravel is found, its pebbles consist of limestone, and with the larger fragments of that formation, they contain the fossils of the calcareous strata at Rama on the north. The

---

* Geo. of the 1st Geological Dist., N. Y.
† Geo. Sur. of Canada.

whole formation consists of the disintegrated rocks of the immediate locality, or those at no great distance north. The grooves and scratches upon the rocks between Niagara and Hamilton, have a north and south direction.

In 1847, W. E. Logan* found on the north shore of Lake Superior, about three miles below the Petits Ecrits, six terraces, in addition to the summit, which, presenting a level surface throughout the whole length, may be considered a seventh. Blocking up the extremity of a deep cone from the rock on one side to that on the other, the accumulation is a barrier to an extensive flat and marshy surface, that spreads out in a valley behind, down to the level of which there is a rapid slope from the summit of the drift, at a distance of about 1,000 yards from the margin of the lake. The height of the ancient beaches as measured by a pocket spirit-level is as follows:

		Above the Lake. Feet.	Above the Sea. Feet.
1st Beach		30	627
2d	"	40	637
3d	"	90	687
4th	"	224	821
5th	"	259	856
6th	"	267	864
7th	" or summit	331	928

The 3d and 4th beaches are the most decidedly marked, the steps, rising behind them, sloping up at an angle of nearly 30°.

Alexander Murray† described the drift on the Kamanitiquia river, which flows into Lake Superior, near Fort William, as consisting between McKay's mountain and the Grand Falls, where the principal display was found, of a light buff-colored clay, covered over by stratified yellow ferruginous sand, both together attaining a thickness of 60 feet above the level of the water. Banks of sand were found on Dog river, at a much higher level than the deposit further down.

In 1848, John L. Leconte‡ described, from a Post-pliocene deposit in a crevice in northern Illinois, *Platygonus compressus* and *Anomodon snyderi.*

Mr. Charles Whittlesey§ designated the different beds of the drift in Ohio and the West as follows :

---

* Geo. Sur. of Canada.
† *Ibid.*
‡ Am. Jour. Sci. and Arts, 2d ser., vol. v.
§ *Ibid.*

1st. " Blue hard pan," resting unconformably on the surface of the stratified rocks. This is a very compact mass of *blue clay, marl* and *sand,* including great numbers of small, partially water-worn, *crushed* and *striated pebbles,* principally fragments of blue limestone and primitive rocks. It contains lime, so much as to effervesce with acids, and to hasten vegetation when applied to land. Beside its strong *blue color,* it is characterized by *imbedded timber,* dirt beds, leaves, sticks, and what are called by well diggers " grape vines." It is so solid as to be almost impervious to water, and is very difficult to excavate.

2d. " Yellow hard pan," resting unconformably on the stratified rocks, and the " blue hard pan." This is a compact material, of a *dull yellow color,* with fewer stony fragments or pebbles, and less calcareous and more aluminous matter than the blue hard pan. It is not quite as solid as the blue, more pervious to water, and contains more and larger pieces of primitive rocks. The clays of the country, used for bricks are principally of this bed. It forms a hard, stiff soil, adapted for grass. The flat regions and savannas of the northwest quarter of the State, are caused by the surface presence of this bed.

3d. " Sand and gravel drift," containing granite bowlders (in small numbers), of large size, and unconformable to Nos. 1 and 2, and the other rocks. It exhibits little regularity of stratification, is composed of inferior patches of coarse sand and gravel, intermingled at all inclinations, evidently the result of long continued and vigorous action of water in rapid motion. The gravel is coarse, but much *worn, rounded* and *smooth,* like the gravel beds of rapid streams. The portion of earthy matter is about one half, of a reddish and yellowish color, showing the presence of oxide of iron, and containing various proportions of sand and clay. Almost every rock in the northern part of America is represented in the gravel; but the greatest part by far is from the underlying and adjacent strata. There are pebbles of quartz, trap, granite, gneiss, conglomerate, limestones of all ages, iron ore, slate, coal and sandstone. In this there has been found timber but very rarely.

4th. The " valley drift," composed principally of debris of the adjacent rocks, and occupying the lower parts of the great valleys of drainage. It is more gravelly and less earthy, and the gravel is more of local origin than in No. 3, while the beds of sand are less common. It is in the " valley drift" or swamp mud that the bones of the mastodon and other large animals are usually found.

5th. "Lacustrine deposits," occupying the basin of the lakes, and for Lake Erie, divided into the "blue marly sand," and the coarse sand and gravel. The "blue marly sand," commonly called the blue clay of Lake Erie, is seen skirting the shore almost everywhere, if the coast is not rocky,—its upper face nearly horizontal, and rising from forty-five to sixty feet above the water. It is of a light blue color, so fine as scarcely to show between the fingers any grit, homogeneous, and in a dry state compact, but brittle. Very rarely, may be seen a primitive pebble, thin layers of leaves and lignite. It is distinctly and horizontally laminated, and at Cleveland is composed of about 75 per cent. impalpable sand, 3 per cent. iron, 6 to 7 per cent, carbonate of lime, 9 per cent. carbonate of magnesia, and of vegetable matter and sulphur. It is impervious to water, and thus causes thousands of springs to appear at its surface, which, passing out over the edges, dissolve and carry it away very fast, forming a quick sand. Its edge is presented to the action of the waves, which dissolve and carry it away rapidly. As it is not tenacious like clay, and not capable of sustaining itself under its own weight, and that of the sand stratum that rests upon it, there are continual breaks and slides along the banks, on both the American and Canadian shores. These avalanches of earth are from one to four rods in width, breaking off in irregular patches, and sometimes sinking, in a night or in a few hours, twenty or thirty feet, leaving huge fissures through which the water of the springs passes, and rapidly washes the earth into the lake.

At the water's edge, the slide frequently raises a bank of about the width of the break, several feet above the surface, driving back for a short time the line of the shore. But the waves acting incessantly dissolve the new barrier, and soon commence their attacks upon the body of the fallen mass, which disappears, and is before long followed by a fresh avalanche from above.

At the city of Cleveland, where the bluff shore rises 70 feet above the lake, the encroachment since the survey of the town in 1796, has been at the foot of Ontario street, 265 feet. The Canadian shore, from Detroit river to Long point, is losing faster than the American. Between Port Stanley and Port Burwell, on the British side, the superior face of the blue marl is about sixty feet, or fifteen feet higher than at Cleveland, and has in the upper part a lighter or more yellow color. In composition the yellowish portion is more argillaceous than the bright blue, and appears to correspond with the yellow clay stratum of Lake Champlain. The greatest thickness of the blue marls can not be computed, as a large part of it lies below the lake level, forming the bed of more than one half of Lake Erie. On the south shore it extends but a short distance into the interior, forming a narrow belt of low country along the lake, and thinning out as the rocks upon which it rests rise to the southward.

The "coarse sand and gravel" of this division, rests conformably on the "blue marly sand," and spreads horizontally over a tract of low, and in general wet land, embracing the western half of Lake Erie, and extending westward into the States of Ohio and Michigan.

On the north, it forms the soil and surface over a large portion of the peninsula, between Lakes Erie and Huron; which seldom rise more than 200 feet above the waters of these lakes. On it, and composed of its coarse water-washed sand and gravel, are seen the "lake ridges," objects of curiosity, and of much utility in a new country, being natural turnpikes that run parallel with the shore. At Cleveland the section is as follows: 1st. Gray, water-washed, coarse sand, resting on the blue marl, 10 feet. 2d. Coarse gravel of the adjacent rocks and sand, 20 to 40 feet. The lake ridges are not precisely horizontal, and are found at *various* elevations, 30, 90, 120 and 140 feet above the water.

There are branches and cross ridges uniting different parallels, that rise and fall several feet in a mile.

6th. Bowlders or "erratic rocks" which he regarded as a "stratum," and the newest of all beds except the alluvium.

The Drift deposits* are very extensive on the southern shore of Lake Superior, and more especially on its southeastern coast. There they not only constitute the only visible formations for nearly 100 miles, but they also attain an astonishing thickness, so as to form, by themselves, ridges and cliffs which exceed in height even those of the Pictured Rocks, being in some places, as at the Grand Sable, not less than 360 feet high. The Drift is less conspicuous along the western portion of the lake shore, although it is not wanting even among the romantic and precipitous cliffs of the Pictured Rocks and the Red Castles.

The Drift of lake Superior may be divided in ascending order, into—

1st. Coarse drift. This is the least conspicuous of all. It is found only in a few places along the southern shore, generally capping the high towering cliffs of sandstone. It is generally a mixture of loam and fragments of rock of different sizes—sometimes worn, but more generally angular. As a leading feature, it is almost exclusively com-

---

* Foster and Whitney's Sur. Lake Sup. Region, 1850.

posed of fragments of the rocks *in situ*, showing that, whatever may have been its origin. it could not have been acted upon by long continued agencies. A few foreign pebbles exist in it, generally trap, and evidently derived from the neighborhood. Greatest thickness 30 feet.

2d. Drift clay, or red clay. It is a mixture of loam and clay, and its color is owing to the decomposition of the red sandstone and trap from which it has been derived. It is mainly composed of very finely comminuted substances, yet there are pebbles interpersed through it, and even bowlders of considerable size, generally rounded and smoothed. Fragments of metallic ores and native copper occur occasionally in it— the latter sometimes weighing several hundred pounds. It is found along the whole southern coast of lake Superior, resting upon the red sandstone, and limited to a certain height, but on the Ontonagon and Carp rivers, it is found in depressions on elevated lands. 500 feet above the lake. At Grand Sable where its base rests on almost horizontal strata of red sandstone, a few feet above the water, and its top is covered by a mass of drift sand, it is 60 feet in thickness, and exhibits lines of stratification disposed with great regularity.

3d. Drift sand and gravel. This is the most widely diffused of the drift deposits on the shores of lake Superior and the northern part of Michigan. The greatest thickness observed is at Grand Sable, where it is 300 feet thick.

4th. Bowlders. These occur of every size and description in great numbers along the whole southern shore. The largest noticed being of hornblende, and measuring 15 feet in length, 11 in width, and $6\frac{1}{2}$ in height. The bowlders have been moved from north to south, but have not come from far, though some of them have been transported from the north shore. It is noticed among the ridges north of Carp river, that the valleys, for the most part, contain bowlders from the next ridge to the north; and there are instances where a ridge did not allow the fragments of the preceding ridge to pass. This limitation prevails only within the hilly portion of the Lake Superior region between the lake shore and the dividing ridge. South of this ridge no barrier occurs.

5th. Drift terraces and ridges. These may be seen both on the north and the south sides of Lake Superior, but they are less striking than around Lakes Erie and Ontario. They are most conspicuous on the south shore, between Saut and Keweenaw point. Their average height is about 100 feet. At a place two miles east of Two-hearted river, the following succession occurs : gravel beach, 5 feet ; sand beach, 12 feet ; 1st drift terrace, 29 feet; 2d drift terrace, 46 feet ; 3d drift terrace, 75 feet ; summit of plateau, 94 feet.

The rocks in many places are grooved, scratched and polished. These phenomena, of course, can be seen only where the drift deposits are absent. The groovings consist generally of parallel furrows, from one to four lines wide—sometimes extending a foot, at others many yards. Where the rock is very hard, they are mere striæ. Hollow spots occur, as if they had been scooped out by a round instrument, and also wide bowl-shaped depressions, known as troughs, which have been caused by the same agency. Grooves and scratches were observed on the road from Eagle river to the Cliff mines running N. 15° E. On an island east of Dead river there are two systems of striæ— one running N. and S., and the other N. 20° E. and S. 20° W. The rock here which is very hard and tough hornblende, is not only grooved and furrowed over its whole extent, but there are, beside, deep trough-like depressions, with perfectly smoothed walls, some 12 to 15 feet long, 4 feet wide, and 2½ feet deep. On Middle Island, east of Granite point, troughs may also be seen 4 feet wide, and 2 feet deep, running like the striæ N. 20° E. On the promontories and islands near Worcester, two miles west of the mouth of Carp river, there are two distinct sets of striæ ; those running N. 55° E. are the most numerous ; those running N. 5° E. the least. The latter cross the former and are therefore more recent. Some of them are, beside, distinctly curved, as if the body which produced them had been deflected in ascending the slope. Each set of striæ extends only about one foot below the water's edge. On the first quartz ridge, one mile from the mouth of Carp river 500 feet high, the striæ run N. 20° E. On the iron ridge south of Teal lake, 750 feet high, the striæ run N. 55° E. At the Jackson forge N. 65° E. A green magnesian rock, with vertical walls, and semi-cylindrical form, on the road leading from Jackson landing to Teal lake is covered with striæ which may be traced along the surface, like hoops around a gigantic cask. On Isle Royal the striæ run N. 50° E. with many local deviations. On the shores of Ackley bay striæ near the water's edge running E. and W., cross others running N. E. and S. W., and others again running S. 75° E. Isle Royale presents but little evidence of drift, though scattered bowlders are found upon it; the surface of the rock s are generally, however, smoothed, as if polished off.

Mr. E. Desor described the superficial deposits on the northern shore of Lake Michigan, the western shore of Green bay, the Big Bay des Noquets, and the valleys of the Menomonee and Manistee. The coarse drift described as occurring beneath the drift proper, at several points along the shore of Lake Superior, seems to be entirely wanting in this district.

Starting from Mackinac westward, the furrows and striæ were noticed at the bottom of St. Martins bay, and two miles north of Pine river, on a point composed of almost horizontal ledges of limestone, having an average direction from E. to W., some running N. 80° E., and others S. 70° and 80° E. At Payment point the direction being from N. 50° to N. 60° E. At the bottom of Big Bay des Noquets, on the west shore of the eastern cove, the direction is E. and W. At the mouth of the Escanaba, in Little Bay des Noquets, the direction is N. E. and S. W. At Oak Orchard, on the west shore of Green bay, the direction is N. 15° to N. 20° E. At the saw mill, near the mouth of the Menomonee, the direction is E. and W.; six miles above Kitson's trading house, E. N. E. and W. S. W.; three miles above Sturgeon's falls, N. 65° E.; foot of the Lower Bukuenesee falls, N. 70° E.; Lower Twin falls, N. 60° to N. 70° E.; and at Upper Twin falls, N. 65° to 70° E. From Green bay, southwestward, they were noticed at Mehoggan point, N. E. by E. and N. N. E.; at Mehoggan falls, N. E. by N.; three miles west of Milwaukee, N. E.; and at Strong's landing on Fox river, N. E. by E.

The true drift seldom approaches the shores of Lake Michigan and Green bay, but it is met with in ascending the rivers at no great distance. Its absence from the coast is the result of subsequent denudation, when the waters of the lake stood at a higher level than at present. It was observed at Pointe aux Chenes, and for a distance of six miles toward Payment point, and on Potawatomee and some of the higher islands. The thickness at Green bay was found on boring to be 108 feet.

Near the junction of the Machigamig and Brule, where the united streams take the name of Menomonee, the river banks are composed of drift, forming bluffs 100 feet or more in height. The drift is composed of sand and layers of gravel more or less interspersed through it, and covered more or less with bowlders. The higher lands adjoining are covered with the same materials. The country adjacent to the Manistee is likewise covered with the drift sand and pebbles. The whole country drained by the White-fish and its branches, and the Escanaba is likewise covered with the drift. The drift clay is well marked, in many places, below the drift sand, especially upon the

Manistee, where it does not generally reach more than 4 or 5 feet above the river, although in one place it was found 10 feet thick. It is very tough, and generally flesh colored, but in one instance it was perfectly white. There were observed, in several localities, rather coarse pebbles of limestone, and even flat stones intermixed with the upper layer of clay, near its contact with the sand.

He described the terraces on the island of Mackinac and the neighboring coasts, on the west coast, and at Pointe St. Ignace and Gros Cap on the north coast of Lake Michigan, which vary in height from 20 to 130 feet. But the terraces are not found farther west on the north shore of Lake Michigan and Green bay, nor in the vicinity of the Menomonee and Manistee.

Mr. Charles Whittlesey,* said of the terraces bordering Lake Erie, that the first ridge, or that nearest the lake, is known as the "North ridge." From Conneaut, in Ashtabula county, to Russelton, Huron county, a distance of 120 miles, the elevation of the ridge above the lake varies from 85 to 145 feet. The second ridge, from Kingsville, in Ashtabula county, to Ridgeville, in Lorain county, varies from 122 to 168 feet above the lake. These ridges consist of coarse, water-washed, yellowish sand, or of fine gravel, principally the comminuted portions of the adjacent rocks. The rocky fragments are not generally worn perfectly round, or oblong, as beach shingle is, but are more flat, with worn edges. There are mingled with the sandstones and shales that compose this gravel, scattered pieces of quartz, flint, granite, trappean rocks, limestone and ironstone. The third and fourth ridges are a little higher, and composed of coarser material.

In 1852, Charles Whittlesey† described the drift in that part of Wisconsin bordering on Lake Superior, and lying between the Michigan boundary and the Brule river, and the sources of the streams flowing into Lake Superior from the south. He divided the drift into—1st, red marly clay; 2d, bowlder drift, coarse sand and gravel.

The red marly clay is a fine-grained, homogeneous marly sand, cemented by argil or clay, with well defined horizontal lines of lamination or deposition; containing, but very rarely, pebbles of granitoid, trappose, sandstone, conglomerate, or slate rocks. This constitutes the shore or lake bluffs most part of the way from the Montreal to the Brule; the red sandstone, on which it rests, showing itself occasionally beneath. It is easily washed away in suspension by the waves, and

---

* Am. Jour. Sci. and Arts, 2d ser., vol. x.
† Owen's Geo. Sur., Wis., Iowa, and Minn.

having little tenacity, falls in slides and avalanches into the water, and is thus cut into deep, narrow gullies by rains. Its surface in the above district is not more than 250 feet above the lake, sloping gradually from the mountains to the shore, as though it formed, at one time, the bed of an ancient sea. On the waters of the St. Louis river on the west, and the Ontonagon on the east, however, the red clay deposits reach to the height of 450 to 500 feet above the lake.

On the " Isle aux Barques" the lime is so abundant in the clay, that it has formed in amorphous concretions throughout the mass. A few leaves and decayed sticks have been seen in the red marly clays, with carbonaceous matter and lignite, but such occurrences are rare. Along the coast there are interstratified beds of sand and gravel of a local character. In the interior, where the clay is visible in bold bluffs, along the water courses, it is more uniform and less intercalated with coarse drift. It rests not only on the sedimentary unaltered rocks, but also on trap and metamorphic and igneous rocks.

The mass of the hills between Chegwomigon bay and the Brule river, is gravel and bowlder drift. It is not very uniform in composition, and is marked by the violent action of water. The central part of this peninsula presents large tracts of barren, water-washed land, and moderately coarse gravel. Both the western and eastern knobs and ridges are of coarse materials; and toward the point or extremity about the " detour," and the adjacent islands, the sand and bowlder deposits are represented.

A section of three miles from the coast to the mountains, four miles southwest of La Pointe, showed red marly clay 95 to 130 feet above the lake, capped by coarse bowlder drift, the top of which is 428 to 509 feet above the lake. This drift is disposed in three very abrupt and well defined terraces. These terraces continue southward around the southern extremity of the mountain, and have the appearance of ancient beaches or shores.

In 1855, Prof. G. C. Swallow* found a fine, pulverulent, absolutely stratified mass of light, grayish buff, silicious and slightly indurated marl, capping nearly all the bluffs of the Missouri and Mississippi within that State, for which he proposed the name Bluff formation. The Bluff above St. Joseph exhibits an exposure 140 feet thick. It is easily penetrated by the roots of trees, which decay and leave encrusting tubes, giving it a peculiar perforated appearance. It extends from Council Bluffs to St. Louis, and below to the mouth of the Ohio.

---

* Geo. Sur. of Missouri.

The greatest development is in the counties on the Missouri, from the Iowa line to Boonville. In some places it is 200 feet thick. At Boonville it is 100 feet thick, and at St. Louis only 50 feet.

The Bluff Group is older than the bottom prairie, and newer than the Drift. It gives character and beauty to nearly all the best landscapes of the Lower Missouri.

He found the drift abounding north of the Missouri river, and existing in small quantities as far south as the Osage and Meramec. Its thickness varies from 1 to 45 feet. The upper part, having the appearance of having been removed and rearranged by aqueous agencies since its first deposit, but before the deposit of the Bluff Group, is described as altered drift. The heterogeneous strata of sand, gravel, and bowlders, is called the bowlder formation; and below this, in some places, a third division exists, which is called the "pipe clay." It contains bowlders more or less dispersed through the upper part of it. It is found in Marion, Boone, Cooper, Moniteau, Howard and Monroe counties, varying in thickness from 1 to 6 feet.

William P. Blake* described the grooving and polishing of hard rocks and minerals by dry sand in the Pass of San Bernardino, California, and on the projecting spurs of San Gorgonia. he said, grains of sand were pouring over the rocks in countless myriads, under the influence of the powerful current of air which seems to sweep constantly through this Pass from the ocean to the interior. Wherever he turned his eyes—on the horizontal tables of rock, or on the vertical faces turned to the wind—the effects of the sand were visible; there was not a point untouched, the grains had engraved their track on every stone. Even quartz was cut away and polished; garnets and tourmaline were also cut and left with polished surfaces. Masses of limestone looked as if they had been partly dissolved, and resembled specimens of rock salt that have been allowed to deliquesce in moist air. These minerals were unequally abraded, and in the order of their hardness; the wear upon the feldspar of the granite being the most rapid, and the garnets being affected least, wherever a garnet or a lump of quartz was imbedded in compact feldspar, and favorably presented to the action of the sand, the feldspar was cut away around the hard mineral, which was thus left standing in relief above the general surface. A portion however, of the feldspar, on the lee side of the garnets, being protected from the action of the sand by the superior hardness of the gem, also stood out in relief, forming an elevated string, osar like, under their

* Am. Jour. Sci. and Arts, 2d ser., vol. xx.

lee. When the surface acted on, was vertical and charged with garnets, a very peculiar result was produced; the garnets were left standing in relief, mounted on the end of a long pedicle of feldspar, which had been protected from action while the surrounding parts were cut away. These little needles of feldspar tipped with garnets, stood out from the body of the rock in horizontal lines, pointing like jeweled fingers in the direction of the prevailing wind.

The effects of driven sand are not confined to the pass; they may be seen on all parts of the desert where there are any hard rocks or minerals to be acted upon. On the upper plain, north of the Sand Hills, where steady and high winds prevail, and the surface is paved with pebbles of various colors, the latter are all polished to such a degree that they glisten in the sun's rays, and seem to be formed by art. The polish is not like that produced by the lapidary, but looks more like laquered ware, or as if the pebbles had been oiled and varnished. On the lower parts of the desert, or wherever there is a specimen of silicified wood, the sand has registered its action. It seems to have been ceaselessly at work, and when no obstacle was encountered on which wear and abrasion could be effected, the grains have acted on each other, and by constantly coming in contact have worn away all their little asperities and become almost perfect spheres. This form is evident whenever the sand is examined by a microscope.

We may regard these results as most interesting examples of the denuding power of loose materials transported by currents in a fluid. If we can have a distinct abrasion and linear grooving of the hardest rocks and minerals, by the mere action of little grains of sand, falling in constant succession, and bounding along on their surface, what may we not expect from the action of pebbles and bowlders of great size and weight, transported by a constant current in the more dense fluid, water? We may conclude that long rectilinear furrows of indefinite depth may be made by loose materials, and that it is not essential to their formation that the rocks and gravel, acting as chisels or gravers, should be pressed down by violence, or imbedded in ice, or moved forward *en masse* under pressure by the action of glaciers or stranded icebergs. Wherever, therefore, we find on the surface of mountains, not covered by glaciers, grooved and polished surfaces with the furrows extending in long parallel lines seeming to indicate the action of a former glacier, we should remember the effects which may be produced during a long period of time by light and loose materials transported in a current of air; and which, consequently, may be produced with greater distinctness, and in a different style, by rocks moved forward in a current of water. The effects produced by glaciers, by drift, or moving sand, are doubtless different and peculiar, so different and characteristic, that the cause may be at once assigned by the experienced observer, who can distinguish between them without difficulty. It is, however, possible that after a sand worn surface, such as has been described, has been for ages covered with moist earth, a decomposition of the surface would take place sufficient to remove the polish from the furrows and leave us in doubt as to their origin.

Alexander Murray* examined a portion of the country between Georgian bay in Lake Huron, and the Ottawa river. He followed the course of the Muskoka river to its head, and by a short portage passed to the source of the Petewahweh, and by its channel descended to the Ottawa. Returning, he ascended the Bonnechere river to Round lake, from which he crossed to Lake Kamaniskiak on the main branch of the Madawaska, and descended the latter stream to the York or southwest branch, from whence he crossed to Balsam lake. He found stratified clays on the Muskoka, between the lake of Bays and Ox-tongue lake, at the height of about 1,200 feet above the level of the sea; the banks expose 10 or 12 feet in thickness, of drab or light buff-colored clays, alternating with very thin layers of fine yellow or grayish sand. At one place, the beds are tilted, showing a westerly dip of about eight degrees, in which they exhibit slight wrinkles or corrugations. Coarse yellow sand overlies the clay, and spreads far and wide over the more level parts, generally forming the bank of the river, where not occupied by hard rock. On the Petewahweh, especially below Cedar lake, the whole of the level parts are covered with sand, which, in some places, is of great thickness. Cedar lake is about 1,050 feet above the sea.

The banks of the Bonnechere display a great accumulation of clay at many parts below the fourth chute, sometimes exposing a vertical thickness of from 70 to 80 feet. Near the mouth of that river, below the first chute, where the clays form the right bank, and are upward of 50 feet high, they are chiefly of a pale bluish-drab color, and are calcareous, while other clays found higher up the stream, are of a yellowish-buff, and do not effervesce with acids. Below the second chute, buff-colored clay is interstratified with beds of sand and gravel, the latter sometimes strongly cemented together by carbonate of lime, the whole being overlaid by a deposit of sand. The gravel is seldom

---

* Geo. Sur. of Can., Rep. of Progr. for 1853.

very coarse, although an individual bowlder may occur here and there amongst it, and it is chiefly derived from the rocks of the Laurentian series. The height of the first chute above the sea, is 265 feet; the second chute, 348 feet;. the fourth chute, including its fall of 39 feet, 432 feet; Round lake, 520 feet, or nearly 60 feet below Lake Huron.

Sand is extensively distributed over the plains of the Bonnechere, and over a large portion of the area between it and the valley of the Madawaska. Most of the valley of the Little Madawaska is covered with sand on either side, and the country between its head waters and Lake Kamaniskiak is one continuous sandy plain. The height of land in passing over the portage to the Madawaska is 968 feet above the sea, and Lake Kamaniskiak is 906 feet above the level of the sea. No organic remains have been detected in any of these drift deposits.

He, afterward,* surveyed the valley of the Meganatawan river and part of the coast or Lake Nipissing. Stratified clay was found on the banks of the Meganatwan, above the second long rapids, east of Doe lake. The color is a brownish drab; it is very tenacious, and does not effervesce with acids. The highest exposure is a little over 1,000 feet above the level of the sea. A fine, strongly tenacious clay occurs on the Nahmanitigong near the main elbow, where the upward course of the river turns to the south at an elevation of 710 feet above the sea. The color of the clay is chiefly pale drab or buff, but bands of reddish clay are interstratified and some of pale blue overlie the whole. The clays of the interior are usually overlaid by a deposit of coarse yellow sand. Among the bowlders on Lake Nipissing, many were observed to be of a slate conglomerate like that of the Huronian series, and they were frequently of very great size.

In the succeeding year† he explored portions of the Huron and western districts of the Province of Canada, and found that the course of the currents which had borne along the drift was from northwest to southeast. This is indicated by the pebbles and bowlders of metamorphic rocks which were clearly derived from the Laurentian and Huronian formations on the north shore of Lake Huron, and by the character of the fossiliferous rocks and pebbles which have been moved a shorter distance, and by the grooves and scratches which invariably have a bearing from the northwest to the southeast.

He, afterward,‡ made a survey north of Lake Huron, where he found

---

* Geo. Sur. of Canada., Rep. of Prog., 1854.
† Rep. of Prog. for 1855.
‡ Rep. of Prog. for 1856.

bowlders derived from the Huronian rocks that had been moved from their source and transported southerly. In the valleys of the Wahnapitae and French rivers, large bowlders of conglomerate rest on the contorted gneiss at various elevations above the mark of the greatest floods, the highest probably over 100 feet. On the Sturgeon and Maskanongi rivers, and on Lake Wahnapitaeping, the course of the grooves and scratches is S. 27° W., with scarcely any deviation, but farther west they seem to alter their course to a more westerly direction, and on Round lake they bear S. 41° W.; while at the long lake, near the outlet of the White-fish river, their direction is S. 49° W. The great deposits of silicious sand, which are spread over the upper valley of the Wahnapitae, above Wahnapitaeping lake, and also the sand in the valley of the Sturgeon river, are probably chiefly derived from the ruins of the Huronian rocks. Lake Huron is 578 feet above the sea; Lake Wahnapitaeping, 938 feet ; Round lake, 775 feet ; Sturgeon river, at the junction of the Maskanongi, 809 feet; and Maskanongiwagaming lake, on the Maskanongi, 862 feet.

In 1859, he described* the drift north of Lake Huron, between the valley of the Thessalon river and the lake coast south of it, and between the valleys of the Thessalon and the Mississagui. A deposit of clay usually of a brownish drab color is spread over a large portion of the region, particularly in the hollows and valleys, and is frequently exposed on the banks of the streams, distinctly stratified, and in considerable thickness. The clay is overlaid with sand which extends far and wide over the highest table lands, and a great part of the country generally. The clay deposits of the Mississagui and Little White rivers, do not appear to attain a height of much more than 160 feet over Lake Huron. Above the Grand Portage at 154 feet above the lake, the clay is replaced by a great accumulation of sand and gravel, the gravel becoming coarser and more prevalent as we ascend the river. On the banks and flats above Salter's base line, 252 feet above the lake, the shingle consists of rounded masses almost all of Syenite, the smallest of which is rarely under the size of a man's fist, and the average as large as a twelve pound canon ball. Many of the masses are much larger, and in addition there are a great number of huge bowlders.

Grooves and scratches on the sides of the lakes, and in the valleys, have the same general bearing of the valleys, and follow the meanderings of the lake depressions. Instances are as follows: On the island

---

* Geo. Sur. of Canada.

south side of Echo lake S. 55° W.; half a mile below S. 70° W.: in a
depression north of Walker lake S. 17° W.; Thessalon river above
Rock lake S. 25° W.; west and south sides of Rock lake S. 15° W.;
east side of bay at Bruce Mines S.; northwest end of Wahbiqueko-
bingsing lake S.; southeast end of same lake S. 12° W.

Instances* of the abraded and polished surfaces of rock are very
numerous on the Canoe route from Lake Superior to Lake Winnipeg.
Near Baril Portage, 143 miles from Lake Superior, and 1,500 feet
above the sea, gneissoid hills and islands are smooth and sometimes
roughly polished on the northerly side, while on the southern side
they are precipitous and abrupt. On Sturgeon lake, 208 miles from
Lake Superior, and 1,156 feet above the sea, the northeastern extremities of hill ranges slope to the water's edge, and when bare are always
found to be smoothed and ground down. The aspect of the south and
southwestern exposures, is that of precipitous escarpments. The
summits of the granite hills near Lake Winnipeg are abraded and
frequently so smooth and polished as to make walking upon them
difficult, if not impossible in moderately steep places.

On the south branch of the Saskatchewan the drift is exposed in
cliffs 50 to 80 feet in altitude at the bends of the river. The drift consists of clay with long lines of bowlders in it at different elevations.
Some of the fragments of shale, slabs of limestone and small bowlders
imbedded in the clay, stand in the drift with the longest axis vertical,
others slanting, and some are placed as it were upon their edges. Long
lines of bowlders lie horizontally from ten to twenty feet below the
surface or top of the cliff, while below, in many places, close to the
water's edge, and rising from it in a slope for a space of 25 to 30 feet,
the bowlders are packed like stones in an artificial pavement, and
often ground down to a uniform level by the action of ice, exhibiting
ice grooves and scratches in the direction of the current. This pavement is shown for many miles in aggregate length at the bends of the
river. Sometimes it resembles fine mosaic work, at other times it is
rugged, where granite bowlders have long resisted the wear of the ice,
and protected those of softer materials lying less exposed.

Two tiers of bowlders, separated by an interval of 20 feet, are often
seen in the clay cliffs. The lower tier contains very large fragments of
water-worn limestone, granite and gneissoid bowlders, above them
is a hard sand containing pebbles; this is followed by an extremely
fine stratified clay, breaking up into excessively thin layers, which

---

* Assiniboine and Saskatchewan Expl. Exped.

envelope detached particles of sand, small pebbles and aggregations of particles of sand. Above the fine stratified clay, yellow clay and unstratified sand occur.

Bowlders are found on the Qu'Appelle and its affluents, below the Moose Woods, and north of the Assiniboine, measuring from 10 to 25 feet or more in diameter.

In Lake Winnipeg, ice every year brings vast bowlders and fragments of rock of the Laurentian series, which occupy its eastern shores, and distributes them in the shallows and on the beaches of the western side. In Lake Manitobah, long lines of bowlders are accumulating in shallows and forming extensive reefs; the same operation is going on in all the lakes of this region, and is instrumental in diminishing the area of the lake in one direction, which is probably compensated by a wearing away of the coast in other places.

A remarkable beach and terrace, showing an ancient coast line between Lake Superior and Lake Winnipeg, separates Great Dog from Little Dog lake on the Kaministiquia canoe route. The Great Dog portage, 55 miles from Lake Superior by the canoe route, rises 490 feet above the level of the Little Dog lake, and the greatest elevation of the ridge can not be less than 500 feet above it. The difference between the level of Little and Great Dog lakes, is 347.81 feet, and the length of the portage between, one mile and 53 chains.

The base of the Great Dog mountain consists of a gneissoid rock, supporting numerous bowlders and fragments of the same material. A level plateau of clay then occurs for about a quarter of a mile, at an altitude of 283 feet above Little Dog lake, from which arises, at a very acute angle, an immense bank or ridge of stratified sand, holding small water-worn pebbles. The bank of sand continues to the summit of the portage, or 185 feet above the clay plateau. East of the portage path the summit is 500 feet above Little Dog lake.

Here we have a terrace 500 feet above Little Dog lake, or 863 feet above Lake Superior, or 1,463 feet above the sea. Another beach or terrace occurs at Prairie portage, 104 miles by the canoe route from Lake Superior, 190 feet above Cold Water lake, or 900 feet above Lake Superior, or over 1,500 feet above the sea.

In the valley of Lake Winnipeg, the first prominent beach or terrace is the Big ridge. Commencing east of Red river, a few miles from the lake, it pursues a southwesterly course until it approaches Red river. within four miles of the Middle settlements; here it is 67½ feet above the prairie; on the opposite side of the river, a beach on Stony mountain corresponds with the Big ridge, and beyond it forms the limit of a former extension of Lake Winnipeg. On the east side of Red river the Big ridge is traced nearly due south from the Middle settlement to where it crosses the Roseau, 46 miles from the mouth of that stream, and on or near the 49th parallel. It is next met with at Pine creek, in the State of Minnesota, and from this point it may be said to form a continuous level gravel road, beautifully arched and about 100 feet broad, to the shores of Lake Winnipeg, 120 miles. On the west side of Red river, north of the 49th parallel, and north of the Assiniboine, from a point near Stony mountain, it extends to near Prairie Portage, where it has been removed by the Prairie Portage river and the waters of the Assiniboine. It may be seen again on White Mud river, about 20 miles west of Lake Manitobah.

In the rear of Dauphin lake, the next ridge in ascending order occurs; it forms an excellent pitching track for Indians on the east flank of the Riding mountain. At Pembina mountain four distinct steps or beaches occur, the summit of which is 210 feet above the prairie.

The lower prairies enclosed by the Big Ridge are everywhere intersected by small subordinate ridges which often die out, and are evidently the remains of shoals formed in the shallow bed of Lake Winnipeg, when its waters were limited by the Big ridge. The long lines of bowlders exposed in two parallel, horizontal rows, about 20 feet apart, in the drift of the south branch of the Saskatchewan above mentioned, are the records of former shallow lakes or seas in that region.

They may represent a coast line, but more probably low ridges formed under water, upon which bowlders were stranded. The fine layers of stratified mud, easily split into thin leaves, which lie just above them, show conclusively that they were deposited in quiet water; their horizontality proves that they occupied an ancient coast or ridge below the comparatively tranquil water of a lake of limited extent; the vast accumulations of sand and clay above them establish the antiquity of the arrangement; and the occurrence of two such layers, parallel to one another, and separated by a considerable accumulation of clay and sand, leads to the inference that the conditions which established the existence of one layer also prevailed during the arrangement of the other. It may be that these are bowlders distributed over the level floor of a former lake or sea, and they may cover a vast area.

The Pembina mountain is par excellence the ancient beach in the valley of Lake Winnipeg. It is not a mountain, nor yet a hill. It is a terrace of table land, the ancient shore of a great body of water, that once filled the whole of the Red river valley. It is only 210 feet above the level of the surrounding prairie, or between 900 and 1,000 feet above the ocean level. High above Pembina mountain the steps and plateaux of the Riding and Duck mountains arise in well defined succession. On the southern and southwestern slopes of these ranges the terraces are distinctly defined, on the northeast and north sides the Riding and Duck mountains present a precipitous escarpment which is elevated fully 1,000 feet above Lake Winnipeg, or more than 1,600 feet above the sea. One of the terraces here is 1,428 feet above the level of the ocean. The denudation of the Cretaceous, in the valley of Lake Winnipeg, has been enormous, because the shales crop out 500 feet above Dauphin lake, where their position is nearly horizontal, and evincing their former extension to the northeast, if not as far as the north shore of Lake Winnipeg. Sand hills and dunes occur on the Assiniboine, Qu'Appelle, South Branch, and north of Touchwood hills.

Prof. E. W. Hilgard† described the drift (he called it the Orange Sand formation) as covering the greater part of the State of Mississippi. It is overlaid by the Bluff Group, and is not, therefore, above Natchez, exposed on the surface, within eight to twelve miles of the Mississippi river ; below Natchez, however, it forms the White cliffs on the Mississippi itself. It does not cover the northeastern part of the State, and is absent from other limited patches. The thickness is quite variable, sometimes reaching 200 feet, though usually not more than 40 to 60 feet. The material is usually silicious sand, colored more or less with hydrated peroxide of iron, or orange-yellow ochre. Sometimes pebbles or shingle, either cemented into puddingstone, or more frequently loose and commingled with sand or clay occur, and at other times limited deposits of clay are found. It contains fossils from the Silurian, Devonian, Carboniferous, and Cretaceous formations which are exposed to the north in Tennessee, Kentucky, Indiana, Illinois and Ohio, and silicified wood from the lignite strata of Mississippi. The character of the surface upon which it rests its own irregular stratification, and the dependence, to a great extent, of the nature of its materials, upon that of the underlying formations, proves, beyond question, that its deposition, preceded and accompanied by extensive denudations, has taken place in flowing water, the effect of whose waves, eddies and counter currents, is plainly recognizable in numerous profiles. Nor can there be any doubt that the general direc-

---

* Geo. Sur. of Miss.

tion of the current was from north to south, although locally changed or directed by the pre existing inequalities of the surface.

The drift is succeeded on the Mississippi by a narrow belt, called the Bluff Group, but in other parts of the State the drift is covered by a yellow loam, which also succeeds the Bluff. The second bottom, or Hommock deposits, and the alluvial, are yet more recent in their character.

Drift materials* are strewn over a great part of the surface of Michigan. At East Saginaw these materials are from 90 to 100 feet thick, and at Detroit 130 feet thick. Wherever large surfaces of the underlying rocks are exposed, they are found to be more or less smoothed and striated. The island of Mackinac shows the most indubitable evidence of the former height of the water, 250 feet above the level of the lake. The trunks of white cedar trees are not uncommon in the drift, and on the north shore of Grand Traverse bay there is a bed of lignite.

In 1862, Prof. J. D. Whitney pointed out,† approximately, the territory in northern Illinois, western Wisconsin, northeastern Iowa, and eastern Minnesota, that is destitute of drift. This tract is several hundred miles in length, and from 100 to 200 miles in width. There is an entire absence of bowlders or pebbles, or any rolled and waterworn materials, which by their nature would indicate that the region in question had been exposed to the action of those causes by which the drift phenomena were produced. The surface of the rock is uneven and irregular, bearing the marks of chemical rather than of mechanical erosion, and there are no furrows, striæ or drift scratches, such as may be observed on many of the rocks over which the drift has been moved.

He concluded:

1st. That there has existed, ever since the period of the deposition of the Upper Silurian, a considerable area, chiefly in Wisconsin and near the Mississippi river, which has never been sunk below the level of the ocean, or covered by any extensive and permanent body of water, and which, consequently, has not only not received any newer deposit than the Upper Silurian, but has also entirely escaped the invasion of the drift, which took place over so vast an extent of the northern hemisphere.

2d. That the extensive denudation, which can be shown to have taken place in this region, as witnessed by the outliers of rock still remaining, and the general outline of the surface, has not been occasioned

---

* Geo. Sur. of Mich., 1861.
† Geo. Sur. of Wisconsin.

by any currents of water sweeping over the surface, under some great general cause, but that it has all been quietly and silently effected by the simple agency of rain and frost, acting uninterruptedly through a vast period of time.

3d.' That a large portion of the superficial detritus of the West, even in those regions where drift bowlders are met with, must have had its origin in the subærial destruction of the rocks, the soluble portion of them having been gradually removed by the percolating water, while that which remains represents the insoluble residuum, the sand and clay, which was originally present in smaller quantities in the strata thus acted on.

Bowlders of Laurentian rocks* are found in considerable numbers scattered over the high table-land of western Canada, south of Georgian bay. A portion of this region attains an elevation of 1,760 feet above the sea. These blocks are generally more angular than those from a similar source found at lower levels, and are associated with many others of local origin.

The stratified drift is separable into two divisions in western Canada, the lower of which, called the Erie clay, had been partially worn away before the deposition of the upper so as to produce unconformability. The Erie clay is commonly more or less calcareous, and always holds bowlders in greater or less abundance. The thickness at any one place does not exceed 200 feet, but clays belonging to this division occur at various levels from 60 feet below the surface of Lake Ontario to 100 feet above Lake Huron, showing differences in level of about 500 feet. It occurs along the north shore of Lake Erie from Long Point westward to the Detroit river, and appears to underlie the whole country between this part of the lake and the main body of Lake Huron. It is found at Owen sound, and along Nottawasaga river, and along the shores of Lake Ontario as far east as Brockville. The upper division is called the Saugeen clay, because it is well exposed along the Saugeen river. It consists of a thinly-bedded, brown calcareous clay, generally containing but few bowlders or pebbles. This division occurs also at all levels from Lake Ontario to 100 feet above Lake Huron, showing differences of level almost equal to that of the lower clay.

At the oil wells, on the 13th and 14th lots of the 10th range of Enniskillen, two beds of gravel, of four and five feet respectively, have been met with in the clay, at depths of ten and forty-four feet from the sur-

---

* Geo. Sur. of Canada, 1863

face, making a total section of clay and gravel of 49 feet. *Unio circulus,
U. gibbosus,* and valves of a *Cyclas* were found in the upper bed of
gravel, and a deer's bone was said to have been found also. Between
the gravel and the overlying 10 feet of clay, a thin layer of impure
mineral pitch, or half dried petroleum, intervenes, inclosing leaves of
land plants, and occasionally insects. Fresh-water shells occur in the
clay on the Detroit river. At Niagara Falls the Silurian limestone is
covered by 120 feet of sandy loam, holding striated pebbles and small
bowlders, and containing near the middle the shells of a species of
Cyclas. It is overlaid by fifteen feet of thinly-bedded, reddish-brown
clay, containing similar pebbles and angular fragments. This deposit,
whose summit is 60 feet above the level of Lake Erie, forms a bank
which continues up to Chippawa. Valves of the *Cyclas* occur in the
upper clay, in calcareous nodules, at a railway cut betwen Kingston
and the Grand Trunk railway station, and leaves of a plant resembling
Vaccinium occur in a laminated brownish clay at Newborough. At
the upper termination of the town plat, on the right bank of the
Goulais river, there is a deposit of the roots and limbs of trees, imbedded in a bluish scaly material, apparently a mass of compressed
leaves and moss, which rests upon a bed of clay, and is overlaid by a
mixture of clay and sand; the whole, with a stratum of sand at the top,
constitutes a bank of from 20 to 24 feet high. The bed of vegetable
matter, which is from one to three feet thick, and about ten feet over
the river at the western end of the exposure, dips gently and evenly up
the stream; while a thin bed of reddish clay, intervening between the
overlying arenaceous clay, and the stratum of sand which forms the
surface, seems to be perfectly horizontal. On the south side of Lake
Superior, between White-fish Point and the Painted Rocks, a great
deposit of sand, interstratified with gravel, is spread over the surface
of the country. At the Grand Sable, a short distance west from the
Grand Marais, it rises here and there almost vertically from the lake
to a height of 300 feet. A bed of vegetable matter occurs below
a layer of mixed sand and clay, and beneath this hill of sand and
gravel, which contains *Thuya occidentalis, Betula paperacea,* and
populus balsamifera.

Behind the Sault Ste Marie, a terrace, varying in its height, but
averaging perhaps 150 feet above Lake Superior, and often composed
of clay in red and drab layers, stretches from the Laurentide hills
southward toward the St. Mary river. About a mile below, and again
about four miles above the foot of the Sault, this terrace comes near

the edge of the river, and recedes in sweeping curves in both directions from each of these points. A bay, two miles and a half in depth, is thus left between them, and is occupied by a barren plain of no great elevation above the river, partly covered with coarse brown sand, and partly strewn with bowlders of northern metamorphic rocks and angular fragments of Silurian sandstone, which are sometimes arranged in small bare ridges parallel to the present direction of the river. The surface has thus the appearance of having formerly been covered with swiftly flowing water.

To the north of Lake Huron, and between the Georgian bay and the Ottawa river, part of the surface of the country consists of bare rock, but where any superficial covering exists, it is almost invariably a yellow sand. A belt of loose gravel, remarkable for its great extent, stretches in a southward direction across the peninsula of western Canada, from near Owen sound to Brantford, a distance of 100 miles. Its average breadth is nearly 23 miles, and its total area more than 2,000 square miles. This great belt of gravel has a general parallelism with the Niagara escarpment, and consists in large part of the ruins of the underlying Guelph and Niagara Groups, though pebbles of the Huronian and Laurentian rocks are everywhere mixed with the others, and fragments of the Hudson River Group occasionally occur.

Beside these clays and sands there are several local accumulations in western Canada, often marked by fresh-water shells. These, together with various ridges and terraces, which are conspicuous features in the surface geology of this region, appear for the most part to have been formed by the waters of the great lakes, when their extent was much greater than at present. The most considerable deposit of this kind is the sandy tract in the county of Simcoe, which extends south-eastward from the head of Nottawasaga bay, and has an area of more than 300 square miles. *Unio complanatus, Cyclas dubia, C. similis, Amnicola porata, Valvata tricarinata, V. piscinalis, Planorbis trivolvis, P. campanulatus, P. bicarinatus, Limnæa palustris,* and *Physa ancillaria,* occur at from 30 to 40 feet above the level of Lake Huron, and twenty miles distant near the Nottawasaga river. *Planorbis trivolvis,* and three species of *Helix,* were found in sand and gravel in a road cutting through a little ridge between 75 and 78 feet above Lake Huron, about a mile south of Collingwood harbor. Two miles west of Cape Rich, worn fragments of bark and wood were met with in digging a cellar on a terrace 155 feet above the lake. There are several terraces of sand and gravel which correspond to ancient water margins on the shores of Owen sound, at 120, 150 and 200 feet above the present level of Lake Huron, and some of the higher terraces continue with great regularity for several miles. Terraces and ancient beaches are found in many places upon Lake Superior. On the north side of the lake, the ancient water margins are frequently marked by the wearing of the solid rock as well as by the loose materials. In a sandy ridge near the western part of Lake Ontario, called Burlington heights, at the height of 70 feet above the lake, several bones of the mammoth were discovered, and in the same excavation, seven feet higher, the horns of the wapiti (*Cervus canadensis*), and the jaw of a beaver (*Castor fiber*), were also found.

The drift in Illinois* is divided into—1st, blue plastic clay, with small pebbles, often containing fragments of wood, and sometimes the trunks of trees of considerable size, which form the lower division of the mass; 2d, buff and yellow clays and gravel, and irregular beds of sand, with bowlders of water-worn rock of various sizes interspersed through the whole; and lastly, reddish-brown clays, generally free from bowlders, and forming the subsoil in those portions of the State remote from the streams, and where the loess is wanting. The scratched and grooved surfaces presented by the underlying limestones, at many localities, and the smoothly worn and polished surfaces that may be seen at others, and the immense size and weight of many of the transported bowlders, which have been carried for hundreds of miles from the nearest outcrop of the metamorphic beds to which they belong, alike preclude the idea that such results have been produced by the action of water alone. Huge masses of moving ice, like the icebergs of the present day, loaded with the mineral detritus of the far northern lands, with angular fragments of hard, metamorphic rock, firmly imbedded in the solid ice to act as a graver upon whatever rock surface they might come in contact, are the only known agencies that seem adequate to the production of the phenomena, characteristic of the drift deposits in this State.

There is an area in the southern part of the State, and another in the northwestern part of the State, over which the drift deposits do not extend. The lead region of Illinois, Iowa and Wisconsin was not invaded by the drift, and is, therefore, entirely free from accumulations of gravel, pebbles and bowlders, that characterize drift areas. The topographical features of the country have been produced by the quiet but

---

* Geo. Sur. of Illinois, vol. i.

ceaseless agency of water, not sweeping over the surface in the mighty currents of the diluvial epoch, bearing the detritus of northern crystalline rocks, and grinding down and bearing away the softer strata, but falling as rain, percolating through the calcareous and magnesian deposits, and gradually carrying them off in solution, leaving the insoluble portion behind, in the form in which we now see it covering the solid rock, as an intimate mixture of the finest argillaceous and silicious particles.

The trunks and branches of coniferous trees, belonging, apparently, to existing species, are quite common in the blue clays at the base of the drift ; and in the brown clays above, the remains of the mammoth, the mastodon, and the peccary are occasionally met with. The fine fragment of a mastodon's jaw, with the teeth, found at Alton, was obtained from a bed of drift, underlying the loess of the bluffs, which, at this point, was about thirty feet thick, and remained *in situ* above the bed from which the fossils were taken. Stone axes and flint spearheads are also found in the same horizon, indicating that the human race was cotemporary with the extinct mammalia of this period. The bones and teeth of a great number of species are found in the crevices of the rocks in the driftless area of the lead region, where they have been washed from the surface, and carried in some instances fifty or sixty feet before finding a lodgment. The most abundant among the remains of animals thus found are those of the mastodon, whose teeth and bones have been procured from a great number of crevices, over the whole area of the lead region ; showing that the species must have lived and flourished in immense numbers, and through a long period of time, since the chances of the preservation of the remains of any one individual by being washed into a crevice, must have been exceedingly small. The remains of both living and extinct species are found in the crevices in such positions, in reference to each other, as to indicate pretty clearly that they were living together. From a crevice, near the Blue Mounds, Prof. Worthen collected the bones and teeth of the mastodon, peccary, buffalo, and wolf—the two former extinct, and the two latter supposed to be identical with the living species.

In 1867, Prof. C. A. White* found drift scratches upon limestone of the Upper Coal Measures, in Mills county, Iowa, near the Missouri river, having a direction S. 20° E., and these crossed by a finer set of scratches, having a direction S. 51° E. And at an exposure of the same limestone, one mile below Omaha, the capital of Nebraska, imme-

---

* Am. Jour. Sci. and Arts, 2d series, vol. xliii.

diately upon the right bank of the Missouri river, and only six or eight feet above the ordinary stage of water, other scratches having a direction S. 41° E.

Prof. F. V. Hayden found erratic bowlders scattered over the country in northeastern Dakota, of all sizes and texture, and especially numerous in the valley of the James river and its tributaries.

In 1869, Dr. E. Andrews* said, that throughout Central Illinois the ancient Pliocene soil still lies undisturbed beneath the bowlder drift. This soil has been met with in excavations at so many independent points, that it may, probably, be considered as the usual floor on which the drift rests. Two of the best observations of it were obtained at Bloomington, Illinois. In sinking two coal shafts, the workmen first passed through 118 feet of unmodified drift clay, whose bowlders and pebbles were all of northern origin, and often scratched by the action of ice. Directly beneath this was a bed of ancient soil, on which logs of wood lay scattered confusedly about, and in which the stump of a tree still stood where it grew. Beneath the soil bed lay various sands, gravels, and clays, and a second dirt bed, but no more northern drift. The stump was of coniferous wood. All of the original drift is clearly stratified.

In 1873, Robert Bell† found the stiff red clay of the Kaministiquia valley, extending westward up the valley of the Mattawa to the outlet of Shebandowan lake, becoming apparently less abundant all the way, and finally disappearing on reaching the lake. Around the shores of this lake, and of nearly all the lakes passed, by way of Lonely lake and the English and Winnipeg rivers to Lake of the Woods, wherever the vegetation is burnt off, the rocky mammillated hills are seen to be strewn with rounded and angular bowlders, from the size of a man's head to a diameter of 30 to 40 feet. Many of these are perched in positions from which they look as if they might be easily rolled into the water below. The striæ on the surface of the rocks occur almost everywhere, and are very general in their course from south to southwest.

In 1875, Prof. George M. Dawson‡ found the striæ on the rocks at Lake of the Woods varying in their course from S. 20° E. to S. 87° W. Bowlders and traveled materials are spread over the country in this vicinity, and especially on the south side of the islands.

---

* Am. Jour. Sci. and Arts, 2d ser., vol. xlviii.
† Geo. Sur. of Canada.
‡ Rep. Geo., 49th Parallel.

The drift deposits cover the second prairie plateau west of the Red river and Turtle mountain, and the eastern front of Pembina escarpment is distinctly terraced and the summit of the plateau thickly covered with drift. The first terrace is about 50 feet above the general prairie level, the second about 260 feet, and the third about 360 feet.

One hundred and twenty miles west of Turtle mountain, the second prairie plateau comes to an end against the foot of the great belt of drift deposits, known as the Missouri Coteau, a tumultuously hilly country, based on a great thickness of drift. The Missouri Coteau is a mass of debris and traveled blocks, with an average breadth of 30 to 40 miles, extending diagonally across the central region of the continent with a length of 800 miles. It appears to have been the work of sea-borne icebergs, and not glacier ice as such.

In 1876, Mr. Robert Bell* said, that in the prairie regions of the northwest territory, loose deposits of Post-Tertiary age cover the surface of the country almost universally, and they are usually of considerable depth. There are immense areas having the same general elevation, or without very great or sudden changes of level, yet, with the exception of the first prairie steppe, there is a remarkable scarcity, or perhaps absence, of extensive stratified deposits of sands and clays, such as occur in the provinces of Ontario and Quebec. The bulk of the superficial deposits is of the nature of bowlder-clay or unmodified drift, which is spread alike over the older rocks from the lowest to the highest levels. The materials of the drift appear to be made up of the debris of the rocks existing *in situ*, immediately beneath or a short distance to the northeastward, together with a greater or less proportion derived from those lying further off in the same direction. As a rule, the softer or more clayey part has come from the underlying strata, while the harder pebbles and bowlders are the furthest transported; still, in washing out the finer ingredients, it is always found that much of the incorporated sand and gravel is of foreign origin. The nature of the transported bowlders and pebbles varies in different localities, but more than half of its bulk, on an average, consists of local material. On the first and second prairie steppes the most abundant constituent of the transported portion is Laurentian gneiss, while the remainder is made up of light-colored unfossiliferous limestones, supposed to be Silurian and Devonian, together with a proportion of Huronian schists, which varies in different localities. On the third steppe, however, smooth pebbles of finely granular quartzite predomi-

---

* Geo. Sur. of Canada, for 1874-75.

nate. These are mostly white, but some are gray, brown, pink, and red, the latter often passing into banded compact sandstone. There are also pebbles of dark, fine-grained diorite, light-colored limestone, and some of dark fine-grained mica schist, and of white translucent quartz, the last mentioned being often rough surfaced. Mr. George M. Dawson thinks this quartzite drift has come eastward from the foothills of the Rocky Mountains, where in the neighborhood of the line (latitude 49°) he found unfossiliferous rocks *in situ*, some of which resemble certain varieties of these quartzite pebbles, but Rev. Pere Petitot collected white saccharine quartzite from the McKenzie river exactly like that of the white pebbles of the third steppe.

While the composition of the bowlder clay of the first and second prairie steppes, and also, to some extent, that of the third steppe, as well as the course of the striæ on the hard rocks on the east side of the prairies, would indicate that the drift had been mainly from the northeastward, the above evidence shows that a large proportion of the transported material on the highest levels has come from the north, or west. A part of what is now found in some localities may have been moved first in one direction and afterward in another, whilst the bulk of the older drift, including, perhaps, even that on the third steppe, has probably come from points between north and east. The quartzite pebbles of the third steppe are all thoroughly water-worn, and appear to be most abundant on and near the surface. The upper 200 feet, or thereabouts, of the south bank of the South Saskatchewan, at the Red Ochre Hills, consists of clayey drift, in which bowlders of Laurentian gneiss occur, while the surfaces of these hills are strewn with smooth quartzite gravel and cobblestones. At the distance of 150 miles to the southeastward, between the Dirt Hills and the Woody Mountain, the proportion of quartzite gravel on the third steppe has diminished considerably, and Laurentian bowlders have become very numerous on the surface.

Between Fort Garry and Fort Ellice. Huronian bowlders and pebbles are scarce, they are, however, abundant in the drift in the banks of the Assineboine for some miles above and below the junction of the Shell river, and in the banks of the Calling river in the neighborhood of the Fishing Lakes. They are noticeable on the surface all the way from these lakes to the Touchwood Hills. Surface bowlders are extremely abundant on the southern and western sides of the gravelly and sandy tract southwest of Fort Ellice, about the head waters of the Calling river, and in many places on the high ground of the third steppe, between the Dirt Hills and the Woody Mountains. By far the greater number of the bowlders in all these localities consists of Laurentian gneiss, many of them are angular, although the majority are pretty well rounded. In each of the above districts, the bowlders are so numerous, over considerable areas, that a man might walk upon them in any direction without touching the ground.

In going from the northwest angle of the Lake of the Woods, toward Fort Garry, the road for long distances, runs upon low ridges of limestone-gravel between swamps, until reaching the drier ground between the White Mouth river and Oak Point, and in this interval, bowlders and pebbles of light-colored limestone are very common. They are also strewn abundantly on the shores around the southwestern part of Lake of the Woods. In the northern part of Lake of the Woods, and in the region of the Winnipeg and English rivers, limestone fragments are extremely rare, so that their sudden appearance, in such abundance, to the West and South of the northwest angle, would appear to indicate the occurrence of this rock *in situ* in the immediate neighborhood.

The magnetic bearings of the striæ in different parts of the country drained by the Winnipeg river, are as follows:

Around Wesaxino lake, S. 10° to 20°W.; two miles South of Sturgeon lake S. 40° W.; southeast shore of Sturgeon lake seven miles from southwestern extremity S. 20° W., and six and a half miles from southwestern extremity, S. 15° W.; North end of Hut lake, S. 25° W.; East end Kitchi-Sagi or Big-Inlet lake, S. 15° W.; inlet of Jarvis lake, S. 10° W.; Minnietakie Falls, S. 35° W.; island on Minnietaka lake, four miles southwest of Abram's chute, S. 45° W.; Abram's chute, at outlet of Minnietaka lake, S. 25° W.; Pelican falls, S. 45° W.; Stormy Point, on North side of Lonely lake, 24 miles from its outlet. S. 60° W.; Shanty Narrows on Lonely lake, 15 miles from outlet, West; outlet of Lonely lake, S. 75° W.; island in Maynard's lake, English river, S. 20° W.; narrows between Tide lake and Ball's lake, English river, S. 70° W.; outlet of Indian lake, English river, S. 30° W.; inlet of Lount's lake, English river, S. 40° W.; outlet of Lount's lake, S. 45° W.; entrance to South arm of Separation lake, English river, S. 50° W.; Winnipeg river, at entrance to Sandy bay, S. 45° W.; northwest shore of Lake of the Woods, seven miles from Rat Portage, S. 25° W.; Manitou Minis, 15 miles southwest of Rat Portage, S. 20° to 30° W.; Hone Point, 18 miles southwest of Rat Portage, S. 45° W.; Dead Oaks Point, 20 miles southwest of Rat Portage, S. 40° W.; and island in Lake of the Woods, 25 miles southeast of entrance to northwest angle, S. 25° W.

In the three prairie steppes there is a marked difference in the general aspect of the surface of the country, and in the character of the river valleys. On the first steppe, the surface is usually level, or undulating in long gentle sweeps, and the beds of the principal streams do not, probably, average more than 30 feet below the level of the surrounding country. On the second steppe the surface is rolling, and the river valleys are usually from 150 to 200 feet in depth, while on the third, the hills are on a larger scale, and either closely crowded together, or they rise here and there to considerable heights overlooking less rugged tracts. The principal river-valleys on this steppe are from 200 to 500 feet deep. The "coulees," as they are termed, form a curious feature of the third prairie steppe. These are valleys, or ravines, with steep sides, often one hundred feet or more in depth, which terminate or close in, rather abruptly, often at both ends, forming a long trough-like depression; or one of the extremities of the coulee may open into the valley of a regular water-course. The coulees sometimes run for miles, and are either quite dry or hold ponds of bitter water, which evaporate in the summer, and leave thin incrustations of snow-white alkaline salts.

The average depth of the river-valleys of the first and second prairie steppes is not affected by the general descent of the country through which they run. From the Little Boggy creek to the Arrow river, the Assineboine must fall 400 or 500 feet, yet the banks of the valley maintain the same general height and the same character throughout the whole distance. Similarly, the fall in the Calling river, from the Sand-Hills lake to its junction with the Assineboine, can not be far from 500 feet, and still its valley-banks have the same average height throughout. The fall in the Red river, from Moorehead to Fort Garry, is upward of 200 feet; but in the whole of the distance, the banks of the river have a nearly uniform height of 20 to 30 feet.

The great valleys of the third steppe cut entirely through the drift and far down into the underlying Tertiary and Cretaceous rocks; those of the second steppe appear to correspond, in a general way, with the depth of the drift, while on the lowest steppe, the streams have merely cut through the modified deposits resting upon the drift, which latter is occasionally exposed at low water at the foot of the banks, or in the bed of the stream at swift places and rapids. The stratified clay, silt, sand, and gravel of the Red river and the lower Assineboine, vary in thickness from almost nothing to 80 or 90 feet, and a variable thickness of bowlder clay is interposed between these deposits and the older rocks, which lie beneath them all. At one place, in sinking a well, after passing through the surface deposits, blue clay was penetrated 70 feet in thickness, followed by 18 feet of sand, gravel, and clay, below which a light-colored limestone was reached. There is ample proof that the Winnipeg basin has been filled with water to the foot of the second prairie steppe, in recent geological times. In digging wells in the city of Winnipeg, wood bark and leaves are sometimes met with, and fresh-water shells occur in the sand deposits between the south end of Lake Manitoba and the Assineboine river, about 50 feet above the former. The level of Lake Winnipeg above the sea is 710 feet, St. Martin's lake 737 feet, Lake Manitoba 752 feet, Lake Winnipegosis and Cedar lake 770 feet, and Lake of the Woods 1,042 feet.

The drift striæ* in the eastern part of Wisconsin are exceedingly variable. Between the Kettle range and Lake Michigan, their course is from S. 4° W. to S. 116° W. Between the Kettle range and the Green Bay and Rock River valley, their course is from S. 12° W. to S. 59° E. In the trough of the Green Bay and Rock River valley, their course is from S. 41° W. to S. 7° E. And on the west slope of the Green Bay and Rock River valley, from S. 94° W. to S. 24° W. The diagram used to illustrate the course of these striæ, resembles the flowing vanes of an ostrich feather, with the shaft pointing to the northeast.

The drift deposits are separated in ascending order, into: 1st—Bowlder clay; 2d—Beach formation; 3d—Lower red clay; 4th—A beach formation; 5th—Upper red clay; 6th—Beach formations. The elevation of the beach ridge which marks the western limit of these deposits above Lake Michigan, near the Illinois line, is 55 feet; farther north, from 40 to 80 feet. North of Milwaukee there is a well-defined terrace, nearly parallel to the lake shore, from 50 to 100 feet high. In the vicinity of Sturgeon Bay, the terrace is replaced by a beach ridge of rather fine yellow sand. Along Green Bay, between Egg Harbor and the mouth of Sturgeon Bay, terraces of rock sustain a relation to the present shore similar to the terraces farther south. These rise, in some cases, almost vertically, to a height of more than 100 feet. The distance between them and the bay varies from a few rods to half a mile or more, and the interval is strewn with water-worn fragments of rock and occasional slight beach ridges.

In Central Wisconsin, the courses of the striæ are not less variable, though but few have been observed. In Dane county, they vary from

---

* Geo. of Wisconsin, vol. ii, 1877.

S. 35° E. to S. 81° W. In Columbia county, from W. to S. 85° W.,
and S. 47° W. In Sauk county, from S. 50° W. to S. 85° W.; and in
Green Lake, S. 68° W.

In southwestern Wisconsin, there is a driftless region of more than
12,000 square miles, or about one-fourth the entire area of the State.
Drift striæ and drift materials are absolutely wanting. The topography of the country shows that it was never invaded by the drift.
Except in the level country of Adams, Juneau, and eastern Jackson
counties, it is everywhere a region of narrow, ramifying valleys, and
narrow, steep-sided, dividing ridges, whose directions are toward every
point of the compass, and whose perfectly coinciding horizontal strata
prove conclusively their subaerial erosion. The ravines are all in direct
proportion to the relative sizes of the streams in them.

The altitude of the country seems to have performed no part in the
causes which kept the drift from this extensive tract of country, for
north of the head of sugar river, the limit crosses high ground, and the
altitudes east of the limit are as great as those to the west; Sauk
prairie is crossed on a level. Where the quartzite range north of Sauk
prairie is crossed by the limit, it is higher (850 feet above Lake Michigan), than any part of the driftless area, except the Blue Mounds,
whilst east a few miles, drift is found at 900 feet in altitude. From
the limit near the east line of Adams county, the country, for 40 miles
to the west, is from 100 to 200 feet lower. From the northwest part of
Adams county, to the Wisconsin river, the limit is in a level country;
whilst from the Wisconsin westward the country north of it is everywhere much higher than that to the south, the rise northward continuing to within 30 miles of Lake Superior. It thus appears that the
driftless area is in a large part lower than the surrounding drift-
covered country. Moreover, there is a scantiness of the drift from 25
to 75 miles north of the driftless area.

Roland D. Irving* said, the lacustrine clays extend inland from
Lakes Michigan and Superior for many miles, and reach elevations of
several hundred feet above the lakes. They are stratified beds of loose
material, chiefly marly clays, with more or less sand, some gravel, and
a few bowlders. They were deposited, evidently, when the lakes were
greatly expanded beyond their present limits. In the Central Wisconsin district, the lacustrine clays have only a small development, most
of the district being either too high to have been reached by the lake
depositions, or else lying behind the dividing ridges. The eastern

---

* Geo. of Wisconsin, vol. ii.

towns of Waushara county, however, are underlaid by a considerable thickness of red clay belonging to this formation. The surface elevation of the country here is 160 to 200 feet above Lake Michigan, and the clays 80 to 100 feet and over in depth, as shown by numerous artesian well borings that yield a flow of water which is obtained from seams of gravel at different horizons in the clay. The clay of eastern Waushara county is part of a large clay area that extends up the Green Bay valley from Lake Michigan, and it is quite significant, that Prof. Irving's map of this lake deposit shows that it extends within about twenty miles of the northeastern part of the driftless area of Wisconsin.

Afterward* he said the lacustrine clays underlie all of the lower levels bordering Lake Superior, above which they rise to altitudes of between 500 and 600 feet. This carries them well up the front slope of the Copper range, and high, also, on the flanks of the Bayfield highland. On the Wisconsin Central, these clays reach to an altitude of 560 feet, and are finally left, on ascending the railroad line from Lake Superior, near where Bad river is first struck.

The clay varies largely in amount of sandy admixture. There is commonly some sand included, though, at times, it seems almost wholly absent, and at others to make up the bulk of the formation. The clayey matter is always of a red color, and always contains a considerable proportion of lime carbonate. The stratification is not always evident, but on the shore bluffs of the Apostle islands, it may be seen in the darker color of the moist sandy layers as compared with the lighter sun-dried clay. In many places, numerous small bowlders, chiefly of some dark greenstone-like rock, are to be seen embedded in the clay, and pebbles of the same, and other crystalline rocks are abundant. On the shores of some of the Apostle islands, and in places along the mainland coast, dark-colored bowlders of large size, presumably washed out from the clay, are very abundant. The entire thickness of these clays can not be less than from 400 to 600 feet, about 100 feet being the greatest thickness seen in any one section.

Mr. E. T. Sweet found a section of the lacustrine sands and clays, with gravel and bowlders, on the north bank of the St. Louis river, about one quarter of a mile from Greeley station, 202 feet in thickness. In the vicinity of Fon du Lac, and southeast of Superior City, along the old St. Paul military road, he found lake terraces at 15, 35, 80 and 120 feet above the present level of the lake, and an indistinct one at the

---

* Geo. of Wisconsin, vol. iii.

height of 300 feet.   Along the Brule river, in the vicinity of the mouth of the Nebagamain, where the river is 300 feet above Lake Superior river terraces are found 30, 80, and 190 feet above the river.   From the top of the highest terrace, or level of the surrounding country, to the corresponding top on the opposite side of the valley, the distance is about a mile.

The lake terraces and lake deposits of sand and clay at these heights in Wisconsin, show that Lake Superior has stood at a height sufficient to have overflowed the highest lands in any of the States south of it. The driftless region in the western half of the State, is alike conclusive against any of the drift phenomena in the eastern part, having been the result of glacial action of any kind, and they both unite in testifying against a continental ice sheet, or glacial period.

In Dakota county, Minnesota, there occurs an outlier of the St. Peter's sandstone, known as "Lone Rock," owing to its standing in a prairie, and forming a conspicuous object for many miles in all directions.   Its summit is about one hundred feet higher than the surrounding country, and from this point a number of outliers and pinnacled rocks of the same sandstone may be seen.   One of these is called "Chimney Rock," from its fancied resemblance to a chimney; and another, standing seventy feet high above the surrounding country, is known as "Castle Rock," the upper twenty feet of which is now so slender that but few centuries will pass before it totters and falls, under the wearing effects of subaerial denudation.   These sandstone outliers are monuments attesting the erosion which has taken place since Silurian times, and yet, in the valleys of this county, the drift prevails and bowlders abound.   In Wabasha county, we have the "Twin Mounds," and in Olmstead county, the "Sugar Loaf Mound" and the "Lone Mound," and numerous isolated bluffs, attesting the erosion for the same period.   In Fillmore county, the Trenton Group forms precipitous bluffs.   It rises perpendicularly from the short talus at the base, which adjoins the creek, forming canons, which widen as we descend the streams, and which, like the monuments of other counties, attest the erosion through long periods of time.   The weathering and erosion have left many scenes in the bluffs of wild and picturesque beauty, as at Weisbeck's dam, in Spring valley, that, standing alone, or considered in their relations to each other, as their bearing is found in all directions of the compass, are convincing proofs of the non-existence of the glacial epoch.   But the strongest proof, it seems, that one could wish against the glacial speculation, may be seen in two lonely towers, in the valley of the south branch of Root river, in this county, known as the "Eagle Rocks." The valley is one of denudation, by the ordinary subærial forces, and it has been excavated out of the Trenton Group; and yet, two lone towers, rising as high as the rocky walls of the valley, are standing to say that no glacial sheet ever moved in this valley.

Indeed, no one having any knowledge of geology, has found any evidence of glacial action in the Mississippi valley, or in the streams that flow into it from Minnesota; but, on the contrary, every geological fact bearing upon the subject is so strongly against it, that we unhesitatingly conclude that no glacier, great or small, ever entered it; and as to the hypothetical continental glacial sheet in this valley, it certainly suggests physical impossibilities. The valley of the Mississippi is one of erosion. At Minisca, the hills are 525 feet high. The slopes are such as are made by ordinary forces, without the intervention of anything extraordinary. The harder layers of rock stand out in bold cliffs on the sides of the valley, while the softer layers form slopes between the harder layers, marking the disintegration and denudation as it takes place under atmospheric influences. Streams enter the valley at right angles, and these are fed by streams flowing into them from the north and from the south in valleys of corresponding depth, and protected by sides of similar slopes and cliffs, and even more rugged bluffs; for, as we recede westerly from the Mississippi river in Southern Minnesota, higher rocks come into view, until the valleys are excavated in the limestones of the Trenton Group, instead of the softer magnesian limestones that abut upon the Mississippi valley. If a sheet of ice were to fill these valleys above the top of the dividing ridges, we may fairly conclude that it would be held so firmly that it could move in no direction; but if it could move either north or south or east or west, the sharp escarpments of magnesian limestone, the rugged bluffs of the Trenton limestone, the bold outliers in the widened valleys, and the pinnacled towers on the level prairies forming the divides between the streams, would be ground down, smoothed off, or entirely torn away.

A trip up the Mississippi river, from Dubuque, Iowa, to St. Paul, Minnesota, or across the country at La Crosse, Minisca, or Lake Pepin, will bring to the view of the observer the incontestible evidences against the existence of a continental glacier, in times so recent as the Pliocene or Post-pliocene. In the absence of the opportunity of taking the trip, turn to Owen's Geological Survey of Wisconsin, Iowa, and Minnesota, and look at the "Natural Section of Hills, Upper Mississippi;" "Cliff of Lower Magnesian Limestone, Plum Creek;" "Alterations of Magnesian Limestone and Sandstone, Kickapoo;" "Lagrange Mountain;" "Castellated appearance of Lower Magnesian Limestone, Upper Iowa;" "Lower Magnesian Limestone, Upper Iowa;" "Cliffs of Lower Magnesian Limestone, Upper Iowa River;" "Outlier of Sandstone, Kinnikinick;" "Outcrop of Upper Magnesian Limestone and Shell Beds, Turkey River," and you will be enabled to form some idea of the bluffs, cliffs, castellated rocks, and pinnacled outliers, that are so utterly inconsistent with the glacial hypothesis.

Such scenes are also presented in the State of Wisconsin, both within what is universally conceded to be the driftless area and without it. Two of these curious isolated eminences are situated in Dark Hollow, north of Wingville, on the head waters of the Blue river, near the junction of Badger Hollow, and composed of the Upper Sandstone, as illustrated in Hall's Geological Survey. Another called the "Stand Rock," in the Dells of the Wisconsin, forms the frontispiece to Vol. ii. of Chamberlin's Survey. But Prof. R. D. Irving informs us that a remarkable feature of all of the palæozoic portion of central Wisconsin, is the occurrence of *isolated ridges and peaks*, rising from 100 to 300 feet abruptly, and often precipitously from the low ground around them, and composed of horizontally stratified sandstone, or of sandstone capped with limestone. Such outlying bluffs lie all along the face of the high limestone country of Columbia and Dane counties, and are, generally, there capped by the same limestone that forms the elevated land, of which they are themselves fragments, others, again, and these are nearly all entirely of sandstone, occur scattered widely over the central plain of Adams and Juneau counties, often covering but a small area, and showing bare rocks from the base to the summit, which not infrequently are worn into jagged pinnacles and towers. He says the driftless area occupies 12,000 square miles (but the map indicates about 13,000 square miles) of the southwestern part of Wisconsin, or nearly one fourth the entire area of the State and that over this area the drift is not merely insignificant, but absolutely wanting. The line of separation of the driftless from the drift area, is thus traced:

Entering the State from the south, on the southern line of Greene county, the drift limit traverses this county centrally from south to north, and continues northward through western Dane and central Sauk; then curving eastward across the southern end of Adams, it follows along the eastern line of that county, passes into Portage, curves westward, and crossing the Wisconsin river again, continues in a nearly westward direction across Wood, Clark, Jackson, Trempealeau, and Buffalo counties, to about the foot of Lake Pepin, on the Mississippi.

He says, that east of this limit, the fragile castellated outliers that abound in the driftless area are wanting, though outliers do occur, though not abundantly, and they are thick and of rounded contour, and more commonly of limestone; but that north of this line the drift is quite insignificant, and all surface irregularities are as purely the result of subærial agencies as in the driftless region itself. And this corresponds with the outliers in Dakota county, Minnesota, mentioned above, which are north of Lake Pepin, and within the drift area.

There are several grand outliers in Jackson county, Illinois; one of these is called the "Back-bone," and another the "Bake-oven." The uplands contain some drift and gravel, but none have been observed south of the dividing ridge that crosses the State through the south part of this county and the north part of Union. The drift clays and gravel do not average more than 20 feet in thickness, and below these there is frequently found a dark blue or black mud, containing branches of trees, and sometimes trees of large size. In Perry county, the drift deposits seldom attain a thickness of more than 30 or 40 feet. But below them, as in Jackson county, there is a layer of blue mud lying on the stratified rocks, which is so full of partly decomposed vegetable matter, consisting of leaves and wood, as to render the water in wells that penetrate it, unfit for use. In Jersey county, the drift consists of about 20 feet of yellowish-brown clay at the top, below which there occurs from 20 to 30 feet of sand and gravel, with bowlders; and this is underlaid by about 15 feet of blue plastic clay, which contains fragments of wood, and even trees of considerable size. In Greene, Calhoun, and Scott counties, there is some evidence of buried channels where the drift is 100 feet or more in thickness. In Cook county, there is abundant evidence of the lake having been 40 feet higher than it is now, and that trees grew upon the surface, at levels lower than the present height of the lake. There is also some evidence here of a buried river channel. In Adams county, below 90 feet of drift clay, with gravel and bowlders, there occur an ancient soil and subordinate clays, without bowlders, or other evidences of drift action. At Sycamore, in DeKalb county, large pieces of wood were met with in the blue clays, at the base of the drift, at 50 feet in depth; and similar instances occur in Kane, Dupage, Richland, Monroe, Morgan, Tazewell, and other counties. Indeed, in nearly every portion of the State, remains of trees are found in the ancient soil in which they grew *in situ* beneath the gravelly clays and hard pan of the drift.

In Martin county, Indiana, near the town of Shoals, on the O. & M. railroad, there are numerous outliers of sandstone of carboniferous age, high and sharp ridges, and much wild and rugged scenery. A high ridge terminates near the east fork of White river, from the top of which there is a projecting mass of conglomerate sandstone, called the "Pinnacle," which stands 170 feet above the level of the stream. On the north side of this ridge, there is a tall outlier, which is called "Jug Rock," from the resemblance which it bears to a jug. It is 42 feet high, and supports, on its top, a flat projecting layer, which is called the "Stopper." A picture of this rock forms the frontispiece to the Second Report on the Geological Survey of that State, by E. T. Cox. The "Knobs," or "Knob stone formation," of Southern Indiana, is so named from outliers of subcarboniferous sandstone that have protected the underlying shaly rocks from denudation during all the ages that have passed since the Carboniferous era. Warren county is situated in the northwestern part of the State, and is deeply covered by the drift, near the base of which, and resting on a broken and irregular floor of Coal Measure rocks, there is generally found a bed of clay somewhat intermixed with quicksand and black muck. In sinking a shaft to the base of this drift, an ancient soil, containing the roots of trees and shrubs *in situ* was discovered, notwithstanding the passage through more than 50 feet of the bowlder drift and clay. And it may be laid down as a rule, in Indiana, that in all cases where the soil was not swept off by the flood of waters in the drift period, it will be found, at the base of the drift, containing the evidences of land vegetation, not materially distinct from that which now prevails on the top of the drift deposits.

There are extensive driftless areas in eastern and southern Ohio. These are marked by outliers, monument rocks, sharp ridges, rugged scenery, and the total absence of the drift sand, gravel, and bowlders, that characterize drift areas. The outspread of the drift materials from the north extends to the sources of the rivers that flow into the Ohio, and over more or less of the land intervening between the minor branches, near where the leading streams arise; but below this, the drift material is found only in the valleys of the principal rivers. It seems that wherever the valley was large enough to carry off the flow of water from the north, the adjacent land was not overflowed, and the height of the water in the valley was marked by river terraces. In eastern Ohio, however, only those rivers which have their sources in the central and northern part of the State, have river terraces, as the Scioto, Hocking, and Muskingum rivers; while the smaller tributaries of the Ohio, such as Raccoon, Shade, Little Muskingum, and Duck creek, have not a vestige of the evidences of the drift from their sources to the Ohio. Some counties are absolutely driftless areas, while others, like Athens and Washington, show that the water passed down the Hocking and Muskingum valleys, but overspread no other part of the country. The same phenomena may be observed in Indiana and Illinois. The water did not cross the great valley of the Ohio until it reached the western part of Kentucky, for the States of Kentucky and Virginia, south and east of the State of Ohio, are absolutely driftless areas.

It is an important fact, that throughout the drift area of Ohio, in all well authenticated cases of excavation, below the drift, where there are no evidences of denudation, at the particular places, there has been found an ancient soil of vegetable mould resting upon the disintegrated stratified rocks in place. The beech, sycamore, hickory and cedar have been found where they grew prior to the existence of the drift period. And beneath this ancient soil, no one has discovered striated or furrowed rocks, such as the glacialists have claimed as an evidence of their theory, and which are not uncommon where the ancient soil does not exist. Wherever a ridge is found having an easterly and westerly direction, the north side and the plains to the north are covered with this ancient soil, reposing on the stratified rocks, beneath the whole mass of the drift. But on the ridges the soil is usually absent, and the rocks are not unfrequently scratched and covered with drift resting upon the abraded surfaces.

A very good illustration of the ancient soil beneath the drift may be seen at the railroad cut north of the tunnel on East Walnut Hills, in the city of Cincinnati. This soil has a thickness in one place of four feet, and consists of a compact mass of very dark, rich, decayed vegetable matter full of roots which are lignitiferous, and still retain the hard woody fibers in a moderately good state of preservation. It reposes on the rocks of the Hudson River Group, and is covered by the sand and gravel of the drift, twenty feet or more in thickness.

The excavation exposed it upon each side, for a distance of about 100 feet, but the masonry will entirely cover it and hide it from view this season.

Commencing in the lower tier of counties in the State of New York where the hills are from 600 to 800 feet above the level of the narrow valleys, as they occur in Cattaraugus, Alleghany, Steuben and Chemung counties, and extending South over all the highlands of Pennsylvania, and over Virginia, West Virginia, the Carolinas, and the eastern parts of Kentucky and Tennessee, and South to the Gulf of Mexico, we have an absolutely driftless area; an area of dry land when the marine clays and sands were strewn over the territory adjacent to the Gulf of St. Lawrence, and over the New England States; and also an area of dry land during the period of the drift, of the central part of the continent, and for untold geological ages antecedent thereto. The elevated hills, precipitous ledges, profound valleys, overhanging rocks, and castellated outliers of the Carboniferous conglomerate in Cattaraugus county, some of which are illustrated in the Geology of McKean county, in the Report of Progress R. of the Second Geological Survey of Pennsylvania, under the name of the Olean conglomerate at Rock City, furnish the most incontestible evidence of the ordinary eroding agents through a period of time, commencing long anterior to the Tertiary epoch, and equally as conclusive evidence that no glacier ever passed over that territory.

During the ages that elapsed from the Carboniferous to the Tertiary, the Ohio river and its tributaries were excavating their valleys, and so also were the streams that flowed through the channel, that drained the northern and central part of the continent, which is now represented by the chain of great lakes. Where the valleys thus eroded remain unaffected by the drift, they are frequently immense chasms. The streams which flowed from the divide into the great drainage system of the north, cut out the valleys precisely as did the tributaries flowing south or east into the Ohio, and to equally as great depth. Could we see northern Ohio stripped of the drift, we would see a country quite as rough and rugged as southeastern Ohio. But there came a time when this drainage system of the north was obstructed in the region of Lake Ontario, and the waters were thrown back over the country, forming an immense lake. From this lake, deposits of clay, sand and gravel were precipitated over the country overflowed, and from the northern shore or sides of the Laurentian mountains, the shore ice transported to the south bowlders and rocky masses, in the same manner that it transports them now from one side of Lake Winnipeg to the other, and thus, much of the country was changed from its broken and hilly aspect into nearly a level plain. And when this lake overflowed the barrier or dividing ridge on the south, and swept over Ohio, Indiana, Illinois, western Kentucky, Tennessee and Mississippi, it transported the material that constitutes the drift deposits of these States, and which extends in the Mississippi valley as far as the Gulf of Mexico. The rush of water was adequate to transport and distribute the finer material, and the shore ice was sufficient to transport the bowlders and larger masses, and distribute them as far south as they occur.

The lake deposits on the hills and mountains near the shores of Lake Superior, occur 600 feet above the present level of the lake, or high enough to overflow all the States to the south. The ancient soil beneath the drift affords evidence that the climate was not materially different from the climate of to-day. The land and fresh-water shells found at different elevations in the drift, and the oft recurring timber transported and buried at all heights within it, show nothing that indicates a change of climate from the time preceding the drift through all its various stages. The ancient beaches prove the different elevations of the lakes, and teach us of long periods of time required for the pebbles and bowlders to be made, that now form these terraces, where they are preserved, and constitute so considerable a part of the drift that was swept southwardly when the lakes overflowed their barriers and carried them away.

The drift is then not only of Post-pliocene age, but much of it dates back through all Tertiary time, and some of it is, probably, much older. But that part of it containing the Mammoth, Mastodon, Dicotyles, Castoroides, and other mammals, with aboriginal man, belongs to the most recent or Post-pliocene era.

The eastern end of Lake Ontario is near a volcanic region, and within the range of the Appalachian system, where there have been important local elevations and depressions, as heretofore shown, by the sinking and rising of the coast from New York to Hudson's bay. The disturbance and elevation has been sufficient to throw the lakes back over the State of New York, and high up on the hills to the north, as shown by the numerous terraces, beaches and lacustrine deposits. This great lake may never have united with the grand body of water which is now represented on a smaller scale by Lake Superior, and again they may have been united at some period, and disunited at others. But all the phenomena presented in this region is to be accounted for by the presence of these lakes at various altitudes.

Lake Superior is in a volcanic region, and near the western end of the Laurentian mountains, and it is not improbable that earthquakes and volcanic energies had something to do with the emptying of these vast bodies of water over the country to the south. The drift deposits, to the west of Lake Superior, which spread over part of Minnesota, and extend as far south as the Missouri river, belong to an overflow of the great central lake of British America, which is evidenced by the terraces and beaches of that extensive region. The overflows have, therefore, not only occurred at different periods of time, but, probably, from three different bodies of water. If then, all the phenomena are to be accounted for by ordinary and well known forces of nature, why call to their aid a glacial period, which will account for none of them.

Taking a broad and general view, we would say that the drift upon the eastern part of the continent, from the mouth of the Hudson river to Hudson bay, is marine, and the striæ upon the rocks were produced under water. The age dates back to the Pliocene era, and probably to the Miocene. When this margin was depressed, a corresponding elevation took place east of Lake Ontario, that blocked up the great river that had drained the central part of the continent, as far west as Lake Superior, during the Triassic, Jurassic, Cretaceous and earlier Tertiary periods. This elevation was more than 500 feet, as proven by the lacustrine clays exceeding that height, which were formed upon the hill and mountain ranges surrounding the great internal lake caused by this back-water, and as further evidenced by the fact, that after the lake had been permitted to stand at this height for so long a period as to form terraces and beaches, that later, it excavated the elevated barrier to a depth of 500 feet, forming a channel, which is now in the bed of Lake Ontario, and when the eastern coast was again elevated, this region was correspondingly depressed.

The drift on other parts of the continent is fresh water or lake drift, and the striæ were produced, except in cases of drifting sand under atmospheric influences, by the action of water forcing harder materials against obstructions, or over barriers, and by floating shore ice having frozen within it, the sand, gravel and bowlders of the place in which it was formed. In the Rocky mountain region, each valley is the limit of its own drift phenomena; but when the northern part of the range was elevated, a very large interior lake was formed in British America, which seems to have covered many valleys, and in times comparatively recent, to have overflowed the country so as to empty itself in part, into the streams that flow south into the Missouri and Mississippi rivers.

Another great overflow took place from the more central lake. This extended over the eastern part of the State of Illinois, over Indiana and the western part of Ohio. The overflow had a width of more than 300 miles, and from its western margin it followed the streams westerly to the Mississippi, and from its eastern margin to the Ohio, so that its greatest width in these States exceeded 500 miles. This overflow may have been produced by volcanic energies in the Lake Superior region, and occurred as late as the Post-pliocene age. It was the great destroyer of the mammoth and the mastodon and other extinct Post-pliocene mammalia. Since that period the lakes have gradually drained themselves to lower levels through the outlet at Lake Ontario, leaving here and there lower lake beaches and terraces. In process of time, Niagara Falls will recede to Lake Erie, and that lake will be drained to its ancient channel, and other beaches and terraces will be left to represent the present height of the lake in the same manner that I have supposed the higher beaches and terraces to represent the former levels. This explanation seems to the author sufficient to account for all the phenomena discovered by the geologists, and it certainly calls to its aid no mythical hypothesis or unknown freaks of nature, but rests upon well-known physical and geological laws.

It is no small tax upon the imagination to believe that a great sheet of ice, having an existence in the north, ascended the Laurentian mountains north of these lakes, and then dipped down into the earth, scooping out Lake Superior 900 feet in depth, pulverizing the material, transforming it into gravel, sand and bowlders, scraping off the soil in some places, and scratching the rocks in others, as it ascended the valleys to the height of the dividing ridge between the waters that flowed to the north and the south, and precipitating itself into the tributaries of the Ohio, Mississippi and Missouri rivers, and depositing behind it in such even and beautiful distribution the sand and gravel that now fills the ancient valleys, and forms a vast, almost level plain over the northern parts of Ohio, Indiana and Illinois, and yet did not sweep off the ancient gravel beaches, in many places, that now mark upon the mountains and hills the ancient shores of vast bodies of water.

To believe in the glacial theory requires all this stretch of the imagination, and to be a real sound stalwart in the faith, there are many other marvelous things which must be accepted. One of these is described by a Pennsylvania geologist, to account for the drift phenomena of New York. He says :

" But when the ice front had been melted back to the southerly crest of the Chautauqua divide, the battle between the elements of heat and cold commenced in earnest. North of the barrier, the ice-king had massed his forces: Lake Erie basin was full of ice, and all the reserves of the north were freely moving down into it. As fast as one skirmish line on the summit was repulsed, another was thrown forward; and thus alternately advancing and retreating, the contest raged for ages before the invading ice was forced back, permanently confined, within the limits of the present lake basin."

The Muse that divulged this information must have been slain in the last glacial engagement, and remained for ages housed up in her little sepulchre, because, otherwise, it is evident that she would have told all about the grand glacial ball which ensued after the victory was complete, when the glaciers danced quadrilles, waltzed and mazourkied, and scratched and furrowed the rocks in all directions, followed by cutting the "pigeonwing" and the great American "hoe-down," when the glaciers shook the gravel, sand and bowlders, which they had collected for war, out of their crests and huge depositories, and covered the earth, which in their great glee they had cut up and striated so beautifully.

In conclusion, the author would seriously call the attention of the reader to the array of facts here collected tending to prove that there is no marine or other deposit which represents a glacial period of time, and, therefore, there is no such geological period; that there is no gap in geological nomenclature into which it can be lodged or injected. That the fossils and animal and vegetable remains teach us of no such period, but quite the contrary. And, finally, that the glacial epoch is a theoretical blunder, without the support of any known facts, and averse to all our geological and palæontological information.

·

www.ingramcontent.com/pod-product-compliance
Lightning Source LLC
Chambersburg PA
CBHW021117270326
41929CB00009B/921